S0-ASN-814

THE INTELSAT GLOBAL SATELLITE SYSTEM

Edited by
Joel Alper
COMSAT Corporation
Washington, D.C.

Joseph N. Pelton
INTELSAT
Washington, D.C.

Volume 93
PROGRESS IN
ASTRONAUTICS AND AERONAUTICS

Martin Summerfield, Series Editor-in-Chief
Princeton Combustion Research Laboratories, Inc.
Monmouth Junction, New Jersey

Published by the American Institute of Aeronautics and Astronautics, Inc.
1633 Broadway, New York, N.Y. 10019

American Institute of Aeronautics and Astronautics, Inc.
New York, New York

Library of Congress Cataloging in Publication Data
Main entry under title:

The INTELSAT global satellite system.

(Progress in astronautics and aeronautics; v. 93)
Includes index
1. Artificial satellite in telecommunication.
I. Alper, Joel. II. Pelton, Joseph N. III. Title:
INTELSAT global satellite system. IV. Series
TL507.P75 vol. 93 [TK5104] 629.1 s [384] 84-18524
ISBN 0-915928-90-6

Second Printing

Table of Contents

Preface vii

Chapter 1 Global Overview of Satellite Communications 1
Santiago Astrain
Bethesda, Maryland

Chapter 2 Voices from the Sky 7
Arthur C. Clarke

Chapter 3 Prologue and Pioneers 21
F. C. Durant III
Astro Associates, Chevy Chase, Maryland

Chapter 4 The Experimental Years 39
Burton I. Edelson
NASA Headquarters, Washington, D.C.

Chapter 5 The INTELSAT System: An Overview 55
Richard R. Colino
INTELSAT, Washington, D.C.

Chapter 6 INTELSAT Satellites 95
Emeric Podraczky and Joseph Pelton
INTELSAT, Washington, D.C.

Chapter 7 The Earth Segment 135
Kunishi Nosaka
Kokusei Denshin Denwa Company, Tokyo, Japan

Chapter 8 Transmission Techniques 191
Giuseppe Quaglione
TELESPAZIO, Rome, Italy

Chapter 9 INTELSAT's Operations 229
H.W. Wood
INTELSAT, Washington, D.C.

Chapter 10 Development of New INTELSAT Services 255
Marcel Perras
INTELSAT, Washington, D.C.

Chapter 11 INTELSAT and the ITU 269
David J. Withers
British Telecom International, London, England
Hans J. Weiss
COMSAT, Washington, D.C.

Chapter 12 INTELSAT System Planning 311
William R. Schnicke
COMSAT, Washington, D.C.

Chapter 13 Research and Development 331
D. K. Sachdev
INTELSAT, Washington, D.C.

Chapter 14 The Next Hundred Years: INTELSAT and the Postinformation Society 363
Joseph N. Pelton
INTELSAT, Washington, D.C.
Joel Alper
COMSAT, Washington, D.C.

Chapter 15 INTELSAT and its Role in the Emerging World Telecommunications System 379
Joseph V. Charyk and Irving Goldstein
COMSAT, Washington, D.C.

Appendix Acronyms and Abbreviations 387
Author Index for Volume 93 419
List of Series Volumes 421

Preface

In two short deacades INTELSAT has revolutionized the world. It has united planet Earth through a complex electronic network that routinely carries two out of three telephone calls—more than one billion conversations a year—across the oceans! It has also become the world's television distribution system and provided the means for 170 countries and territories to be linked together instantaneously, effectively, and inexpensively through a new communications technology that sets aside politics and other artificial barriers separating races, cultures, and nations. Through the INTELSAT global satellite system, international cooperation is being achieved on the high frontier of space.

Dramatic increases in performance have been achieved in an incredibly short period of time. The INTELSAT VI satellite will have a capacity that is 160 times greater than that of the diminutive Early Bird, which started it all in 1965. The volume of traffic carried has increased from 75 satellite circuits as of year-end 1965 to nearly 35,000 circuits today for international traffic, plus nearly 40 transponders for domestic service in some 25 countries. This miracle of modern technology and its effective employment to benefit world trade and world peace have not come easily. It has required the dedicated efforts of men and women from more than 100 countries. INTELSAT currently employs scientists and engineers, accountants, lawyers, and other specialists from around the world to operate the system and to design the networks of the future.

Some of the key people who played a dominant role in the development of INTELSAT have kindly contributed to this book—published during the 20th anniversary year—thus preserving INTELSAT's past, describing its present, and predicting its future. This book is at once a history, a description of INTELSAT's current technological and operational features, an analysis of the key policy issues that face the Organization in an increasingly complex international telecommunications environment, and a long-range look ahead.

As editors, we have been fortunate to receive the support and cooperation of many people in making this project a reality. We are especially indebted to Arthur C. Clarke, the father of the communications satellite and, hence, the grandfather of INTELSAT, for granting permission to reprint one of his earliest technical articles on satellite communications. We also wish to thank the past and present Directors General of INTELSAT, Santiago Astrain and Richard R. Colino, without whose support this project could not have come to fruition, as well as Joseph V. Charyk, the Chairman of the Board and Chief Executive Officer of the Communications Satellite Corporation (COMSAT), who played a key role in INTELSAT's development.

We also would like to express our special appreciation to the editors of the AIAA's *Progress in Astronautics and Aeronautics* series, particularly Martin Summerfield and Burton Edelson, who initially conceived this project and continually encouraged it during its 18-month gestation period. We particularly thank editors Camille S. Koorey of AIAA, Ellen D. Hoff of COMSAT, and Mary Z. Weisend of INTELSAT, who shepherded the volume to publication. Special tribute is due all of the authors who contributed material to make this book possible: Santiago Astrain, Joseph V. Charyk, Arthur C. Clarke, Richard R. Colino, Frederick C. Durant III, Burton I. Edelson, Irving Goldstein, Kunishi Nosaka, Marcel Perras, Emeric Podraczky, Giuseppe Quaglione, D.K. Sachdev, William R. Schnicke, Hans J. Weiss, David J. Withers, and the late H. William Wood.

Finally, we would like to thank everyone who played an important behind-the-scenes role in the production of this book, particularly Pep Ruddiman, Janet M. Tingley, Sherri O'Bannon, and Warkat Widaya, without whose support and assistance this book would never have been completed.

No volume, however long and exhaustive, can completely cover the technical, operational, and human detail encompassed by the INTELSAT global system. Nevertheless, we hope that as a reference book, as a "readable" history, and as an educational and technical data base this volume will be a worthy record of what INTELSAT has accomplished in the development of a global telecommunications network linking all mankind.

The editors would like to consider this book a memorial to a very special individual, H. William Wood, INTELSAT's Deputy Director General, Operations and Development, who died March 18, 1984. His outstanding contribution to INTELSAT was recognized in the following tribute of the INTELSAT Board of Governors: "Mr. Wood played a leading role in the formation of the

INTELSAT satellite system, both during his tenure as Vice President, INTELSAT Management Division of COMSAT, and as Deputy Director General, Operations and Development, of INTELSAT. He was known throughout the international telecommunications world as the key operations person for the INTELSAT system, and he will be sorely missed by all of his friends and colleagues in the satellite communications field.''

Joel R. Alper
Joseph N. Pelton
June 1984

1. Global Overview of Satellite Communications

Santiago Astrain*
Bethesda, Maryland

Man is unique on the planet Earth--and perhaps in the solar system and galaxy as well. He is set apart by his intelligence, sense of humor, and drive to communicate. For thousands of years now, his way of life has been in constant flux as the transition from hunter/nomad to farmer and town dweller to industrialized creature has caused sweeping social, economic, political, and cultural change.

Today, modern man is increasingly tied to a global culture and economy. Like his predecessors, he faces difficult problems, but he has greatly benefitted from the progress of the last two centuries. Advances in health care, education, and communications have improved the quality of life around the globe. And there are many who believe that, within the next two centuries, the tools of a postindustrial society (namely, computers, advanced communications systems, robotics, artificial intelligence, and advanced energy sources) can largely eliminate poverty and make possible an acceptable standard of living for all of humanity.

One of the characteristics of man's intellectual and social development has been the use of tools. The first tools were largely physical extensions of his anatomy. However, with the passage of time, they increasingly became intellectual prosthetic devices that extended the ability of his mind, as well as his body. Today, there are computers that, when vertically integrated with other processing units, can achieve an effective power of one billion calculations per second. Soon there will be a new central processing unit capable of two billion calculations per second, and the end is nowhere in sight. To give you some idea of the awesome ability of these "mechanical men," using one billion calculations per second as a convenient benchmark and assuming that a human could make and record the same calculation in four seconds' time, then the computer would equal the total mathematical speed and power of all humanity working in concert.

Invited paper received October 11, 1983.

*Former Director General, INTELSAT (retired Dec. 31, 1983).

Electronic communications have also made tremendous strides and resulted in dramatic breakthroughs at all levels of society. As recently as 1964, there were only a few high-quality telephone channels linking continents. But, in 1965, the first INTELSAT satellite, Early Bird, was launched and its capacity of 240 voice circuits immediately tripled the communications capacity available across the Atlantic Ocean region and carried the first live transoceanic television transmissions. Yet, today, Early Bird is merely a toy when compared with satellites such as those of the INTELSAT VI series, the first of which is scheduled to be launched in 1986. INTELSAT VI will have more than 150 times the communications capability of Early Bird and seven times the lifetime--an incredible net "efficiency" increase of 1000 times. The INTELSAT VI satellite will be capable of relaying three billion bits of information per second--the equivalent of transmitting the Encyclopedia Britannica between the United States and Europe some 20 times a minute.

A look at Table 1, which compares the world of global communications in 1964, 1983, and 2000 (projected), shows how remarkable the change in communications around the world has been since the advent of satellites. It should not be imagined that the remarkable growth in global communications during the era of satellites is coincidental. Indeed, there is every evidence that satellite communications has, in many instances, been the primary stimulus to rapid communications development regardless of whether a country was "developing" or "developed."

However, when INTELSAT was created in 1964, it was generally regarded as nothing more than the beginning of another scientific experiment. Few people realized then that the creation of INTELSAT was actually the start of a global communications revolution that later would kick off a global economic revolution as well. In order to understand how this came about, a bit of historical review is necessary.

The rapid explosion of experimental satellite communications technology in the late 1950's and early 1960's--satellites such as SCORE, COURIER, ECHO, RELAY, Telstar, and SYNCOM--occurred in a world that was an institutional vacuum as far as practical arrangements for the exploitation of space communications were concerned. Quite simply, no one had a practical use for that "new fangled" technology. In due course, however, discussions by heads of state and the passing of Resolution 1721 by the United Nations General Assembly--proclaiming, "Communications by means of satellites should be available to the nations of the world as soon as practicable on a global and nondiscriminatory basis"--led to the negotiations for the establishment of INTELSAT.

Not surprisingly, the INTELSAT Agreement and the INTELSAT Special Agreement, which were signed into effect on an interim basis on Aug. 20, 1964, reflected a compromise between divergent national viewpoints, and, indeed, the Agreements were structured

Table 1 Global communications, 1964-2000.

INDICATOR	1964	1982	2000 (PROJECTED)
Population (World)	3,100 Million	4,300 Million	6,000 Million
Population (U.S.)	190 Million (6.0%)	235 Million (5.5%)	290 Million (4.8%)
Electrical Power (World)	3,000 Billion Kwh	8,500 Million	24,000 Billion Kwh
Electrical Power (U.S.)	1,100 Billion Kwh (36.1%)	2,250 Billion Kwh (26.4%)	4,600 Billion Kwh (19.1%)
Telephones (World)	175 Million	472 Million	1,630 Million
Telephones (U.S.)	80 Million (45.7%)	175 Million (37.1%)	422 Million (25.9%)
World Transoceanic Telephony Service	678 Voice Circuits	40,000 Voice Circuits	Equivalent of 2,000,000 Voice Circuits
U. S. Transoceanic Telephony Service*	330 Voice Circuits (48.7%)	9,200 Voice Circuits (23%)	175,000 Voice Circuits (21.8%)
Individual Submarine Cable Capacity	128 Telephone Circuits	4,000 Telephone Circuits	150,000 - 200,00 Telephone Circuits
Individual Satellite Capacity (International)	240 Telephone Circuits (1965)	12,000 Telephone Circuits Plus 2 Color TV Channels	100,000 - 300,000 Telephone Circuits

*In operational full-time service.

to allow participation either by governments or private enterprises designated by their governments. INTELSAT was created on the basis of commercial operating procedures, but, just as significantly, it was premised on the principle that all charges would be nondiscriminatory and the same charges would apply to all users for the same service, regardless of their traffic requirements.

The first four years of INTELSAT operations after the launch of Early Bird were years of dramatic progress. The number of Earth station-to-Earth station pathways grew from 1 to 20. The number of half-circuits in operation grew from 150 to 1142, representing, in effect, an eightfold increase. The number of hours of television transmissions increased from 80 to 1372, and, perhaps most significantly, the total communications capacity increased more than tenfold. At the end of 1968, INTELSAT had a system capability of some 2400 two-way voice circuits and four television channels. This dramatic shift in only four years' time enabled INTELSAT to reduce its utilization charges by some 35%. By the end of 1968, therefore, INTELSAT had already established itself as the predominant international overseas telecommunications facility, providing two-thirds of the world's overseas service.

Despite those dramatic gains, however, there remained serious constraints upon INTELSAT's capabilities. It was, for instance, providing service to just 25 countries. It was still providing services only in the Pacific and Atlantic Ocean regions, and even though, by that time, the participants in the INTELSAT system represented nearly 90% of the world's overseas telecommunications requirements, there was no significant offering of service to the developing countries.

In 1969, however, when INTELSAT extended its service offerings to include the Indian Ocean region--just a few weeks before it televised the first moon landing around the world--the second phase of its development began, with a startling effect. Membership in the Organization surged to 80 countries, as developing countries recognized that the benefits of satellite communications were particularly suited to their needs. Earth stations in developing countries seemed to mushroom overnight in places like Balcarce (Argentina), Ras Abu Jarjur (Bahrain), Tangua (Brazil), Longovilo (Chile), Hong Kong, Jatiluhur (Indonesia), Shahid (Iran), Umm Al-Aish (Kuwait), Sehouls (Morocco), Utibe (Panama), Pinguay (Philippines), Si Racha (Thailand), and Pulantat (Guam). Global satellite communications through the INTELSAT system increased dramatically not only between developed countries, but between countries at all levels of economic development.

The leapfrog forward to modern communications for most developing countries has been remarkable--not only because it happened, but because it came about so suddenly. By the end of 1970, INTELSAT was providing service not to 25 but to 60 countries. Earth station-to-Earth station pathways had jumped from 20 to 131, and voice half-circuits in operation had nearly quadrupled--from 1150 at the end of 1968 to 4250. Net investment in INTELSAT had increased to nearly $200 million, and satellites had become the backbone of overseas communications.

Rapid development of the INTELSAT global system was to continue unabated through 1975. By then reliability of service had stabilized at 99.9%; the number of countries and territories served had increased to more than 100; and the number of voice channels in operation was close to 15,000. System capacity as of year-end 1975 was an amazing 16,000 two-way voice circuits plus eight television channels, while annual space segment utilization charges had plunged to $8460 ($U.S.). Growth in demand for overseas service in excess of 30% per year had become commonplace.

By the mid-1970's, however, INTELSAT could begin to catch its breath. Experience gained during the previous decade made possible diversification of services that enabled it to serve its customers even better than before. In 1973, came the demand-assignment system, known as SPADE, designed for countries with thin routes of traffic. In 1976, came the 10/11-m (32.8/36-ft) Standard B Earth station to offer a cost-efficient alternative to the 30-m (98-ft) Standard A terminal, particularly for countries with small traffic routes. Perhaps most significant was the introduction in 1975 of domestic services on the INTELSAT system, starting with Algeria, but quickly followed by Brazil and Norway. The leasing of spare capacity on the INTELSAT system for domestic purposes has been an especially important innovation. Today, some 24 countries--ranging alphabetically from Algeria to Zaire--use INTELSAT as an integral part of their domestic communications network. By the close of the 1980's, perhaps as many as 60 countries will use INTELSAT in this manner.

By the end of the 1970's, INTELSAT had won global acceptance as the world's overseas telecommunications network, and demands for a vast array of services challenged it to respond to the forthcoming "information revolution." It moved quickly to meet that challenge in a number of ways, most notably by providing mobile communications capacity to the International Maritime Satellite Organization (INMARSAT), and now, in the 1980's, by approving Standard E and F Earth station performance characteristics for international overseas business communications, including the option of direct-to-the-customer's-premises services via a 3.3-m (10.8-ft) terminal, as well as urban gateway stations in the 3.5- to 8.0-m (11.5- to 26-ft) class.

INTELSAT is also introducing digital communications services--first with time-division multiple-access (TDMA) equipment that will be followed by space-switched time-division multiple-access services soon after the INTELSAT VI begins operation in 1986. This will be followed by advanced modulation and coding techniques, such as 32- and 16-kbit/voice processing in the late 1980's and early 1990's. Planning is currently underway to explore how INTELSAT might provide aeronautical mobile services, high-definition television, business-optimized, low-cost video-conferencing services, remote location data collection and transfer, and possibly other services--if and when it evolves as the owner and operator of space platforms or satellite clusters in the 21st century.

Today, the INTELSAT system must serve the needs of a dramatically different global telecommunications environment than existed when satellite services began via Early Bird. Now it is the global network of submarine cables and INTELSAT communications satellites that makes possible today's global economy. Each year, millions of airline reservations, trillions of dollars in electronic funds transfer, and billions of telephone conversations--those essential elements of a global economy--are totally dependent upon the INTELSAT system. Today, this system carries two out of three overseas telephone calls and virtually all live overseas television broadcasts. During the 1980's, thousands of new Earth stations will be added to the INTELSAT international and domestic networks. Thousands of international and domestic pathways will carry vital messages upon which national and international economies depend. System concepts that will produce millions of voice circuits, thousands of video channels, and rates approaching a trillion bits of information a second will be developed in research laboratories around the world for implementation in the 1990's.

The political, social, and economic environment of satellite communications has changed perhaps even more radically than has the demand for new service. Fiber optic technologies promise highly reliable and low-cost trunking capabilities that will challenge INTELSAT to ever-higher performance levels, while separate domestic and regional systems are seeking a niche in the overall global telecommunications network. The proliferation of satellite systems, particularly those of the United States, the Soviet Union, Australia, Canada, Japan, and the European nations, as well as those of industrializing countries such as Brazil, India, Mexico, and Indonesia, has precipitated new concerns about the geosynchronous orbit. If the trend of proliferation continues, it will mean the waste of the precious orbital arc and must jeopardize effective and cost-efficient global interconnectivity. Further complicating matters is the request of some developing countries for a "guaranteed right" of access to the geosynchronous orbit. Thus, in response to this new environment, INTELSAT must, of necessity, find innovative and pragmatic ways to introduce needed new services and to achieve improved economies of scale while, at the same time, ensuring that its service offerings remain at the highest level of reliability.

System planning now under way, and described in detail in later chapters of this book, explains how the challenges facing INTELSAT in the 1980's and 1990's can be met and how, even in the 21st century, INTELSAT will continue to serve as the global bridge in space that connects the nations of the world.

2. Voices from the Sky*

Arthur C. Clarke†
1968

I must confess I have forgotten how it all began, in detail, but the events of the past few years have made me go back through my records and pick the brains of my friends; so I have been reconstructing, rather like an archaeologist, the events of 1944–45. During the last months of the war, we were trying to restart the British Interplanetary Society (B.I.S.). Chiefly involved in this were Eric Burgess, Val Cleaver, Ralph Smith, Harry Ross and our President, Dr. L.R. Shepherd. Towards the end of 1944, the V-2 arrived and its impact, in more ways than one, was considerable. It convinced people that there was something in big rockets. Incidentally, the only V-2 I ever heard was during a meeting of this B.I.S. committee. We had just heard Val Cleaver, who had recently returned from America, telling us about a distinguished American rocket expert who said the V-2 was all a propaganda rumour. At that point there was a bang in the distance which ended one rumour!

Although the rocket had arrived, space travel still seemed a very long way off and we were trying then, as I recollect, to think of commercial uses for rockets that would fill in the rather long period before anyone could do anything about space flights. One of the members of our group was Ralph Slazenger. His name is familiar to all sportsmen and he was a glider enthusiast. So one of the things we investigated was rocket assisted take-off for gliders. I think we imagined that this was one of the best commercial prospects for rockets. Yet, as far as I know, no gliders have taken off by rocket; it's easier to use a winch and a cable. Another idea was meteorological sounding rockets. That was quite a good one but only recently, in fact, have rockets been made cheap enough for wide-scale use for this purpose.

Communications satellites came in as a rather long range afterthought. The idea seems to have gelled in my mind in the Spring of 1945. I had spent the war years as the technical officer in charge of the prototype, and later the first operational ground

---

*Reprinted from "Spaceflight," March 1968, by permission of the British Interplanetary Society.

†Reprinted by permission of the author and his agents, Scott Meredith Literary Agency, Inc., 845 Third Avenue, New York, New York 10022.

control approach (GCA) radar system in the world. So I was equally immersed in electronics and astronautics; and in May 1945 I wrote a memorandum, called "The Space Station: Its Radio Applications." (Fig. 1)

This memorandum contained practically all the points in the article which subsequently appeared in Wireless World, but there were one or two additional considerations. In reading this document recently, I was interested to see that one of the points I made concerned the important services communications satellites would render to aircraft over the great oceans--services that could be obtained in no other way. This extremely important use of communications satellites has only recently been demonstrated. It is the only way to provide really reliable communications for aircraft on trans-Pacific flights or even over the mid-Atlantic.

In July 1945, I expanded this memorandum into the article which I submitted to the R.A.F. censors. It was very promptly passed (I'd love to know what they thought about it!). Two weeks later it had been accepted by Wireless World and it appeared the following October.

Fig. 1 Arthur C. Clarke autographing commemorative posters of his 1945 memorandum, "The Space Station: Its Radio Applications," at TELECOM '79, Geneva, Switzerland.

I should like to make just a few points regarding this article.

In the first case, it only discussed the synchronous or 24-hour orbit. The advantages of this orbit, which gives you a satellite that maintains station above a fixed point on the East at the equator, seem so obvious that alternatives did not appear worth mentioning. However, Dr. W.F. Hilton (another member of this Society) developed the very interesting idea of the 12-hour orbit, which the Russians in fact used, and which has some advantages for countries in high latitudes. I am only sorry I did not think of this as well as the 24-hour orbit!

I must confess I never thought of the time delay problem, partly because I was thinking almost entirely in terms of broadcast services where, of course, time delay does not matter. I also thought in terms of manned space stations because in 1945 there was no alternative in the available technology. It was a good day on our unit if only one tube blew in the day's operations. The idea that you could have complex electronic equipment, without hoards of servicemen and spare parts to replace the burnt out components, then seemed almost as fantastic as space travel itself. As far as I can remember, the immediate reaction to the appearance of the Wireless World article was precisely zero. There was no criticism, no comment; no one thought it was crazy--at that time I don't think many people would, in fact, have done so.

The only press reaction--and this was rather interesting--was in America. "The Los Angeles Times," 3 February 1946 did eventually pick it up, I think through the Rand Corporation, and wrote quite a good article.

I am often asked: "Why didn't I patent communication satellites?" It is a question I have often asked myself, and there has been some speculation on the subject by patent lawyers as to whether it could have been done. As far as I can remember, there are three reasons why I did nothing then. First, I frankly did not believe it would happen in my lifetime. Second, I lost interest in taking it much farther once the details had been worked out. And third, I was busy on more important(?) matters--namely my first novel. However, I didn't forget the subject completely and I publicised it wherever possible. For example in The Exploration of Space, 1951, the end plates show the complete three-station communication satellite network, and in Prelude to Space, which appeared in 1950, there was also a reference to synchronous satellites. I had forgotten this until I came across a reference to it in the NASA Chronology of Communication Satellites. It was a summary of the novel in which the NASA historian makes the plaintive and totally irrelevant remark: "The hero of this novel is an historian."

From 1950 onwards, the idea was pretty much in the public domain and was generally taken for granted by everybody. It was obviously only a matter of time before the electronic engineers updated it in the light of contemporary technology. Indeed, within ten years of my 1945 paper, three inventions had completely transformed the situation. The first was the solar cell, which converted sunlight directly into electricity. This, in fact, was predicted in the paper itself. Secondly was the transistor and its various solid state relatives. Third was the travelling wave tube, the special wide-band amplifier, which alone made possible the amplification of another B.I.S. Member, Dr. John Pierce, and R. Kompfner. John Pierce was Director of Communication Research at Bell Telephone Laboratories--and, incidentally, an occasional science fiction writer himself.

John has just written his memoirs of the era of satellite communications which has been published by the San Francisco Press of the University of California. It gives the whole story of his early work, and as a friend of longstanding it was my privilege to write an introduction to this important work. Pierce instigated the Echo project, the great balloon which is still one of the most spectacular objects in the sky. He heard that NASA was going to launch this large metallized balloon for the determination of upper air density. Incidentally, this idea also originated, so far as I know, with the B.I.S. It was put forward by Kenneth Gatland, Anthony Kunesch and Alan Dickson in their paper, "Minimum Satellite Vehicles", at the London I.A.F. Congress in 1951. After the success of Echo, John Pierce was then largely responsible for starting the far more ambitious and significant Telstar project which resulted in the first relay of television. Telstar was also the first, and certainly not the last, privately owned satellite, being built and paid for by the American Telephone and Telegraph Company, who hired a NASA launch vehicle to put it into orbit. Most people will agree, I think, that the progress has been somewhat spectacular since that time and much of the credit for this is due to a team at the Hughes Aircraft Company led by Dr. Harold Rosen, who developed the first synchronous satellites, Syncom, Early Bird, and their successors.

When Telstar was orbited in 1962, the launch vehicles available could put only very limited payloads into the 36,000 km (23,320 miles) high synchronous orbit. And synchronous satellites also require more power and involve more complexities than those that operate at lower altitudes, since they have to incorporate a control-jet system to keep them from drifting off station. John Pierce and the AT&T team avoided this difficulty with Telstar by using a low altitude orbit of only 950 to 5600 km (589 to 3472 miles). As a result, of course, the rapidly moving satellite was available for service only a few minutes each day. With great courage and ingenuity, Rosen and his colleagues at Hughes decided to shoot for synchronous orbit in 1963, even though they could put only a very small payload into it with available launch vehicles. Actually, it is harder to put a payload into synchronised orbit than it is to hit the Moon, although the Moon is ten times farther away. This is a hard fact of celestial mechanics.

Syncom was what you might call the minimum possible synchronous communication satellite--the smallest one that could do the job; and it triumphantly demonstrated the principles involved. It led to Syncom 3 which relayed the Tokyo Olympics in 1964. It is hard to believe that we now take TV coverage of the Olympics completely for granted.

In quick succession the Syncoms led to the more advanced Early Bird--the world's first commercial communications satellite--launched on 6 April 1965. I was at Comsat headquarters the night it went up and I certainly won't forget that evening. Early Bird led to the more advanced Atlantic and Pacific satellites which, in turn, led to the establishment of a global system in 1969. Rosen and his team were dedicated men, fully aware of the social implications of what they were doing. One of the engineers in the trio who started this work was Don Williams, who offered Hughes his life savings to start the project, because at the time there was so little interest. Later he became so worried about the state of the world that he killed himself. I'm sorry I never had the chance of meeting him. Rosen himself said once: "What we are doing is trying to save the world;" which is a theme I'll return to later.

So much for the past and the present. What of the future? It is certainly difficult to predict the impact that any invention will have on society at large. I have to thank Mr. Wedgwood Benn for sending me this example of the difficulties of prediction. About 80 years ago there was a Parliamentary commission which confronted the Engineer-in-Chief of the Post Office with this new-fangled invention, the telephone. He was asked if it was likely to have any considerable use and he replied: "No, it may be valuable to the Americans but we in England don't need telephones. We have plenty of messenger boys."

I also recall some of the early predictions (not personally) about the use of motorcars. When these were first introduced it was pointed out that these vehicles would obviously only be used in the cities; there were no roads outside the cities on which they could run.

I feel that many of the predictions about the impact of communications satellites will be of the same short-sighted nature. If a revolutionary invention does not fit into the pattern of society, that is just too bad for the pattern of society, and what we have seen already is only a faint intimation of the changes to come. These probably cannot be fully appreciated in such semi-developed countries as the United States and the United Kingdom. We already have TV, radio, newspapers and communications media; most of the World hasn't! The social changes to come over the whole planet are not incomparable to those brought about in the West by electronics since 1920. And they will happen quickly--not in the remote future.

First of all, I should like to say a few words about radio, not TV-communications satellites, which have been partly overlooked in the general glamour of television. Reliable domestic radio services are not available over most of the world. Long distance services are of poor quality, particularly in the tropics and especially towards sunspot maximum. By 1970, there will be 130 million VHF sets in the world, many of which could pick up direct radio broadcasts from satellites which an Atlas-Agena rocket could launch into an 8-hour 12,872 kilometers (8000 mile high) orbit. Radio has a tremendous advantage over TV for direct satellite broadcasting. It needs only a fraction of the power and it does not require a synchronous orbit or directional antennae. Yet, incredibly, Congress has cut NASA's request for funds for broadcast radio satellites. I should like to ask the State Department or the Foreign Office, "how useful would you find a satellite that could broadcast directly into China so that anyone with a transistor radio could listen in?" If it had been pushed, this facility might be available now.

The Telephone and the Global Village

The telephone is perhaps the best example of a communications explosion brought about by electronics. Today's satellites carry tens of thousands of telephone circuits, in contrast to Early Bird's 240, but we have yet to scratch the surface.

At an American Astronautical Society's Conference on the commercial uses of space, a Rand Corporation scientist discussed the "ultimate" communications satellite which might be in use 10 to 20 years from now. This would have 10 million two-way voice channels available without interference. By spacing these around the Equator, one could easily have a hundred of such satellites without interference, making possible a thousand million telephone circuits; all people awake at any one time could be on the phone! You may say that a thousand million telephone circuits is excessive but is it? Computers can be more talkative than people!

In effect this could shrink the world to the scale of a "global village," or at least a world community which is psychologically like a village. The phrase "the global village" is due to Marshall McLuhan of whom you will be hearing much more shortly, if you haven't already.

As well as telephone circuits for speech, all sorts of facilities will be made instantly available between all points on the Earth. Satellites have been used for decades to transmit the vast amounts of documentation needed for transoceanic flights, so that the documentation gets to, say, New York, before the aircraft does; this alone would probably save enough payload to take another passenger. This is just one of the things that can be done.

It will soon be possible for anyone anywhere to consult a video display from which any form of information can be obtained from anywhere on Earth. So that anybody can effectively jump to anywhere in the world by pressing the right buttons, can see anybody, get any information. One result of this will be a vast reduction in travel for work as opposed to travel for recreation.

As well as a reduction in commuting, there might even be, ultimately, a dispersal of the cities. We will never solve our traffic problems by covering the world with concrete super highways; we will solve them by making it unnecessary to have so much traffic. And this is one of the likely results of communications satellites. We will never rebuild all our slums; we have got to try, of course, but by the time our cities are fit to live in, no one will have to live in them. Anyway, we probably will not need them so badly because the population explosion of the twentieth century will be followed by the inevitable population crash of the twenty-first. Now, when we all live in one "village," even though it is a village spread over the entire world, and we may want to talk with each other at any moment, it's obviously intolerable that half of us will be asleep while the rest are awake. This is going to cause some social problems. Anyone who has friends or contacts in other Continents or even in the Americas, won't want to be woken up at odd hours for discussions. What can we do about this? Can we abolish sleep? I think maybe we will have to. Perhaps this is a little drastic. It may be theoretically possible no doubt to prove that sleep is not necessary, and of course there are many creatures that don't sleep. Perhaps the best solution was put forward by Sidney Sternberg, in a very funny paper given to the AIAA in 1966. He said: "A century ago, people set their watches, if they had them, or their clocks, more or less by the Sun and no one bothered about agreeing with anyone else's time. But when the railroads came in clocks had to be synchronized and over countries like the United States, they had to bring in zone times." Mr. Sternberg suggests that the next stage to be brought in by communications satellites will not be zone time, but common time. Everybody in the world would have to have their watches set at the same time, and he uses the slogan "One world, One bedtime."

This also involves problems, because if we set our watches now at 7 o'clock on the present 24-hour clock, it means that half the world would have to get up when the Sun was shining and half the world would have to get up in the middle of the night, so they would see perpetual darkness. For the solution to this, let us switch to Sidereal time, which has 23 hour 56 min. in a day. And then you see, during the course of the year, the clock ticks right round, so if we only woke up in England say at 6 o'clock by our watches in the Spring, 6 months later, by 6 o'clock by our watches, it would be the other half of the day. You might say that you woke up when the Sun arose in the spring, and when the Sun was vertically beneath us in the winter. But the effect of this would be, of course, that everybody on Earth would have at least equal time in the Sun, even though they would drift round the daylight and night hours. As Sternberg points out, some of the professions would be rather hard hit; he mentions farming and burglary. But you can never please all people all of the time.

Although perhaps we should not take this suggestion too seriously, it does hint at the kind of social changes that communications satellites may introduce. After all, could the Victorians have imagined our present telephone-based society? And the pre-communications satellite has been used for long distance and inter-continental communications, and it can provide services, especially TV, not available in any other way. But this is a short-sighted view and the situation is going to start changing soon. A recent study by Francis Gicca of Raytheon's Space Communications System, points out that today's satellites are cheaper than any other system for a moderate number of messages for distances of over 2414 kilometers (1500 miles). By the end of the century, they could be cheaper even for single channels over only 161 kilometers (100 miles). When this occurs, most homes are likely to be equipped with simple satellite communications terminals. Such terminals will be only slightly more complex then present colour TV sets and less expensive to mass-produce. Incidentally, they will probably operate in the millimetre region. This is the great frequency band for "comsats," because of the enormous band width and high gain possible with small antennae. I think, for instance, one of the effects of this will be to get rid of today's unsightly TV antennae which straggle all over the houses and make the city look hideous. The antennae needed for this kind of system might be small flat discs you can lay in the roof.

John Pierce's memoir mentioned earlier comes to some important conclusions. "There is and probably always will be far more communications within a continent than between continents. Suddenly I realized that domestic communications will be the big future of satellite communications. Within a very few years, it will be possible to orbit a satellite with a communication capacity of tens or hundreds of thousands of telephone circuits. The satellite would incorporate switching equipment which would transfer blocks of circuits from one pair of cities to another to meet fluctuations in demand. Such a capacity does not change former communications concepts--it makes them obsolete." I hope the Postmaster-General is aware of this. And if John Pierce is right, and he hasn't done so badly in the past, communications satellites will affect the man in the street even more closely than is generally imagined. But they won't be only used for trans-Atlantic calls and telegraphs; they will be used for ordinary trunk calls.

Television Broadcast Satellites

Now I would like to say something about direct television broadcast from satellites straight into the home, something that hasn't been done yet, but which was really what I was discussing in 1945. This is an enormously complex subject and I can mention only a few facets. Contrary to suggestions that have been made I believe that the main application of direct broadcasting will be to the undeveloped countries, which have neither ground stations nor TV. They need it and we don't; we already have TV. Also it is going to be easy to provide in undeveloped countries and in rural regions generally. We must appreciate the noise factor. In cities the level of RF noise is so high that you'd need a powerful signal but I've seen studies that show how to adapt a television home receiver to pick up signals from a satellite. This kind of satellite will be available in 5 to 10 years.

If you live in the City, your adaptor might cost $2000, but if you lived in the country, it would cost only $100; and that is a tremendous difference.

There have been many studies of education TV from satellites, and I would like to mention one made by Philip Rubin, then of Hughes Aircraft. This concerns an educational TV system from Mexico and the astonishing thing is its cheapness. It works out at about $2.00 per person per year for the cost of the system. And if there were more than 20 million people in the area, it came down to about $.50 per person per year. This includes the cost of the satellite, the ground stations, the receiver--everything. And it has been suggested that isolated schools in remote areas will be fitted with simple 3 meter antennae. This service would provide high quality colour on 12 channels at a cost of about $2.00 per person per year.

Rubin has also sent me a study relating to India, which has even greater potentiality. I did not know, I must confess, until he sent it to me that the United Nations had already built a centre for research and training in satellite communications at Ahmedbad in North West India. I'd like to quote a letter sent to me recently about the use of educational TV satellites. "The Indian response was enthusiastic--the two most motivated people were Dr. Vikram Sarabhai of India's Atomic Energy Commission and Dr. Sripati Chandrasekhar, Minister of Family Planning. Dr. Chandrasekhar's problem is to prevent 350,000,000 births in the next 20 years. The thought of immediate communication with the great masses of people had led him to believe he could really do his job."

I must say it gives me a rather strange feeling to realize I may stop millions of people being born, but I would stress again the astonishing cheapness of this communications satellite distribution system. The question is not, "can poorer countries afford satellite television?" It is, "can they afford anything else?" Communications satellites are the cheapest way of educating a backward community; for spreading information on agricultural techniques, more efficient food production, farming and hygiene as well as all the things which are necessary in order to have a stable, healthy society.

Incidentally, there is a good reason why India is an excellent place for communications satellites and I discovered this only recently. The equatorial belt is not circular. It has two high and two low points. One "high" is at 20° W over the Atlantic; as a result Atlantic satellites are on the top of a gravitational hill and tend to slide off it. They need control jets to push them back to the top from time to time, and can no longer stay in place when they run out of hydrogen peroxide to keep them on station. Desmond King-Hele in Observing Earth Satellites observes sarcastically that it is almost as if Nature in her wisdom has decided that Europe ought not to be exposed to the influence of American television! However, when they finally slide off the gravitational hill, synchronous satellites drift around the equator and come to rest at the low point. I've been delighted to discover that retired "comsats" go to Ceylon (or Sri Lanka, if you prefer), as I also retired there some time ago. In my old age I hope to watch the orbiting junkyard immediately overhead through my telescope--and to watch the occasional flare of the rockets as the repair and maintenance crews shuttle around the most valuable piece of space real-estate in existence, the 35,800-kilometer (22,300 mile-high) strip above the equator.

The Future Comsat

In 1945 I took it for granted that communications satellites would be large manned space stations assembled in orbit and staffed with servicing crews. Thanks to the incredible advance in electronics miniaturization, the first ones have actually turned out to be automatic communications satellites barrels. However, the ultimate communications satellites will be large manned structures associated with smaller unmanned ones, which will be regularly visited for maintenance and the replacement of fuel supplies for station-keeping. The importance of this from the viewpoint of long-term economics cannot be over-estimated. Today's communications satellites cost several times as much as they should, because they have to be made as reliable as humanly possible. Imagine what today's telephone system would cost if the exchanges had to be designed to last five years without servicing, and it was impossible to get at them if they went wrong. Several communications satellites have failed to go into the correct synchronous orbit because of apogee motor malfunction. There are millions of dollars worth of perfectly serviceable satellites up there that can only be used for a few hours a day. A rendezvous with a manned spacecraft could put that right and boost it up to the correct orbit. And I would remind you, although it isn't exactly a communications satellite, of the heartbreaking demise of the complex and expensive satellite known as the Orbiting Astronomical Observatory. It is up there in a perfect orbit, but its power supply failed soon after launch. A man with a screwdriver might have been able to fix it. In fact, had it been designed for servicing in the first place, it might never have failed. This is just one of the reasons why we need man in space; it may be the least important but it is going to be worth billions a year in hard cash.

Britain's Space Role

I would like to end by taking both a wider and a narrower view of the space potential. And say something about the position of the United Kingdom in the space age.

More than a century ago, this country led the world into the age of global telecommunications, when it spun a web of deep sea cables round the planet. In the summer of 1867 Brunel's masterpiece, the Great Eastern, laid the first operational trans-Atlantic telegraph cable, linking the new world with the old. That, incidentally, was the creation of an Anglo-American consortium; so also was the first Atlantic telephone cable, which began service in 1956.

What part, I wonder, will the United Kingdom play in the age of communications satellites? This country certainly cannot do everything in space. There are some that can say that it should do nothing. We should argue with them patiently. But what we should not tolerate is the apparently invincible ignorance of those who think that nothing in space is worth doing. I've been listening with mounting incredulity to some of the attacks on space exploration that have been made in this country, and it is only the Law of Libel which prevents me from quoting from the literary weeklies and the serious Sundays, not to mention the British Association. Reading these effusions, I'm reminded of the remark someone once made about the commentators on 'Hamlet': "Are they mad, or only pretending to be mad?" Regrettably, some of the silliest and most ill-formed criticisms of space exploration have come from scientists who should know better. But unfortunately they are no more immune to prejudices than other people. I would not care to say whether it is short-sightedness, cowardice or mere jealousy which has prompted the more intemporate of these attacks.

Time and history will dispose of these squeaking pygmies, but they may do grave damage before they are silenced. The position is well summed up by Dr. Frederick Seitz, President of the U.S. National Academy of Science, who obviously should be able to see both sides of the question. "Our children," he says, "will wonder what manner of people we were, that we ever seriously questioned the value of space exploration."

The communications satellite is only the first, and in the long run it may not even be the greatest, of the benefits of space. It is just one of the tools that will be needed by those who, in Dr. Rosen's phrase, are "trying to save the world."

And the world has got to be saved on two levels. We can call them the material and the intellectual or, if you like, the physical and the spiritual. Communications satellites are heavily involved at both levels.

I have already mentioned how they may be used--how they must be used--to improve the standards of education which alone can drag the have-not nations out of misery and starvation. They can contribute even more directly through meteorological satellites and environmental satellites with which they are inextricably linked.

Meteorological satellites indeed have already added much to the convenience and safety of people in the Western World, and incidentally have saved millions through the efficient application of the processed data to farming and other human pursuits. But this is trivial to what they could do elsewhere. For example, at Christmas 1966, a hurricane moving in across the Bay of Bengal struck between India and Ceylon. It wiped out whole communities, including a train going across the ferry to India, and destroyed entire shipping fleets. The approaching storm was photographed by a weather satellite the day before, but there was no active warning system to get the information to the people involved. This of course has now changed for ever.

In the East, life depends on the Monsoon. I saw the first satellite photograph that covered the whole of Asia. There were a billion people in this picture and their lives depended on the distribution of certain white patches--the pre-Monsoon clouds. For the first time, we're beginning to understand how the weather system works. The environmental and earth sensing satellites, which survey the ground for natural resources, are going to be equally important.

Let me just give two examples of this. It has been discovered that from satellites, one can detect certain types of fungus infection on plants and trees. From the satellite this can be detected several days before a man on the ground walking through the same forest becomes aware of it. When I was visiting NASA headquarters, they showed me some photographs taken from the Gemini flights of the Gulf of Mexico showing incredible detail of underwater geological formation, of the shores, and fishing--particularly shrimp--areas. They had shown them to one of the men who work in the fisheries in the Gulf; he said the photograph told him more in 5 minutes than he had learned in 20 years of fishing in those same areas.

That is why it makes me mad to hear such statements as: "Why did you spend the money on space instead of on the starving millions?" That money one day is going to stop them starving. In some areas it will be the only way. I refer to a statement by a recent clerical gentleman who made this very point. Surely, he must realize perfectly well that even if the space programme were stopped the money would not go to relieve human misery. So why talk such dangerous and misleading nonsense? I would have a far better case if I said: "Why don't you stop building your churches and give all the money to famine relief?"

Incidentally, why can't the critics of the space programme even get their elementary facts right? I have seen it reported that the non-stick frying pan was one result of the space programme. But even that isn't correct because it's based on Teflon, and that goes back to the atomic energy programme 20 years before. If they must pick up sticks to beat us, at least they should be able to reach for the right sticks. The same debate criticized the space programme for draining so many brains away from this country to the United States. The simple answer is that we didn't encourage these people. They were ready to go because they realized that the most important and interesting work lay elsewhere--it isn't just the money that attracts them.

I admit that many of the criticisms of the space programme are aimed at the Apollo manned lunar programme. That I am sure will justify itself in due course. But almost anything that has been attempted in the lunar programme would still have to be done even if the Moon was not there, because of the need to develop the technology to open up near space. Without this, none of these things I have mentioned can be exploited properly. Robots have done a wonderful job showing what is possible. But only men on the spot can really complete this job. If you try to imagine the problems of building and operating purely remote controlled physics laboratories, factories and observatories, you will see how ludicrous it is to imagine that we can obtain more than a small portion of the benefits of space endeavour without having men on the spot. The record of the Mercury and Gemini flights established this beyond doubt.

Let me make it clear that I'm not defending everything that the U.S. and the U.S.S.R. has done in space. We all know that there has been waste, confusion, politics, even corruption involved. Many things could have been done better; many should not have been done at all. The entire Apollo programme may turn out to be a case of doing the right thing for the wrong reasons and perhaps not in the best possible way. But this, I'm afraid, is typical of all human endeavours in this imperfect world.

Two years ago the astronautical fraternity met at the I.A.F. Conference in Athens. I'm sure Pericles had the same problem with people who very sensibly wanted to use money for slum clearance and the sad fact is that the genius who created the Parthenon died in prison awaiting trial for misappropriation of funds. Yet the Parthenon is one of the chief glories by which Greece's Golden Age is remembered. Our space achievements will be our greatest legacy to the future. Indeed they will create that future. They will make it possible to have a future; which finally brings me back to communications satellites.

Do not judge them merely in terms of the drivel you see on television every other night of the week. They are the next giant stride towards what Arnold Toynbee once called, in a different context, the "unification of the world." Whether we like it or not, the world of the communications satellite will be one World in which all men speak, or at least understand, a common language. Now, at last, we have the means to turn back the clock to the moment when, as recorded in Genesis, confusion fell upon mankind. The legend of the Tower of Babel has an uncanny relevance for our times. Higher than the wildest dreams of its builders, 35,800 kilometers (22,300 miles) above the Earth, we may regain what was once lost, when the Lord said, "Behold they are one people, and they have all one language, and this is only the beginning of what they will do; and nothing that they propose to do will now be impossible for them."

3. Prologue and Pioneers

F. C. Durant III*
Astro Associates, Chevy Chase, Maryland

Before communications satellites, before any satellites, there were the enthusiasts and the visionaries. These were men and women who believed that space flight was feasible, if only there were a program and money--at least a few million dollars or marks or rubles anyway. Some of these believers became leaders.

The world war that took place between 1939 and 1945 was important to the development of rocket technology. But even before the war, there were those who believed in rocket technology. They banded together in groups called "rocket" or "interplanetary" societies. Sometimes they experimented with solid- and liquid-fueled rocket motors, and some even launched a few. These self-styled rocketeers looked to three scientists as their inspiration and true leaders. These three great pioneers worked independently to develop rocket theory and conceived of step, or stage, rockets. The first was a Russian, the second an American, and the third--still living--was born in Transylvania (now Romania) and is now a German citizen.

But the true start of satellite communications came earlier still, even before the three great space pioneers. There was a time when technology was insufficient to build rocket motors, and before that, there was a time when it was not even recognized that rocket motors were needed to achieve space flight. Let's start at the beginning.

The dream of space flight is an ancient one. However, for centuries no one had even a vaguely scientific notion concerning what would be required to travel to the moon or to those moving lights in the night sky. Cannons and firecracker power were conceived, but were not suited to the task.

Modern understanding of the solar system can be dated to the works of Nicolaus Copernicus (1473–1543), famous as the proponent of the heliocentric solar system, and Galileo Galilei (1564–1642), who developed the astronomical telescope. Isaac Newton

---

Invited paper received October 11, 1983.

*Former Assistant Director of National Air and Space Museum, Washington, D.C. (1964–1980).

(1643–1717) formulated the three fundamental laws of mechanics and the law of universal gravitation. Yet, despite these theoretical and empirical breakthroughs in the fields of physics and astronomy, the nature of the limited distance of the Earth's atmosphere (blending into the near-total vacuum of space) and extremes of temperature were still unappreciated.

Flight in the atmosphere began with balloons, in 1783. This was 120 years before the Wright brothers learned enough about the aerodynamics of heavier-than-air, controlled flight to initiate man's development of aviation.

By the turn of this century, it was recognized that reaction propulsion alone would make Earth satellites and space flight possible. In fact, as A. V. Cleaver (1917-77) remarked, "Ninety percent of the science required to go to the moon was available by 1900." Yet, the technology was lacking. Important engineering principles (mechanical, chemical, and electrical) had yet to be discovered, and materials (metal and nonmetal) had yet to be developed.

Science fiction in the late 19th century was increasingly prolific. It played an influential role in preparing the general public of industrialized nations for space travel as a technical possibility. The novels of Jules Verne and H. G. Wells influenced the great pioneers, as well as those dedicated leaders who would emerge in the 1920's through the 1950's.

Quite apart from ideas on its use in space, rocket propulsion had developed for several hundred years as a war weapon and as a propulsive element for pyrotechnics, or fireworks.

No one knows who invented a "rocket motor" by tamping black powder in a cylinder, closed at the forward end and restricted to a short neck at the aft end. Certainly black powder--a mixture composed of potassium, nitrate, charcoal, and sulfur--all finely ground (separately!!!) must have been available.

Most scholars credit the Chinese with this invention. In the 13th century, "arrows of flying fire" are deduced by some historians as possibly having been propelled by rocket jets. In any event, rocket propulsion for warfare, as well as for fireworks displays and signaling, evolved slowly over the next five centuries.

Near the end of the 18th century, the use of war rockets by Indian troops against the British sparked research in England. William Congreve (the younger) developed a series of metal-bodied rockets with incendiary, or shrapnel-loaded, warheads. Congreve not only experimented and standardized his production, he developed a "rocket system." The rockets were transported unassembled and attached to their 2.44-m (8-ft) guide sticks at the battle site. They were launched either along the ground and/or in a ballistic trajectory against enemy troops and cavalry.

From 1810 through the 1870's, war rockets were used in most European battles, as well as (occasionally) in the United States, Mexico, South America, and Africa. The phrase "the rockets' red glare" in the American national anthem immortalizes the Congreve rockets used in the Battle of Fort McHenry at Baltimore, Md., in 1814.

By the late 19th century, rifled guns made of alloyed steel generally surpassed rockets as a battlefield weapon. Some use was made of modified rockets to throw lines from shore to sailing vessels in distress and, in the whaling trade, to propel harpoons.

Rockets were not used in World War I except as launched from aircraft against hydrogen-filled observation balloons and for signaling.

So much for the mundane, terrestrial uses of rockets in the early 20th century. By then, our three great space flight pioneers were alive and busy with their dreams and calculations as to how extraterrestrial flight might become possible. All three were modest, independent, hard-working scientists, as well as fine mathematicians. Only the youngest would actually witness man in space, on the moon, and see the Space Shuttle become a routine "trucking" operation.

Konstantin Eduardovich Tsiolkovsky (Fig. 1) (1857-1935) was born in Kaluga, USSR. A mathematics teacher, he formulated basic calculations for rocket thrust and staged rocket vehicles. He recognized the greater value and utility of liquid propellants (including liquid hydrogen and oxygen), conceived of a closed biological cycle (using plant life to replace oxygen on long voyages), and considered the problems of high acceleration on launching to Earth satellite velocity and the zero-g condition of free-fall. All of these points were contained in his A Rocket Into Cosmic Space[1] and subsequent writings.

Fig. 1 Konstantin Eduardvich Tsiolkovsky (1857-1935), Russian astronautical pioneer, shown in 1924. From N. A. Rynin, Interplanetary Flight and Communication, Vol. III, No. 7 (NASA TT F-646), p. vii.

In the United States, Robert Hutchings Goddard (Fig. 2) (1882–1945) was growing up in Worcester, Mass. Goddard developed rocket theory and experimented with steel rockets, tapered, and with carefully narrowed nozzles to achieve supersonic exhaust flow. By 1919, he was a professor of physics at Clark University in Worcester, Mass. The Smithsonian Institution sponsored (modestly) his research from 1917 to 1929. Goddard not only derived the formulas for staged rocket flight but was a remarkable inventor. In his lifetime he was issued 48 patents. Eventually, 166 additional patents were issued, bringing the total to 214. In 1958, the U. S. Government settled $1 million for rights of use by the U. S. Department of Defense and the National Aeronautics and Space Administration (NASA).

Experimenting initially with double-base powder, Goddard shifted to liquid oxygen and gasoline in 1920. His research effort was directed toward high-altitude sounding rockets for scientific research. Privately, as his notebooks and archives eventually revealed, he dreamed of flights to the moon and interplanetary exploration. A Method of Reaching Extreme Altitudes[2] set forth the theory of rockets and the results of his laboratory tests.

On March 16, 1926, before three witnesses (and assistants), Goddard launched the world's first liquid-propellant rocket at Auburn, Mass. The motor had a thrust of less than 4.53 kg (10 lb). It rose slowly to a height of 12.5 m (41 ft), curving over to land 56 m (184 ft) away. Not very far, but a longer flight than that of the Wright brothers at Kitty Hawk, N.C., 23 years before!

Between 1926 and 1941, Goddard conducted static tests of rocket motors. Three more rockets were launched at Auburn (1926, 1928, and 1929). The last one carried a barometer and a camera to record data. With the support of Charles A. Lindbergh,

Fig. 2 Dr. Robert Hutchings Goddard (1882-1945), "Father of American Rocketry," looking through telescope at control gages of control shack 1000 ft. from launching tower. (NASA photograph.)

funds were provided by the Daniel and Florence Guggenheim family, who were philanthropic aviation enthusiasts. Goddard moved to Roswell, N.M., and continued the development and launching of larger, ever more complex, sounding rockets. During those years, 31 more rockets were launched. His designs included gyro stabilization using aerodynamic and jet deflector vanes; turbopumps driven by a gas generator; gimbal-mounted, clustered rocket motors; automatic, sequencing launch systems; flight trajectory tracking and recording devices; and parachute recovery. These and other techniques were later developed and refined by others. But Goddard did them first.

Our third pioneer is Herman Julius Oberth (Fig. 3) (b. 1894). He was born in Sibiu, Transylvania, a town located in what is now Romania. In 1894, it was part of the Austro-Hungarian Empire. As a boy, Oberth was captivated by the thought of space flight. He was thrilled by Jules Verne's From the Earth to the Moon,[3] although he appreciated that a cannon was not the way to put men into space. During World War I, he proposed to the German War Department a long-range missile that would use liquid air and alcohol. The concept included gyro control. Ballistic range was controlled by continuous measurement of acceleration, which, he wrote, triggered "the closing of the propellant valves at the proper moment."

After the war, Oberth studied for his doctorate at Heidelberg. His first book, The Rocket Into Interplanetary Space,[4] was written as a challenge to other scientists to defend his concepts of rocket propulsion and its use in space flight. This was in 1923. In 1929, his classic, Means for Space Travel,[5] was published.

These three men shared certain fundamental characteristics: independent thought, great self-confidence, and enormous perseverance. All were ahead of their time. In each case,

Fig. 3 Prof. Hermann Julius Oberth (b. 1894), Romanian-born rocketry and astronautical pioneer. (Photo ca. 1928 by Ufa motion picture company, Berlin.)

acknowledgment of their efforts was modest and slow in coming from their peers at home. Language barriers restricted dissemination of their writings outside their own countries.

By the late 1920's, however, science fiction had sparked an interest--albeit of small conflagration--in the rocket and interplanetary societies. These small groups of enthusiasts met and discussed the elements of space flight. Some conducted experiments. They gave lectures to attract attention and new members. Their funds were minimal--or nonexistent. They were ardent disciples of Tsiolkovsky, Goddard, and Oberth, whose interest and patronage they sought.

Between 1927 and 1933, such groups were formed in Germany, the Soviet Union, the United States, Great Britain, and Austria. The activities of these groups varied greatly. The interesting point, though, is that many individuals who would later play key leadership roles in early space programs were active members of these societies (Figs. 4 and 5). With evangelistic zeal, they publicized the concept that space flight was possible in our lifetime.

The outbreak of World War II slowed or stopped such society activities, particularly in Germany. There, a young engineer, Wernher von Braun (1912-77), had to pigeonhole his dreams of space flight to become technical leader of a burgeoning missile development program. By 1945, Germany had been defeated, but

Fig. 4 Photo taken on Aug. 5, 1930, after a successful test of Prof. Hermann Oberth's Kegelduse rocket at the Chemisch-Technische Reichsanstalt, Berlin-Plentzensee. Left to right: Rudolf Nebel; Dr. Ritter (of the Chemisch-Technische Reichsanstalt); Mr. Barmuller; Kurt Heinisch; unidentified; Prof. Hermann Oberth; unidentified; Klaus Riedel (Riedel II) in white coat; Wernher von Braun; unidentified.

Fig. 5 American Rocket Society ARS No. 4, Staten Island, New York, Sept. 9, 1934. Preparing for launch of liquid oxygen/gasoline rocket. L. to R.: John Shesta (designer of rocket); Carl Ahrens; and ARS president, G. Edward Pendray.

the whole world was aware of rocket propulsion, if only for weapons. The best-known weapon was the V-2 (Fig. 6). Standing 14-m (46-ft) tall, with a 1.98-m (6 1/2-ft) diameter, the V-2 could carry a 1000-kg (2205-lb) warhead nearly 322 km (200 miles). Postwar examination of the V-2 rocket by American engineers brought detailed knowledge of large rocket motors. Von Braun and about 200 of his top technical team came to the United States and helped boost the United States' knowledge and understanding of this rapidly developing field. Within a year, they were assembling V-2's at the Army's White Sands Proving Ground in New Mexico. V-2's were used to teach launch techniques and as sounding rockets to conduct high-altitude research. By 1955, von Braun and his team had become American citizens and were developing the United States Army Redstone ballistic missiles at Huntsville, Ala.

In the Soviet Union, Sergei Pavlovich Korolev (1906-66) had been the leader of the Moscow Group for Study of Reactive Motion (MosGIRD) (Fig. 7) in 1931. With a small fund supplied by the government, experimental work was accomplished over the next few years, and some liquid-propellant rockets were launched successfully. During World War II, he designed rocket propulsion units as boosters for combat aircraft. Korolev is credited generally as the guiding force and team leader of Soviet postwar programs of sounding rockets, ballistic missiles, and as the developer of Soviet space launch vehicles.

Arthur C. Clarke (b. 1917) was born at Minehead, near the southeastern coast of England. As a young charter member of the British Interplanetary Society (BIS), he worked on studies for space ships before World War II. An anachronistic British law prevented active experimentation by the BIS. The law, prohibiting fireworks, stems from the attempt by Guy Fawkes to blow up the House of Parliament--in 1606!

An avid reader of science fiction, Clarke started writing science fact as well as fiction. By 1951, when his income from writing exceeded his government pay as an auditor, he made writing his full-time profession.

Fig. 6 V-2 (A-4) rocket without launch tower, White Sands, N.M., ca. 1946. From the George P. Sutton collection.

In 1944, when V-2's were raining on London, Clarke appreciated that the Germans had indeed developed a large reliable rocket engine--just the thing to make possible satellite flight! In October 1945, Wireless World[6] published his "Extraterrestrial Relays." In this brief paper, Clarke pointed out that a spacecraft in Earth orbit, above the Equator at an altitude of 35,800 km (22,300 miles) would circle the Earth in one day. Put another way, the angular velocity of the satellite was the same at any point on the Earth's surface. Moreover, three "space stations," spaced equally around this equatorial orbit, would be splendid for long-distance and transoceanic microwave television and telephone links. Transmissions beamed up from one spot on the Earth could be relayed to a receiver station thousands of miles distant. Electrical energy for operation could be obtained from the sun by means of a heat engine driving an electric generator. Clarke also pointed out that more direct use of the sun's energy might become possible through development of the solid-state solar cell.

Of course, the "extraterrestrial relays" Clarke wrote about were "fairly large structures with their own maintenance and operating crews." In 1945, no one envisioned the developments in electronics (namely, transistors) that would make unmanned communications satellites possible. This lack of foresight is not surprising when one considers that all the thinking and paper planning for space exploration was an extension of aviation. After

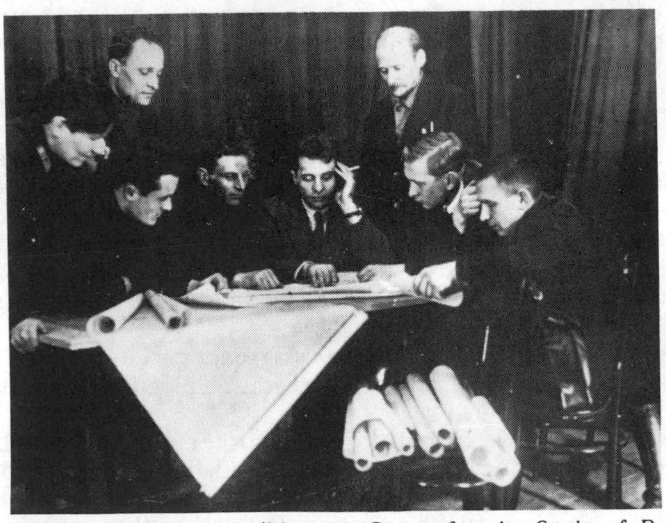

Fig. 7 Members of MosGIRD (Moscow Group for the Study of Reactive Motion), 1931. Shown are: (right, standing) Fridrikh Tsander and second from left (seated) Sergei Korolev. Others are, first on left, Sumarokva, Sumarokov (standing), Yuri Pobedonostev, B. I. Cheranovsky, Zaborin and A. Levits. From the K. E. Tsiolkovsy State Museum of Cosmonautics, Kaluga, USSR.

all, that was the latent thrill of the dream--man traveling in space, looking down at the Earth, traveling to the moon and the planets.

The specific orbit that Clarke defined was later to be known as the geostationary, or "Clarke Orbit." A few years ago, an attempt was made to locate an earlier recognition of this useful orbit. Only two were found. One was in the classic 1928 study of a space station for astronomical studies entitled Problem of Space Travel[7] by Hermann Noordung. Noordung was the pen name of an Austrian named Potocnik, about whom little is known. His study not only specified a geostationary orbit, but even his living quarters were in the shape of a torus. Rotation of this wheel-like spacecraft, he theorized, would produce artificial gravity.

This rotating wheel concept influenced many rocketeers, including von Braun and his team. When Collier's[8] magazine published an influential series of articles on space flight between 1952 and 1954, von Braun used this concept. The Arthur Clarke/Stanley Kubrick collaboration in the film "2001--A Space Odyssey" (1968) featured a more sophisticated variation of the rotating wheel to provide a measure of artificial gravity.

To find the only other known reference to the "stationary" orbit, we must go back to the original pioneer, Tsiolkovsky. He did not note the advantage of this orbit for communications or observation (reconnaissance) purposes, but he did point out that at that altitude above the Equator an object would be traveling at the same angular velocity as the Earth.

This observation was made in a paper published by that remarkable man in 1895! Tsiolkovsky was examining gravitational loads--high gravitational forces on Jupiter, weightlessness in free-fall, etc. He imagined a "high tower." As one climbed the tower (noted to be "impossible to construct"), gravity slowly lessened because of the Earth's rotational velocity, creating centripetal--outward (balancing)--force. He wrote that at "a distance of 34,000 versts (a Russian measurement of distance, equivalent to 35,800 km or 22,300 miles) (gravity) would be annihilated." Tsiolkovsky went on to suggest that an object released at this (geostationary) altitude would be weightless and circle the Earth in orbit.

This appreciation by Tsiolkovsky was uncovered during studies in the Soviet Union, Sri Lanka, and the United States on the feasibility of a "space elevator."

In September 1949, Eric Burgess, another charter member of the British Interplanetary Society, also published a paper[9] on the details of an active geosynchronous communications satellite. A few years later another science fiction fan and author who became a leader was John R. Pierce. "I am sure that science fiction helped to attract me ... to engineering," he writes in a memoir. Pierce received his doctorate at the California Institute of Technology in 1936 and joined Bell Telephone Laboratories in New Jersey. Interplanetary communication intrigued Pierce. He recognized that it would be easier to transmit a microwave signal to the moon than across the country. The latter required beaming a signal to a receiver every 48 km (30 miles) or so, amplifying it and transmitting it to another receiver. Atmospheric fading, he realized, would not exist in space. Pierce had not known of Clarke's Wireless World article but deduced and calculated how passive reflector satellites as well as active repeater satellites might be valuable some day.

Pierce correctly recognized the value of unattended simple satellites, as well as the significance of the geostationary orbit. He presented such a paper at a meeting of the Institute of Radio Engineers in October 1954.[10]

A comment on science fiction might be appropriate here. All three of the great rocket pioneers acknowledged the influence of the works of Jules Verne (1828-1905). Goddard also read and reread The First Men in the Moon,[11] by H. G. Wells (1866-1946), with whom he corresponded on occasion. The young disciples of the rocket pioneers read not only the imaginative works of Verne and Wells, but new science fiction as well. The best examples of this genre extrapolated contemporary technology and broke no known physical laws.

In the Soviet Union, Tsiolkovsky wrote several small books devoted to the problems of organization for interplanetary travel and its prospects. He argued that, after establishment of the first artificial Earth satellite, a logical progression would follow: flight to the planets, interplanetary stations, and cities and great communities in space adapted "to the needs of expanding humanity."

In Germany, Kurd Lasswitz's On Two Planets,[12] and books by Otto Willy Gail were read widely. Willy Ley, a founder of the German Rocket Society, wrote for prestigious Berlin newspapers. In 1934, Ley emigrated to the United States, where he became a leading popularizer of space flight, writing and lecturing until his death in 1969. Ley's Rockets, Missiles and Men in Space[13] has gone through 20 printings and updatings since 1944.

In the United States, science fiction proliferated in pulp magazines such as the Amazing Stories and Wonder Stories series of Hugo Gernsback's publishing empire. Authors of Science Wonder Stories were among the handful of founding members of the American Interplanetary Society in 1931.

By 1954, a number of events had occurred which focused attention on the increasing feasibility of building a satellite. Von Braun had met United States Naval Commander George W. Hoover, an inventive pilot and engineer with a "can do" philosophy. Hoover was in the Air Branch of the Office of Naval Research (ONR) in Washington, D.C. Within a few months of their meeting in 1954, Hoover was actively seeking top-level approval of a joint venture between the United States Army and the Office of Naval Research, known as Project Orbiter. This project envisaged the use of a specially modified Redstone launch vehicle to launch a satellite. Redstone had been used successfully in ballistic missile tests.

The American Rocket Society (the erstwhile American Interplanetary Society) had formed a Space Flight Committee in 1953. In 1954, a report was presented to the National Academy of Sciences. The Academy, and various other scientific organizations in the United States, were planning the American role in an International Geophysical Year (IGY). The IGY was to run from July 1957 through 1958. It was a cooperative effort, with 70 nations participating in global geophysical atmospheric and oceanic data gathering projects. President Eisenhower announced on July 29, 1956 that the United States would launch an Earth satellite as part of the United States program for the IGY. However, the Army/ONR project was not chosen for funding. Rather, a project proposed by the Naval Research Laboratory, called Project Vanguard, was selected. The Vanguard project required updated design rocket engines and a totally new solid-propellant rocket motor. One reason given for the decision by the various committees was that "military rockets would not be used." Suffice it to say that service rivalries, personal jealousies, industrial lobbying, and power politics were among the factors in the choice of the Vanguard proposal. Furthermore, the project was never accorded sufficient funds to insure a successful launch early in the 18-month IGY.

The Soviet Union announced that they, too, would include an Earth satellite in their IGY program. The United States officials seemed to assume that, despite the technical obstacles still to be overcome with Project Vanguard, their launch vehicle and satellite would be superior to the Soviet's effort--about which no details were known.

On Oct. 4, 1957, the USSR launched the 85-kg (185-lb) Sputnik I into orbit. On November 3rd, Sputnik II was launched, carrying an instrumented dog. With Sputnik II, it became clear that there was strong Soviet interest in manned space flight. It was at this point that von Braun received a go-ahead to proceed with the Army satellite launch vehicle, Jupiter-C.

On Dec. 6, 1957, Vanguard blew up on the first attempt to launch an American satellite. The Army launched the first United States satellite, the diminutive Explorer I, weighing only 8.3 kg (17.6 lbs), on Jan. 31, 1958. A Vanguard satellite did get off on March 17, 1958.

The Space Race was on. The stakes were international prestige and a lead in the Cold War. Nevertheless, many space enthusiasts still hoped and dreamed of peaceful uses of outer space and of a united mankind that might travel to the moon, the planets, and, ultimately, the stars.

The history of launch vehicles and space exploration, however, tells us only a part of the story and only of some of the early pioneers. Parallel to the development of rocketry was that of electronics and radio communications. Again, the story is that of a few great pioneers: Alexander Graham Bell (1847-1922); Thomas Alva Edison (1847-1931); Heinrich Hertz (1857-94); James Clerk Maxwell (1831-79); Samuel F. Breese Morse (1791-1872); and last but certainly not least, Guglielmo Marconi (1874-1937), who led the way to modern electronic communications.

The development of the telegraph in 1832 and the telephone in 1876 was, in both cases, related to the transmission of electrical impulses through wire--something understood to be possible since Benjamin Franklin's experiments with a kite and key during an electrical storm. The telegraph and telephone were, in many ways, a logical progression of ideas. The key breakthrough that led to the idea of an active communications satellite, a concept put forth by Arthur Clarke, was, of course, radiocommunications. This was an idea that ultimately required a leap of faith. Transoceanic radiocommunications, in fact, was developed without knowing the true scientific reason why it worked.

There are many people alive today who can remember when communication by radio was little more than a scientific novelty, still in the experimental stage--when ships moved across the oceans in lonely silence, having no means of making contact with the shore once they passed beyond the range of visual signaling.

That was the situation, as far as most people could have known it, when the 20th century began. Yet, the new science had actually emerged from the confines of the laboratory and made its debut in the commercial world in the closing years of the 19th century.

The interest of the entire international scientific fraternity had been stirred by a demonstration performed by German physicist Heinrich Hertz at the Karlsruhe Polytechnic Institute, described in a paper published in 1888,[14] in which electromagnetic waves were generated artificially and intercepted by a receiver. Professor Hertz's apparatus was copied and the demonstration repeated in lecture rooms all around the world. Students as far away as Sydney University, for example, saw the method demonstrated in Australia in the same year the paper was published. However, progress beyond this point had to await the guiding hand of another inspired individual. Five years before the turn of the century, most experimenters still found themselves unable to extend the transmission and reception range of the "Hertzian waves" farther than the length of a laboratory bench.

Now, out of the shadows of obscurity, stepped a young Italian amateur scientist, one of those rare beings who, through outstanding qualities of mind and character, is given the power to alter the course of human history. His name was Guglielmo Marconi.

In 1895, the year of his 21st birthday, Marconi still lived with his parents in Pontecchio, near Bologna. He had already been conducting experiments for some years, alone, in a spare room of the family home. Starting from the Hertzian model and utilizing refinements developed by other leading researchers, he had succeeded, through a painstaking program of experimentation and modification, in progressively improving the range over which signals could be sent. He had also incorporated Morse telegraphy instruments into his apparatus so that he was actually transmitting and receiving signals in Morse code rather than merely creating waves of energy. In this, Marconi revealed his practicality of outlook. He was, in fact, fired with an ambition to create a system of communication that would travel where wires could not reach.

The electric telegraph had then been in existence for more than 50 years and formed a worldwide system, with submarine cables providing links between continents. The people of Queen Victoria's England and their civilized contemporaries probably considered they had reached the point of ultimate sophistication in rapid communications. The telephone, by this time 20 years old, seemed unlikely to usurp the telegraph from its preeminence in long-distance communications. Not until the second half of the 20th century would submarine cables be developed that were capable of carrying speech circuits effectively across the oceans. Meanwhile, the young Marconi adopted the familiar, well-tried technique of telegraphy to aid him in his work.

At the villa in Pontecchio, his experiments progressed from the spare room to the garden. And now Marconi made a great technical advance. He introduced aerials and Earth connections to both transmitter and receiver. This inspired innovation enabled him to extend the signal range dramatically. Using metal cylinders attached to the tops of poles, he found that the range was directly related to the dimensions of the cylinders and their height above the ground. Soon his apparatus was able to send and receive Morse signals over a distance of 2400 m (about 1.5 miles). The time had come, he decided, to start putting his findings to practical commercial use.

The world was not disposed to accept his claims too readily. He offered the Italian Government a demonstration. The Government declined. In 1896, he moved to London, then the hub of a vast empire and the capital and business center of the world's leading industrial and commercial nation. He was not entirely a stranger, for he had attended school in Bedford as a small child and he had relatives in England, his mother being a member of a well-known Irish family. In London, on June 2, 1896, Marconi applied for the world's first patent for wireless telegraphy, which was duly granted. The following year saw the formation of his first company, The Wireless Telegraph and Signal Company Ltd., with his cousin, Jameson Davis, as managing director.

Despite the difficulties of gaining financial support for a venture so utterly speculative, Marconi gathered a team of highly able scientists and engineers around him and continued to work tirelessly on research and development. In 1899, the first wireless messages across the English Channel were passed between the company's demonstration stations.

In 1900, Guglielmo Marconi decided to commit his company's resources to an experiment which, if successful, would be the most dramatic demonstration of wireless telegraphy yet seen. He proposed the construction of high-powered stations on either side of the Atlantic Ocean, so that he might attempt to transmit signals between Britain and North America.

The move represented an act of sheer faith on the part of Marconi and his board of directors. Not only was there no experience to show that the plan would work, but the prevailing scientific theory about the behavior of electromagnetic waves insisted that it could not. The waves were known to travel in straight lines. As no one yet knew of the existence of the ionosphere, or of the ionospheric reflection effect, the general assumption was that the transmitted waves, being unable to follow the curvature of the Earth's surface, would leave the Earth tangentially and shoot off into space.

Another argument insisted that the "thunder" of the high-power stations would disturb or drown out the traffic between ships and the existing shore stations. Marconi was confident he had already solved this problem. As to the "curvature" theory, he had nothing to cling to but his own conviction that the theorists would be proved wrong.

The company built two entirely new stations for the experiment: at Poldhu, Cornwall, and at South Wellfleet, Cape Cod, Mass. The antenna system at each station consisted of about 400 wires suspended in a cone formation from twenty 60.7 m-(200-ft) masts. After 11 months of feverish planning and hard work, in September 1901 all 20 masts at Poldhu collapsed in a gale. A month later, the same fate befell the masts Cape Cod. Lesser mortals would have abandoned or postponed the project in the face of such misfortune. Within days of the first disaster, however, a temporary replacement antenna had been erected at Poldhu. Marconi, impatient to get results (and to convince his board that the company's expenditure of £50,000 had not been in vain), took ship for Newfoundland with two assistants on November 27th, having decided to make his initial trans-Atlantic tests in an east-west direction only, using a temporary kite-borne receiving aerial.

So it came about that, on the 12th of December 1901, Marconi and his assistant, George Kemp, in a room of a deserted military hospital at St. John's, Newfoundland, each in turn pressed a telephone earpiece to his ear and heard, faintly but repeatedly, the three dots of the Morse "S" being transmitted across 3218 km (2000 miles) of ocean.

Marconi had now satisfied himself that trans-Atlantic wireless telegraphy was feasible. The rest of the world, however, remained skeptical. In a bold effort to provide unassailable proof, he staged a more public demonstration two months later. During the voyage of the liner Philadelphia from Southampton to New York, with

Marconi onboard as a passenger, messages were received from Poldhu and recorded on Morse inker tape as the ship steamed westward. Daylight reception was excellent up to a distance of 1126 km (700 miles), then the signals became too faint to operate the tape machine. However, after dark, the signal strengthened again, and good reception continued, under these conditions, up to 2494 km (1550 miles). In darkness, single test letters were detected as far away as 3377 km (2099 miles).

With financial assistance from the Canadian Government, the Marconi Company now built a new high-power station at Glace Bay, Nova Scotia, and the first test wireless signals across the Atlantic from west to east were transmitted by this station in November 1902 and received at Poldhu. In December, messages of greeting were sent to the people of England and Italy by a correspondent of The London Times; to H.M. King Edward VII by the Governor-General of Canada; and to the English and Italian monarchs by Marconi himself. In January, the President of the United States, Theodore Roosevelt, addressed a special message of greeting to the King and the people of the British Empire, which the newly completed Cape Cod station transmitted. Intended for retransmission by the Glace Bay station, this message was in fact picked up by Poldhu, so becoming the first wireless communication received across the Atlantic from a station in the United States of America.

Much work remained to be done before the company found itself technically ready to offer a reasonably reliable commercial service between England and North America. Reception proved to be erratic in quality over such long distances, and the reasons for this were not yet understood. The British and American physicists, Heaviside and Kennelly, each put forward the hypothesis, in 1902, of a layer of ionized atmosphere at a certain height above the Earth's surface that reflected wireless waves back to Earth; but these ideas were not fully investigated until the 1920's. Meanwhile, the pioneers proceeded on a trial-and-error basis. No one had any reason to suppose that the use of short wavelengths might offer any advantages and, in fact, experience so far had led to the adoption of increasingly longer wavelengths, coupled with higher power input, for long-range working. Wireless, it seemed, was an unpredictable medium, tantalizingly effective at some times and bafflingly useless at others.

Experimentation and development continued for another five years before a commercial trans-Atlantic service opened. This was inaugurated at last on Oct. 15, 1907, operating through two new stations incorporating all the latest devices known to the Marconi engineers, including large directional antenna systems. The essential electronic portion of the communications satellites had now been born.

It was thus that radio communications and, ultimately, microwave communications came to be!

The story of the development of satellite communications is, in truth, the story of a handful of pioneers who had the vision to foresee a totally different world. The pioneers of space and the pioneers of electronics were separated in time, schooling, culture, and nationality; but they all shared the vision.

Acknowledgments

Photographs in this chapter courtesy of Astro Associates, Chevy Chase, Md.

The author is greatly indebted to Dr. Joseph Pelton for his contribution to the latter portion of this chapter dealing with the history of radio communications.

References

[1]Tsiolkovsky, K. E., "A Rocket into Cosmic Space," Moscow, 1903. English translation of the article in Works on Rockets Technology, NASA, Washington, D. C. 1965. NASA Technical Translation TT F-243, Nov. 1965.

[2]Goddard, R. H., A Method of Reaching Extreme Altitudes, Smithsonian Miscellaneous Collections, Vol. 71, No. 2, Smithsonian Institution, Washington, D.C., 1919.

[3]Verne, J., De la terre à la lune (From the Earth to the Moon), J. Hetzel, Paris, 1865.

[4]Oberth, H., Die Rakete zu den Planetenräumen (The Rocket into Interplanetary Space), R. Oldenbourg, Munich, 1923.

[5]Oberth, H., Wege zur Raumschiffahrt (Means for Space Travel), R. Oldenbourg, Munich, 1929.

[6]Clarke, A. C., "Extra-Terrestrial Relays," Wireless World, London, Oct. 1945.

[7]Noordung, H., Das Problem der Befahrung des Weltraums: der Raketenmotor (Problem of Space Flight), R. C. Schmidt & Co., Berlin, 1929.

[8]Collier's magazine, New York, 1952-1954.

[9]Burgess, E., "The Establishment and Use of Artificial Satellites," Aeronautics, London, Sept. 1949.

[10]Pierce, J., "Orbital Radio Relays,": address delivered at the Institute of Radio Engineers, Princeton, N. J., Oct. 14, 1954, published in Jet Propulsion, Vol. 25, Apr. 1955.

[11]Wells, H. G., The First Men in the Moon, George Newnes, London, 1901.

[12]Lasswitz, K., Auf Zwei Planeten (On Two Planets), 2 vol., Verlag B. Bleischer Nachfolger, Leipzig, 1897.

[13]Ley, W., Rockets, Missiles and Men in Space, Viking Press, New York, 1968.

[14]Hertz, H., Karlsruhe Polytechnic Institute, paper published in 1888.

4. The Experimental Years

Burton I. Edelson*
NASA Headquarters, Washington, D.C.

"It may seem premature, if not ludicrous to talk about the commercial possibilities of satellites," wrote Arthur Clarke in 1957. "Yet the aeroplane became of commercial importance within 30 years of its birth, and there are good reasons for thinking that this time scale may be shortened in the case of the satellite, because of its immense value in the field of communications."[1]

Good reasons indeed! Only a few weeks later, the Soviet Union launched Sputnik I, and the United States soon followed with Explorer I and a number of other scientific and applications spacecraft--many involving communications experiments. Technical and economic feasibility was soon demonstrated. As a result, not 30, but only eight years elapsed before the first commercial satellite, Early Bird, entered service; and in just 12 years, commercial satellite service extended around the globe and became profitable.

How was it that the communications satellite became of commercial value in such a short time? Three ingredients were necessary for this to happen, and, fortunately, all three were (or shortly became) available: technology to create the system, communications requirements to form a market, and a management structure to implement the system. This chapter treats the technology; following chapters will show how the market formed and how INTELSAT became the implementing organization.

Many different technologies are needed to create a communications satellite system. These flowed to INTELSAT from diverse sources. Launch vehicle and spacecraft technologies came from work supported over many years by the U.S. Department of Defense (DOD) and, for a shorter but more intense time, by the National Aeronautics and Space Administration (NASA). The communications and electronics technologies, on the other hand, came mostly from civil and commercial sources, as did the systems design and engineering. The first INTELSAT satellites were

---

Invited paper received October 11, 1983.

*Associate Administrator for Space Science and Applications.

launched on NASA DELTA launch vehicles, which were derived from earlier military rockets. The spacecraft technologies for the first INTELSAT satellites came from NASA experimental spacecraft, and the communications technologies were inherited largely from commercial microwave systems. The first Earth stations used on both sides of the Atlantic in the INTELSAT system can trace their roots to radiotelescopes, radars, and microwave transmission systems. The technologies of INTELSAT are truly eclectic, and their origins and development paths interesting indeed.

Forming the Concept

As mentioned in Chap. 3, although there is a long history of space technology, it is to Arthur C. Clarke that we are principally indebted for the concept of using Earth orbiting satellites for telecommunications. In 1945, he recognized the unique nature and usefulness of the geostationary orbit[2] and suggested "space stations" in that orbit as a means "to link TV services to many parts of the globe." He foresaw (incorrectly) a rather limited future for terrestrial radio links and cables as means for distributing communications services. "A relay chain several thousand miles long would cost millions," he wrote, "and transoceanic services would still be impossible."

A few years later in the United States, John R. Pierce of Bell Laboratories, writing without knowledge of Clarke's words, independently suggested several promising system configurations employing passive as well as active satellites in low-altitude as well as geostationary orbits.[3] He raised a point that has since proved the very basis for the ascendance of the satellite over the cable for transoceanic telephony. Looking at the first 36-channel submarine cable just then (1955) being laid under the Atlantic by the British Post Office and American Telephone & Telegraph Company, and costing some $35 million, Pierce asked, "Would a channel 30 times as wide, which would accommodate 1080 phone conversations or one TV signal, be worth 30x35 million dollars?" He recognized this as an absurd thought and predicted (correctly) that a technical solution, could be found before the commercial demand reached that point. Satellites, of course, have proved the solution and we can now provide a thousand channels and more--not at a cost equivalent to many cables, but for only a fraction of the cost of one cable.

With the concepts of Clarke and Pierce in place, the launch capability demonstrated by several Sputniks and Explorers, and electronics technologies becoming available, it was possible in the early 1960's to envision an operating satellite communications system. Still, the means and organization for developing and operating a system were not apparent.

Then, in 1961, President Kennedy took two important steps to accelerate the pace of the United States space program. One, quite famous now, was his announced goal to send an astronaut to the moon in that decade; the second, not so well known, was his "Policy Statement on Communications Satellites."[4] His statement recognized the potential value of satellites to provide communications services, established government policy to coordinate activities and carry out research and development, called for implementation by the private sector, and, most importantly, invoked an international effort with these words: "I invite all nations to participate in a communications satellite system, in the interest of world peace and closer brotherhood among peoples of the world."

President Kennedy's bold and prescient statement led to passage by the United States Congress of the Communications Satellite Act of 1962,[5] which, in turn, set in motion a series of events that resulted in the formation of the Communications Satellite Corporation (COMSAT) in 1963, INTELSAT in 1964, and the initiation of commercial satellite service in 1965, and the establishment of full global coverage in 1969.

It is interesting to realize now that the 1961 statement was based largely on prediction and promise--very little on the results of development or on demonstrated technology. At the time, no active satellite had flown to test voice or video transmission; no spacecraft had ever been put in geostationary orbit; no data on reliable operation of electronic devices or rotating mechanisms in space were available; and some considerable doubt existed as to the acceptability of long-delay transmission paths for commercial service. Still, engineers and managers pushed onward; and as each technical, economic, operational, political, and organizational problem unfolded, they seemed somehow to find a solution.

As the stage was being set for the initiation of commercial satellite communications, a number of particularly knotty technical and operational questions arose. Those engineers and planners in DOD and NASA involved in developing technology and launching experimental satellites, and those members of COMSAT's technical staff responsible for designing and specifying the initial system, searched diligently for the solutions to these key crucial problems:

Should the satellites be active or passive? Previous experimental work, using both the Earth's natural satellite, the moon, and the metallized plastic balloon, ECHO, had shown that signals could be bounced successfully off a passive reflector in orbit and received on Earth. But a very high level of transmitted power would be required and the aperture of the receiving station would have to be quite large. An active electronic receiver, amplifier, and transmitter on the satellite would provide much system gain. But could this electronic instrumentation withstand the trauma of a rocket launch and survive in orbit for the months or years necessary to make the system economically viable?

What orbit should the satellites be in? With relatively low-powered rockets, rudimentary guidance systems, and limited orbital control capabilities, the geostationary orbit suggested by Clarke, though desirable, looked very difficult to attain in the early 1960's. Test flights had been at altitudes of a few hundred kilometers (125 to 185 miles). Then, too, the question was raised (largely by experienced system engineers at AT&T) as to whether the long delay time involved in transmission through a satellite in geostationary orbit (over 1/2 s for a two-way trip) would be tolerated in commercial telephone service. There was concern both for the inherent delay and for the ability to suppress "echo" on the circuit. However, since a geostationary satellite system could provide global coverage with three satellites, whereas a medium altitude [say, 3000 to 5000 km (1860 to 3100 miles)] would require 20 or more satellites, and since the former would require just one Earth antenna at each location, while the latter would need two steerable antennas, there were powerful technical and economic reasons for going to a geostationary satellite system if it proved technically possible.

What frequency band should be used? Various experimental military and civil satellites had used frequencies in the vhf and uhf bands for tracking, telemetry, and command functions; but, for telecommunications, it would be more efficient to use higher frequencies in the "microwave" region (above 1 GHz), where a great deal more bandwidth would be available and, therefore, a higher information transmission capability attained. Microwave transmitting and receiving equipment was then available from radar systems for ground use, but not for space. Could it be made available? Particularly, could light, efficient, and reliable power output tubes be developed for spacecraft use? If so, which frequencies in the 1 to 10 GHz band should be used for transmission up? and down? Because of the power output tube challenge, it seemed desirable to use the lower frequency on the down path.

What services should be provided? The obvious advantages of satellite communications over other forms lie in their wide area coverage, broad bandwidth, and direct access to small terminals. These advantages would seem to make satellite communications particularly useful for television broadcast and mobile services. Indeed, television broadcast was originally suggested by Clarke. However, the commercial infrastructure already existed for transoceanic point-to-point telephone service, and a market existed in the early 1960's, at least in the Atlantic Region.

Satellite Experiments

The years from 1958 to 1964 were the true "experimental years" for satellite communications. During this crucial period, technology developed rapidly. At the beginning of the period, little technology existed; no service had been tested in orbit; and not one of the questions raised above could be answered. But by the end, enough technology was available to provide confident answers to those questions and to create the INTELSAT System.

Passive Satellites

Even before Sputnik, design and experimental work had been conducted on passive satellite communications. The United States Navy began testing moon relay communications in 1954 and started an operational service between Washington, D.C., and Hawaii in 1959.

A metallized balloon satellite was first suggested for tracking purposes at the London Congress of the International Astronautics Federation (IAF) in 1951. In 1956, William J. O'Sullivan, at the Langley Research Laboratory of the National Advisory Committee for Aeronautics (NACA--later to become NASA), proposed to develop a 3.66-m (12-ft) diam version of such a balloon. His concept came to the attention of John Pierce of Bell Laboratories and William H. Pickering, Director of the Jet Propulsion Laboratory of CalTech, who suggested it as a passive communications satellite. First, the DOD turned down the opportunity to sponsor this project, preferring to concentrate on active satellites, which seemed more likely to meet their requirements. But then, NASA, in one of its earliest decisions, decided to support the concept, and this resulted in the ECHO program.

Leonard Jaffe, an engineer who had just come from the Lewis Research Center in Cleveland to be Chief of Communications Satellite Programs at NASA Headquarters, Washington, D.C., saw the potential value in both passive and active communications satellites. He was responsible for directing the ECHO program and all the active satellite development work which NASA was later to pursue and which developed the many technologies that eventually became the foundation for INTELSAT and many other systems. Jaffe's book, Communications in Space,[6] published in 1966, is the best source of detailed information on the early development of communications satellites.

Passive satellites were touted as "simple, reliable, and long-lived." The ECHO spacecraft were the simplest type of isotropic reflectors. There was no expectation that they would ever develop into operational systems, but their purpose was to test propagation through the atmosphere and ionosphere and to develop transmission techniques. Not so incidentally, they turned out to be valuable in the development of Earth station equipment and technology.

ECHO I was a 30-m (98-ft) diam metal and plastic balloon, launched in August 1960 into an inclined 1600-km (992-mile) orbit from Cape Canaveral on a Thor-DELTA vehicle. ECHO I provided the first live, two-way voice communications via satellite. Many other optical and radio tests were made using stations around the world. Within a week of its launching, the first transoceanic satellite signal transmitted from Bell Laboratories in New Jersey was received in Paris by an Earth station of the Centre National d'Etudes de Telecommunications (CNET).

ECHO II, (Fig. 1) somewhat larger and more rigid, was launched in January 1964, by a Thor-Agena. Tests with ECHO II included transmission between a station at the University of Manchester, England, and the State University at Gorky, USSR. The ECHO satellites proved the feasibility of radio transmission via satellite, measured propagation characteristics, and demonstrated the effectiveness of various items of transmitting, receiving, coding, and modulation equipment. Perhaps because they were so easily visible to observers on Earth, the ECHO satellites excited the general public and stimulated enormous interest not only in satellite communications, but in all the practical applications that space systems had to offer.

Whereas the ECHO balloons were discrete reflectors, Project West Ford, sponsored by the United States Air Force, used an orbital "belt" of small wire filaments as dispersed reflectors. This project was carried out by the Lincoln Laboratories under the direction of Thomas F. Rogers and Walter E. Morrow Jr. The advantages of a dispersed reflector system over a discrete system were continuous availability and large reflection area, whereas the disadvantages included frequency limitation and interface with

Fig. 1 The ECHO II passive reflector satellite.

other systems. Successful tests of West Ford were conducted in 1963. Although no attempt was made to implement an operational system, the project contributed in important ways to developing ground terminal equipment and transmission techniques. Also, it brought the efforts and talents of people like Rogers and Morrow and others at Lincoln Labs working on military research to bear on the field of satellite communications, with many future beneficial results to civil systems.

Active Systems

The first active communications satellite was SCORE, a 60-kg (132-lb) payload built in just a few months by the DOD and carried into orbit on the side of an Atlas rocket in December 1958. As a "delayed repeater," it received a message from Earth, stored it on tape, and later, in another part of its orbit, transmitted the message to ground. SCORE transmitted President Eisenhower's Christmas message to the world and became known as the first "voice from space." Its 8-W vhf transmitter lasted less than two weeks on battery power and died on New Year's Eve.

The second DOD effort was COURIER. Like SCORE, a delayed repeater, but this time a separate spacecraft intended to develop an operational capability. COURIER was spherical, 130 cm (4.27 ft) in diameter, with a mass of 230 kg (508 lb). It was a highly complex, solar cell powered satellite with four receivers, four transmitters, and five tape recorders, processing and repeating several kilobits per second of digital data. It was launched on a Thor-Able-STAR vehicle into a 1000-km (620-mile) orbit in October 1960. SCORE and COURIER, together, proved that delicate and complex electronic equipment could be made to survive the trauma of rocket launch and function in orbit. COURIER demonstrated all essential subsystems of an active satellite--communications, power, telemetry, and command. It also demonstrated the unfortunate and inevitable tradeoff between complexity and reliability. COURIER operated perfectly for 18 days, then failed, due to a command system fault.

Encouraged by the success of COURIER and SCORE, the DOD, in late 1960, embarked upon a most ambitious project--Advent--aimed at placing a 1000-kg (2210-lb) body-stabilized, high-powered microwave satellite in geostationary orbit. Two years were scheduled for this task. The Air Force assumed responsibility for building the spacecraft; the Army and the Navy for the ground and shipboard terminals. Advent was terminated in 1962, after the expenditure of $170 million. The reason was simple: Requirements were set well beyond technological capability to meet them--a lesson to be learned over and over again in aerospace systems development.

Telstar was the most famous experimental communications satellite of all. Its technical contributions were so significant and its impact on the public so great that its name for a while became generic for "communications satellite." Telstar was developed by Bell Laboratories of AT&T, under the guidance of John Pierce, and, with AT&T corporate funding, it became the first significant undertaking in space by private enterprise.

Telstar was an 88-cm (2.89-ft) diam faceted sphere, weighing 80 kg (176 lb), covered by solar cells, with an antenna belt around its waist. Significantly, Telstar received at the 6-GHz and transmitted at 4-GHz bands that later were assigned to commercial service and used exclusively by INTELSAT and other "fixed service" systems during the 1960's and 70's. Telstar was the first satellite to use a traveling wave tube (TWT) power output amplifier. Its output was 3 W. A complete description of Telstar can be found in the Bell System Technical Journal.[7]

Telstar I (Fig. 2) was launched on July 10, 1962, on a Thor-DELTA vehicle, into an elliptical inclined orbit with a perigee of 950 km (589 mi) and an apogee of 5600 km (3472 miles). Its period was 158 min. For months, many types of communications tests were conducted between Earth stations especially built or modified for the project at Andover, Me. (United States), Goonhilly Downs (England), and Pleumeur-Bodou (France). These Earth stations were coordinated by the NASA ground station committee formed and chaired by Len Jaffe.

The day Telstar I was launched, live television was transmitted from the United States and received in England and France, and transmitted from France to the United States. This spectacular demonstration was followed by many other communications tests over its four months of operation. Telstar I demonstrated the specific communications techniques and equipment that would be used in commercial systems. It also showed that a medium-altitude satellite system could be used effectively, with antennas tracking the many satellites and communicating simultaneously.

Telstar II was launched in May 1963 into a considerably higher orbit [10,800 km (6696 miles)], giving it a period of 225 min, with relatively long periods of visibility across the Atlantic. Telstar II served long and well enough to prove to all that a medium-altitude commercial system was feasible and, to many, that such a system was preferable.

NASA's medium-altitude communications satellite, named RELAY, followed Telstar by several months. Although similar in concept, RELAY, built by the Radio Corporation of America (RCA), differed from Telstar in several ways. RELAY had two transponders and used 10-W TWTs. RELAY I was launched in December 1962, and RELAY II in January 1964, both into inclined orbits with apogees of about 7400 km (4588 miles). Both performed very well over several years, demonstrating live television transmission around the world, including "firsts" to Germany, Italy, Brazil, Japan, and the Soviet Union.[8] Several Earth stations, later to become part of the INTELSAT system, including those at

Fig. 2 The Telstar satellite.

Fucino, Italy, and Raisting, West Germany, first operated with RELAY.

Far and away the most important experimental communications satellite was SYNCOM (Fig. 3)--the first to be placed in geostationary orbit. Its design became the model for INTELSAT's first four generations of satellites and for satellites in many domestic systems. The concept of a spin-stabilized spacecraft for use in geostationary orbit was created by Harold A.

Rosen and Donald Williams, of the Hughes Aircraft Company, in 1959–60. After the DOD turned the mission down, NASA awarded Hughes a contract, in August 1961, to design and build a satellite following their concept. SYNCOM was planned to fly on a Thor–DELTA rocket which could put about 60 kg (132 lb) into an elliptical transfer orbit with an apogee of 36,000 km (22,320 miles). The spacecraft was to be designed with its own apogee kick engine capable of circularizing the orbit and placing about half the weight, or 35 kg (77 lb), into final orbit. The SYNCOM satellite was packed full of spin-stabilization and control equipment. Its communication package had rather limited capacity in two transponders, each with 2-W TWT output stages. Because the SYNCOM system inherited ground and ship terminals from the defunct Advent program, it had to use their frequencies in the 7/2 GHz bands.[9]

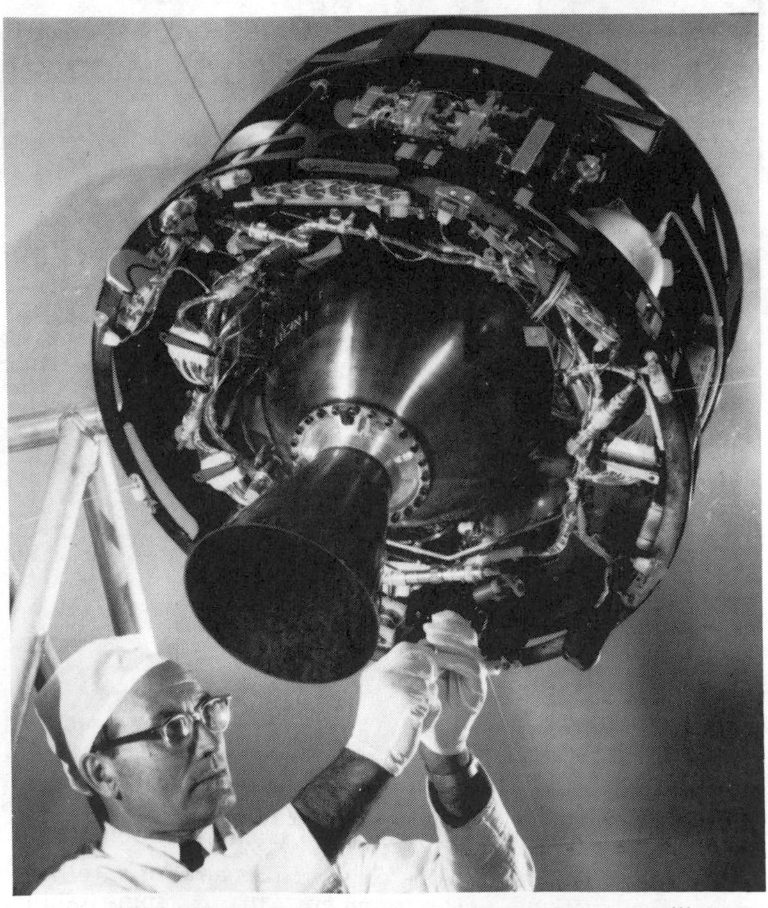

Fig. 3 SYNCOM, the world's first geostationary satellite.

SYNCOM I, launched in February 1963, failed to survive the apogee motor firing; but SYNCOM II, in July 1963, and SYNCOM III, in August 1964, succeeded admirably and demonstrated the great utility of the geostationary orbit for almost all satellite communications services.

Actually, due to limited propulsion capability, the orbit of SYNCOM II was not in the equatorial plane, but was inclined 33 deg so that it traced a figure-eight pattern in the sky. Still, it clearly demonstrated the great advantage for ground stations of minimal tracking and continuous coverage. Communications links, once established, could be maintained for days or months without interruption. Its obvious use for full-time telephone and live television was quickly accepted. SYNCOM III, using a thrust-augmented version of Thor-DELTA, a true geostationary satellite, clinched the argument for most system engineers.

Tests and demonstrations with the SYNCOM satellites went a long way toward gaining acceptance of the use of geostationary orbit satellites for telephone service. NASA, Stanford Research, AT&T, and the Federal Communications Commission (FCC) all participated in evaluating the tolerance of users for the 1/2-s round-trip delay along the 150,000-km (93,000-mile) path from ground station to satellite to ground station, and back. Electronic echo suppressors were developed to limit the amount of original signal on the return path. The result of these tests was that, while critical engineers could not agree, the general public found the circuits quite acceptable.

SYNCOM set the stage for INTELSAT. It proved out the rocket, spacecraft, and communications technologies needed for geostationary satellites, and it introduced a superb design and manufacturing team from Hughes Aircraft Company to the communications satellite business. Harold Rosen, a co-author of the concept, went on to become the world's leading designer of communications satellites for over two decades and was a principal technical contributor to the INTELSAT system.

It should be noted that the network of ground stations developed for use with the NASA experimental satellites--Andover, Me. (Fig. 4) (United States), Goonhilly Downs (England), Pleumeur-Bodou (France), and Raisting (West Germany)--provided the early base for commercial service in the INTELSAT system.

Development Satellites

The INTELSAT system, as we shall learn in later chapters, benefited greatly from the technologies developed during the experimental years between SCORE and SYNCOM (1958-64). However, it was still not clear in the years 1963 and 1964--when COMSAT was forming and making initial technical decisions--and even into the late 1960's, what the ultimate INTELSAT system configuration should be.

Some questions were settled early. Active satellites were deemed more effective than passive ones for commercial service. The 6/4 GHz band, tested by Telstar and allocated by the International Telecommunication Union (ITU), would be employed, but the orbit question was still open. Although SYNCOM had proved that the technology would work in geostationary orbit, the question of whether the geostationary orbit would be acceptable for commercial telephone service was not settled. Also, concern was expressed as to whether the electronics, stabilization, orientation, and electrical power systems would be suitable for long life in orbit.

INTELSAT considered both medium and geostationary orbit satellites for its initial system. The COMSAT technical staff, working then as "manager" for INTELSAT, observed carefully and critically as NASA and the DOD continued their experimental efforts to test and demonstrate new technologies; to improve spacecraft performance and reliability; and to develop more efficient communications techniques. In doing this, COMSAT was fortunate to have on its early technical staff a number of engineers who had previous first-hand experience with satellite experiments. To mention only a very few: Siegfried H. Reiger, a brilliant systems analyst from the Rand Corporation, headed their technical team; Sidney Metzger, from RCA, who had participated in projects

Fig. 4 Original Earth station at Andover, Me. (United States).

SCORE and RELAY, became COMSAT's chief engineer and made major contributions to Earth station technology; and Martin J. Votaw, who had built experimental satellites at the NRL, became project manager for satellite development. Wilbur L. Pritchard, of Aerospace Corporation, who had extensive experience developing military systems, was appointed the first director of COMSAT Laboratories and led it to become a principal research center for satellite communications technology. This team created the INTELSAT system, first using inherited technologies from experimental projects and, later, technologies developed at COMSAT Labs as part of INTELSAT's own R&D program.

During its first decade of operational service, INTELSAT continued to benefit greatly from the resultant flow of spacecraft and communications technologies from experimental satellites. Many of these came from the NASA Applications Technology Satellite (ATS) program. Six satellites in this series, each designed to test different new technologies, were built and launched by NASA during the ATS program in the years 1967-74.[10] The technologies included electronically and mechanically despun antennas (ATS-1 and ATS-3, respectively), hydrazine propulsion for stationkeeping (ATS-3), and the use of large antennas (ATS-6). All six demonstrated the use of wide-band multiple-access microwave communications; ATS-5 introduced K-band, and ATS-6, L-band communications. It is interesting that ATS-2 and ATS-4, both intended to be body-stabilized spacecraft, failed to attain proper orbit, and thus were prevented from demonstrating the advantage of this technique. ATS-1 and ATS-3, both spin-stabilized spacecraft, performed very successfully and are still operating. No doubt this comparison of stabilization systems influenced the future course of satellite design. However, ATS-6, a body-stabilized spacecraft, did achieve orbit in 1974, and with its multifrequency 9.1-m (30-ft) diameter antenna, very successfully demonstrated the use of satellite broadcasting, networking, and data collection with small antennas. The NASA ATS satellites, all of which were built by Hughes Aircraft Company, were extremely useful in developing and stockpiling technologies that were later to be used not only by INTELSAT, but by domestic, maritime, broadcast, and military systems as well.

The SYMPHONIE satellite, sponsored by France and Germany, also contributed to communications technology development and demonstrations during this period. SYMPHONIE I, a body-stabilized spacecraft launched in 1974, was very helpful in demonstrating the advantages of satellite communications for regional and domestic services and helped greatly to spread knowledge and capability in this burgeoning field to Europe and around the world.

The Lincoln Laboratories of MIT were very active in developing technology specifically for the military services. Their work on the Lincoln Experimental Satellites (LES) series was done at X-band and at vhf, but involved advances important to commercial communications, such as transmission efficiency, spacecraft power, high gain antennas, stabilization and stationkeeping, and the use of small ground terminals.

The LES developments led to the Tacsat program, which was designed to provide both experimental tests and operational communications services for the United States military forces. An important technology advance in Tacsat was the "gyrostat" stabilization concept, which allowed a cylindrical spacecraft to be spun about its long central axis. This concept found immediate application in the INTELSAT IV spacecraft, which served successfully throughout the 1970's and into the 1980's.

One other development satellite program made important technological contributions during this period. This was the Canadian Communications Technology Satellite (CTS). Launched in 1976, it used an advanced body-stabilized spacecraft design with lightweight, foldout solar arrays. It carried a high-power 200-W 12-GHz traveling wave tube power amplifier and demonstrated broadcast and "thin route" communications services to very small ground terminals.

Conclusion

INTELSAT went into operation in 1965, having benefited from the very exciting and productive period of development and tests of experimental communications satellites from 1958-64. During those six years, all of the spacecraft subsystems (structure, stabilization, control, propulsion, and electrical power) and all of the communications subsystems (antennas, wide-band receivers, filters, and power amplifiers) were developed, tested, and demonstrated to the extent that INTELSAT could proceed directly into operational service. Then, during its first decade of operations, INTELSAT continued to receive the technological output of many development satellite projects. Those technologies carried INTELSAT through the fourth generation of satellites. Then, starting with INTELSAT V, more dependence was placed upon technologies resulting from its own R&D efforts than from outside sources. INTELSAT was on its own!

References

[1]Clarke, A. C., The Making of a Moon, F. Muller, London: Harper, New York, 1957.

[2]Clarke, A. C., "V-2 for Ionosphere Research and Extraterrestrial Relays," Wireless World, Feb. 1945, pp. 58.

[3]Pierce, J. R., The Beginnings of Satellite Communications, San Francisco Press, San Francisco, Calif., 1968.

[4]J. F. Kennedy Presidential Statement, "Policy Statement on Communications Satellites," July 24, 1961.

[5]"Communications Satellite Act of 1962," U.S. Congress, Public Law 87-624, Aug. 31, 1962.

[6]Jaffee, L., Communications in Space, Holt, Reinhart, and Winston, New York, 1966.

[7]Dickieson, A. C.,"The Telstar Experiment," Bell System Technical Journal, Vol. 42, No. 4, July 1963, pp. 739.

[8]Pickard, R. H., and Metzger, S., "Relay I--A Communication Satellite," Astronautics and Aerospace Engineering, Vol. 1, Sept. 1963, pp. 64-67.

[9]"Syncom Engineering Report," Vol. I, Syncom Projects Office, Goddard Space Flight Center; see also NASA TR R-233, March 1966.

[10]"Applications Technology Satellite and Communications Technology Satellite User Experiments for 1967-1978," prepared for NASA Lewis Research Center by University of Dayton Research Institute, Dayton, Ohio, June 1979.

5. The INTELSAT System: An Overview

Richard R. Colino*
INTELSAT, Washington, D.C.

Introduction

In a stimulating book discussing trends in American society, entitled Megatrends,[1] author John Naisbitt scrutinizes the transition from an industrial to an information society and states that "the real importance of Sputnik is not that it began the space age, but that it introduced the era of global satellite communications." And, in a curious manner, this is true. Satellites are seen as the missing catalyst in the growing information society of the 20th century, exploding in the 1980's not just in the United States but throughout the world. Communications satellites have, Naisbitt writes, "transformed the Earth ... into a global village" and, instead of turning the world outward toward outer space, "the satellite era turned the globe inward upon itself." The communications satellite, rather than television, has produced this "global village." Again, as Naisbitt and many others have observed, the globe is a truly interdependent place inhabited by people of various races, backgrounds, and cultures in which "it is especially critical that the industrial countries forge a new relationship with the Third World." The INTELSAT global satellite system and organization have been eminently successful in linking the world together not only in the communication of intelligence and information but also as an outstanding example of the establishment and nurturing of a meaningful and mutually beneficial relationship between the more highly industrialized nations and the lesser developed ones so as to accelerate the rate of development of lesser developed nations and to foster in a concrete fashion the much discussed phenomenon of global interdependence.

It is, of course, ironic that the 20-year-old competition between the two major space powers--the Soviet Union and the

---

Invited paper received October 11, 1983.

*Director General.

United States--should provide the underpinning for the establishment of an international organization of more than a hundred member-nations and a single global system of communication satellites. But the "Space Race" has resulted in a joining together of virtually all mankind in a common experience and mutual awareness. One of the most important consequences of the successful launching of Sputnik by the Soviet Union was the rapid emergence of communications satellite technology in the United States and a willingness to share its application with other nations in a cooperative manner intended to benefit all participants and users but, above all, developing nations and areas.

The origins of the INTELSAT Organization are thus, in a sense, rooted in the worldwide rivalry between the Soviet Union and the United States in all its dimensions, but particularly in the contest that emerged in the 1950's and early 1960's to demonstrate a superior prowess in outer space. The technology developed rapidly, both in terms of placing objects in outer space through launcher development programs--a derivative of missile technology--and in terms of scientific exploration and application of outer space. Both nations sought to send unmanned probes into the deepest recesses of outer space and to develop manned flight programs for outer space and lunar landings. Both were also interested in the application of the technology for scientific satellite programs relating to weather, communications, navigation, reconnaissance, education, environmental resource surveying, and other applications. In the United States, however, through the efforts of the Federal Government and the private sector, much of the early effort and attention was focused on communications satellite development.

The rapid development of the launcher vehicle and radio wave (microwave) communications in the United States provided the two essential components for the successful application of technology to the design and development of satellites to carry telecommunications services for practical purposes. As early as the 1940's, experts envisaged the use of satellites in Earth orbit as microwave repeater stations to carry telecommunications signals between widely separated points on the Earth's surface, and the emergence of high-quality and reliable transistor repeaters hastened the actual development of what had earlier appeared to many to be sheer science fiction! Early studies by Arthur C. Clarke, in 1945, and John R. Pierce of Bell Laboratories, in the early 1950's, however, required additional efforts and encouragement from both the Federal Government and the private sector in the United States. With the creation, in 1958, of the National Aeronautics and Space Administration (NASA), the United States embarked on dedicated efforts to explore outer space. Within a few months of its creation, NASA was assigned manned space programs in addition to its unmanned activities. Of major importance and relevance was the NASA program to develop active experimental communications satellites. With RELAY and SYNCOM, NASA made significant progress toward establishing an operational satellite communications system. There was a need under these programs to test the proposition that active space-based repeaters would work and provide acceptable service, despite the inherent time delay. This, however, required the active participation of other countries from the outset. Thus the NASA program helped to stimulate the construction and use of wide-band experimental Earth stations in France, Germany, and the United Kingdom as well as the United States, and various narrow-band stations, from time to time, in Brazil, Italy, Japan, Spain, and even a ship station anchored in the port of Lagos, Nigeria, to communicate with United States' Earth stations. Close technical cooperation in these programs, and the pattern of international participation in the early development of the use of an object in outer space for communicating among and between points in different countries, proved to be a harbinger of the interest and dedication later evidenced for commercial communications satellites. Moreover, the SYNCOM satellite, developed during this early experimental period, proved to be the direct ancestor of the world's first commercial communications satellite, INTELSAT I--better known as Early Bird.

Parallel with the scientific and developmental programs, the United States, through domestic policy and legislative actions, and other members of the United Nations were endeavoring to establish an international framework that would lead to the establishment and operation of a global commercial system of communications satellites. Although few satellites had been launched into outer space for scientific and other purposes, the United Nations had begun to establish norms of behavior for the exploration and use of outer space, including principles contained in several resolutions adopted by the General Assembly in 1962 and 1963.[2]

These resolutions stated that the rules of international law applied to outer space; that outer space was to be free from national appropriation; that nations were responsible for the use of outer space whether by them or by nongovernmental agencies authorized by them; and that ownership was not affected by the passage of objects through space. In 1963, a specialized agency of the United Nations, the International Telecommunication Union (ITU), allocated certain radio frequencies for use in the communications satellite service and for experimental purposes.[3] These actions helped greatly in setting the stage for the introduction of commercial communications satellite services around the world.

Most important, however, and most impressive with 20 years' hindsight, were the actions taken by the executive and legislative branches of the United States Government to articulate the principles and establish a structure for exploitation of the new technologies. At the very time that initial experimental satellites were being constructed, President Eisenhower, in one of his last statements in office, announced that the early establishment of a communications satellite system was an important national objective. This was followed, on July 24, 1961, by President Kennedy's declaration that private ownership and operation of the "U.S. portion of the system is favored" and that satellite communications should be developed through United States leadership "for global benefit at the earliest practicable date."[4] The United States Government issued an invitation to all nations to participate in a communications satellite system in the interest of world peace and brotherhood. In 1962, again prior to the first transoceanic television broadcast via Telstar, bills were introduced in both Houses of Congress and a message delivered by President Kennedy advocating American participation in a worldwide system that was to be organized along the lines of a private corporation: a Communications Satellite Corporation.[5] After lengthy debate in both Houses, the Communications Satellite Act of 1962 was passed and signed into law on August 31, 1962, when President Kennedy stated:

> The benefits which a satellite system should make possible within a few years will stem largely from a vastly increased capacity to exchange information cheaply and reliably with all parts of the world by telephone, telegraph, radio and television. The ultimate result will be to encourage and facilitate world trade, education, entertainment and new kinds of professional, political and personal discourse which are essential to healthy human relationships and international understanding.[6]

The Satellite Act established a set of principles that was remarkably prescient in anticipating the structure of the organization later to be known as INTELSAT. These principles were well received by other nations around the world because they contained a sense of sharing and generosity. This American willingness to share their space technology was seen at the time, and may still be seen today, as providing the basis for a truly cooperative undertaking. Among the declarations to be found in Section 102 of the Satellite Act were:

> ...it is the policy of the United States to establish, in conjunction and in cooperation with other countries, as expeditiously as practicable a commercial communications network, which will be responsive to public needs and national objectives, which will serve the communications needs of the United States and other countries, and which will contribute to world peace and understanding.

> The new and expanded telecommunications services are to be made available as promptly as possible and are to be extended to provide global coverage at the earliest practicable date. In effectuating this program, care and attention will be directed toward providing such services to economically less developed countries and areas as well as those more highly developed, toward efficient and economical use of the electromagnetic frequency spectrum, and toward the reflection of the benefits of this new technology in both quality of services and charges for such service.

The Satellite Act did not specify the structure of the arrangements to be entered into in order to achieve these objectives. The private company authorized to be created under this legislation was a unique corporation, established for profit-making with public and corporate shareholders, and with no government investment. This corporation was, nevertheless, given a mission to achieve certain public policy goals. The newly formed Communications Satellite Corporation (COMSAT) was charged with establishing as quickly as possible the commercial communications satellite system and harnessing and applying a new technology for the betterment of all nations. COMSAT and representatives of the United States Government together entered into discussions and negotiations with officials from telecommunications organizations and other governments to consider the arrangements under which a global commercial system of communications satellites could be established and operated.

Clearly, the United States "team" negotiating the arrangements to establish a satellite system had some clear guidelines: a sense of urgency in getting a satellite system established; an emphasis on the commercial aspects of the venture to be undertaken; a sense of avoiding the pitfalls of "typical" international organizations; the importance of sharing with developing countries this new and largely American technology; the desire to achieve the broadest participation possible while endeavoring to avoid political-type compromises; a commitment to establish a single system in order to avoid waste of resources; and the need to achieve economies of scale in what was still an unknown area of activity with respect to the commercial acceptability and financial success of the technology.

Preliminary discussions were held in 1962 with the United Kingdom and Canada, and they were followed, in 1963, by meetings between representatives of the United States Government and the European Conference of Postal and Telecommunications Administrations (CEPT). The results of those meetings were communicated to officials of COMSAT, and further discussions took place between representatives of COMSAT and various European nations in the latter part of 1963.

Those early contacts made clear the interest on the part of other countries not only in using the satellite system that would be established, but also in having ownership rights that carried with them an active participation in the design, development, production, establishment, management, and operation of the satellites. In short, European countries clearly indicated that they wished to see arrangements entered into in which they would be "partners" from the outset with the United States. Moreover, unlike other arrangements for the establishment and use of telecommunications facilities, including transoceanic submarine cables, which were generally left for telecommunications administrations to conclude and implement, it was made certain that "governments" would assume some role in the global satellite venture. This was true for several reasons: The use of outer space was felt to be a responsibility of governments; the technology (both launch vehicle and satellite) was governmentally developed and largely controlled by a United States agency; and, finally, the governments of developing and developed countries alike perceived that there were industrial and economic processes in space communications that transcended the fundamental concerns of telecommunications organizations. In short, there was a broad appreciation, in the wake of Sputnik, of the enormous impact of space on the public and the strategic interests of the individual nations. Thus, foreign ministries of the various European countries determined to play a role in this matter established a single agency--the European Conference on Satellite Telecommunications (CETS)--to participate in the discussions and negotiations for establishing the satellite system. It also was clear that the agreements governing the early phases of the system would be in two parts: one intergovernmental and the other dealing with issues normally determined among telecommunications administrations.

Serious negotiations were begun in Rome in February 1964. The participants included representatives of foreign ministries and telecommunications organizations from Western Europe, the United States and Canada, and, shortly thereafter, Australia and Japan. At that time, these countries accounted for some 85% of the world's long-distance international telephone traffic. Thus, it was clear that they would form the backbone of the system and the organization as it conducted early experiments and evolved into an operating organization. In the remarkably brief period of just over six months, two agreements entered into force and the new international entity was created.

These agreements were: an intergovernmental Agreement Establishing Interim Arrangements for a Global Commercial Communications Satellite System, and a Special Agreement and Supplementary Agreement on Arbitration among telecommunications entities designated by the participating governments. These agreements, constituting the "interim arrangements," entered into force on the date they were opened for signature: the 20th of August 1964.† The agreements were interim in nature for a variety of reasons, including the yet unproved technology for commercial and operating purposes.

It is quite impressive, in retrospect, that the institutional framework for application of this technology was in place prior to actual proof of the benefits of the satellites. It is also interesting that, despite the preferences of some of the participating nations, it was seen to be acceptable to form the organization as an unincorporated joint venture involving both public and private telecommunications entities as Signatories of the Special Agreement. Those interim arrangements allowed the Signatories to focus their attention and efforts on a number of specific tasks and objectives before addressing the permanent arrangements for the organization and system. They also permitted the shaping of the organization and the allocation of responsibilities and powers in a manner designed to enhance the attainment of specific goals and objectives. INTELSAT, from the first, was thus success and action oriented. It was to become an international organization that worked.

The purpose of the new undertaking was to design, develop, construct, establish, maintain, and operate the space segment of a single global commercial communications satellite system. That was to be accomplished in several phases. First, INTELSAT launched an experimental/operational satellite into synchronous orbit in 1965. This was followed by successive launches of increasingly sophisticated satellites that were configured to achieve global coverage in midyear 1969. Thereafter, emphasis has been placed on improving the system and extending its services as appropriate.

The interim arrangements were to remain in force until superseded by definitive arrangements. As the first step toward permanent arrangements, the governing body of INTELSAT was to issue a report no later than Jan. 1, 1969 to all governments that were Parties to the Agreement. The report was to contain recommendations for definitive arrangements. The governing body did so, and the definitive arrangements were discussed and negotiated for over two years (from 1969 to 1971) and, after formal ratification, entered into effect on February 12, 1973. Thus the interim arrangements remained in force from 1965 until early 1973, governing the many actions and programs of the INTELSAT effort for almost nine years.

†More precisely, the Agreement and the Interim Agreement entered into force on Aug. 20, 1964 and the Arbitration Agreement on June 4, 1965. 15 UST 1705, TIAS No. 5646. The name INTELSAT was adopted on Oct. 28, 1965, and appears in later editions of TIAS. In its original form, INTELSAT was created as an international consortium, without legal personality. This was changed in 1973, when INTELSAT became a full-fledged intergovernmental international organization. It is thus incorrect to refer to INTELSAT as a consortium.

The interim arrangements were quite simple with respect to organizational structure. They specified a governing body called the Interim Communications Satellite Committee (ICSC) and a Manager (COMSAT) to execute and implement the decisions of the ICSC. Investment shares in the organization were determined on the basis of projections of long-distance traffic likely to be carried by satellites. This resulted in the United States (which included domestic traffic, i.e., Mainland-Hawaii, Mainland-Alaska, and Mainland-Puerto Rico) having an initial share and vote of some 61%. The European countries had a 30% share, and the remaining 9% was divided among Australia, Canada, and Japan. Voting was based upon the same investment shares for those represented on the ICSC. It required a total investment share of a Signatory or Group of Signatories of 1.5% to have a seat on the ICSC. If unanimity could not be achieved, then decisions were reached on substantive matters by a vote of COMSAT plus 12-1/2%, regardless of the COMSAT investment share of voting right at the time. (All investment shares, called "quotas," of Signatories were reduced pro rata as new countries joined the venture, provided that the COMSAT share never was reduced below 52.5%.)[δ]

The dual role of COMSAT as the United States participant in INTELSAT on the ICSC and as Manager of the joint venture assured efficiency and speed in meeting the objectives of early placement of Early Bird in orbit, as an experimental and operational-phase system. COMSAT's resources also proved critical to attracting interest on the part of developing countries in joining the enterprise and in using satellites to advance their development. It also clearly posed problems of "American domination" and the "conflict of interests" between international considerations and the domestic interests of COMSAT and the United States Government from time to time.

Several ways were developed to ameliorate such problems, including establishment of "expert" subcommittees of the ICSC that reviewed and evaluated virtually everything the Manager submitted to the ICSC, whether financial, technical, contractual, or involving patent and data rights. Also, because of the keen technical interest on the part of the other INTELSAT Signatories, the Manager adopted a foreign "assignee" program to permit qualified people from other countries to be assigned to the Manager's staff for varying periods of time; initially, this was limited to technical people, but eventually it encompassed other elements of the staff functions. Finally, the Manager's staff was, to an extent, sequestered. COMSAT management separated staff functions and responsibilities so as to make a meaningful distinction between obligations running to the ICSC directly and

---

δA total of 17% of the quota was reserved for new members to take up as they joined. Since the original membership reflected some 85% of total global telecommunications services, this seemed a reasonable figure.

those to the COMSAT Corporation as a whole. In the same light, the ICSC for a time hired its own Executive Secretary with responsibility for the language and conference services staff.

Of course, basic policies governing the rights and obligations of the Parties and Signatories with respect to the single global system were set forth in the Agreements.π Also set forth in the Agreements were policies with respect to procurement matters; financial obligations; use of the system by nonmembers; criteria for joining the venture (it was required that a nation be a member of the International Telecommunication Union); ownership and use of patents, data, and information; payment of monies due INTELSAT by Signatories and the consequences of failure to meet such obligations; withdrawal; minimal shares for those wishing to participate; amendment of the Agreement and Special Agreement; charging principles and procedures for the use of the system; and a host of other important subjects.

The estimated cost of establishing the organization and achieving the successive phases of space segment envisaged during the period when the interim arrangements were expected to be in force was set at $200 million (U.S.) in the Agreement (Article VI). The ICSC had the power to determine that as much as $300 million could be invested, but that a special conference of all Signatories to the Special Agreement was required to raise the sum above that figure. (There was some ambiguity concerning the firmness of the requirement for Signatories to contribute above the $200 million sum, but the issue never arose.)

Certainly, at the time the interim arrangements were negotiated, the total cost of satellites to be manufactured and launched on other than an occasional "experimental" basis was unknown. Likewise, it was not even clear whether the satellite system would consist of low-, medium-, or synchronous-altitude satellites. There was not even firm evidence that telephone service by means of geosynchronous satellites located some 35,800 km (22,300 miles) above the Earth's surface, with its small but perceptible time delay, would be found acceptable to the general public. Thus the interim arrangements were almost "articles of faith" that gambled on the assumption that the technology somehow would work.

The interim arrangements echoed many of the principles and concepts to be found in the United States COMSAT Act of 1962. Perhaps the key element was the desire to have the single global system be part of an improved worldwide network that would provide service to all areas of the world, make efficient use of the radio frequency spectrum, and offer communications via satellite to all nations on a nondiscriminatory basis. In addition, it was hoped this would contribute to world peace and understanding. These principles proved so central to INTELSAT global cooperation

π"Single" appeared only in the Preamble but it was clearly the intent of the founders that there be but one "global system."

that they were carried over to form the basic principles of the organization's definitive arrangements that are in effect today.

The Interim Agreements entered into force on Aug. 20, 1964 for some 11 countries (14 signed them). The membership had grown to 68 by the time the first Plenipotentiary Conference met in Washington, D.C., in 1969 to consider definitive arrangements, and to 83 by the time the definitive arrangements came into force in 1973.

On August 20, 1971--almost two and a half years after the opening of the first Plenipotentiary Conference--two interrelated Agreements were opened for signature. Thus, seven years after the start of INTELSAT under interim arrangements and after a period of incredibly successful operations, INTELSAT was on the road to having a permanent charter and organization. About one and a half years later, on Feb. 12, 1973, having received the necessary ratifications, these two new Agreements entered into force.

One is called "Agreement Relating to the International Telecommunications Satellite Organization 'INTELSAT,'" an agreement, with four annexes, among States; the other, the "Operating Agreement Relating to the International Telecommunications Satellite Organization 'INTELSAT,'" with one annex, is among States party to the Agreement or telecommunications entities public or private, designated by a Party to the Agreement.[7] Together, they provide the charter, mission, structure, policies, and powers for INTELSAT to carry on its business of operating an ever-expanding and improving worldwide system of satellite facilities and services. At the time these Agreements entered into force, they did so for some 80 nations. Today, INTELSAT membership totals well over 100 countries and is still growing.

These constitutional instruments for man's first and most successful cooperative venture in the use of outer space are complex and detailed in comparison with many other international agreements and, indeed, even in comparison with the interim arrangements. At the time of their negotiation and entry into force, they appeared to be unique; however, since then the International Maritime Satellite Organization (INMARSAT) has been established and has agreements that are closely patterned after INTELSAT's. The INTELSAT definitive arrangements reflect a legacy of the experience gained under the interim arrangements and of the issues not completely resolved in 1964 which continued to be of importance in the negotiations years later. As an example, in 1964, many European countries wanted INTELSAT to have an organizational structure and status more akin to traditional international organizations, without regard to its special operating nature; and some were interested in seeing INTELSAT as an independent legal entity not dependent on another entity to implement decisions and oversee its management activities. Instead, a loose joint venture structure was adopted.

Thus, in the definitive arrangements negotiations, there was a continuing desire to confer at least an independent juridical status upon the INTELSAT organization and thus allow it the right to be seen as an independent organization with its own in-house staff.

The question remained in the 1969-71 negotiations as to the balance between this desire and the practical need to maintain the expertise of the COMSAT management and staff, which had successfully deployed and operated the early global system of satellites. The interim arrangements had provided all participants able to sit on the ICSC, alone or in a group, with an opportunity to be involved in the decision-making process and the formulation of policies for the organization. Those countries not seated on the ICSC, however, had no direct access to such participation. The Interim Agreement had taken this condition into account and had pledged to provide an opportunity to all parties in the future to contribute to the determination of general policy.[8]

The question of the rights and obligations of Parties and Signatories under the new definitive Arrangements with respect to the establishment and use of other satellite systems was also anticipated as a key issue for discussion. Many (especially from Europe) desired the resolution of this issue in a manner perhaps different from the interim accords. This, at least in part, was due to the development of space technology capabilities in Europe and elsewhere and the inexorable wish to apply new industrial capabilities in practical ways. Similarly, the emergence of industrial know-how outside of the United States in the fabrication of satellites signaled detailed discussions of revised procurement provisions of the new agreements. As more countries had joined the venture, their governments had become interested and involved, and they, too, tended to prefer an organization in the more "traditional" mold, at least in terms of structure, roles of governments, voting and like matters, in contrast to the highly pragmatic and functional approach of the pioneering INTELSAT negotiators. To some, the INTELSAT organization required emphasis only in the application of the technology for economic purpose in the context of broad international cooperation, while others strongly desired for INTELSAT to be similar to other international organizations. The final arrangements were, necessarily, a balance of fundamentally differing points of view and a result of tradeoff and compromise. Yet, the pragmatic, as well as the technically and economically optimum approach, generally prevailed.

INTELSAT, under its definitive charter, is thus not surprisingly, a "hybrid" organization. It is somewhat "traditional" in appearance and possesses similarities to other international organizations created by intergovernmental agreements, but its purpose, power, structure, and decision-making processes are oriented toward practical and commercial objectives. In comparison to the interim arrangements, it is not as "lean" an organization, from the vantage point of commercial decision making. This feature does not, however, appear to have had adverse consequences in terms of economic or technical performance thus far.

The mission and charter of the organization under the definitive agreements are similar to those under the interim arrangements. The focus is still the space segment, which is defined to include the satellites and related terrestrial support equipment and facilities to run the global system.

As noted previously, there are two interrelated Agreements, as was the case with the interim accords, although the two Agreements are very detailed and complex in comparison with the earlier instruments. The primary objective of INTELSAT under the definitive arrangements is to continue and carry forward the space segment under the interim arrangements and to provide this space segment on a commercial basis, with the primary purpose being the provision of international public telecommunications services, under the INTELSAT definition of services. This means, essentially, all types of point-to-point services, public or private, capable of being provided by satellite.** Unlike the Interim Agreements, the scope of activities is defined in both functional and geographical terms, and "public telecommunications services" is defined and contrasted with "specialized telecommunications services" in Article I of the Agreement. INTELSAT may provide both types of service (i.e., public or specialized) under different provisions of the Agreements and with different financial and approval requirements, but its basic purpose is "public" telecommunications services, that can be "fixed" or "mobile," and which are available for use by the public, such as telephony, telegraphy, telex, facsimile, data transmission, transmission of radio and television programs for further transmission to the public, as well as leased circuits for all of these purposes. (Thus "public" means, in effect, all users--general public, business, governmental agencies, private data networks, etc.)

Membership in INTELSAT is open to any State that is a member of the ITU, and all nations and areas of the world may have access to the system regardless of membership. As was the situation under the prior accords, Earth stations are not owned or operated by INTELSAT but are regulated in accordance with domestic laws and policies. INTELSAT does, however, set general standards for Earth station performance and performance acceptance testing.

As suggested earlier, the most radical changes in the organization were in its structure--not in INTELSAT's scope or

---

**Article II of the Agreement. The Article refers to "the global commercial telecommunications satellite system" and the Preamble refers to the desire to continue the development of the system with the aim of achieving a "single global commercial telecommunications satellite system."

commercial objectives. The role of governments is defined more directly and actively than under the Interim Agreements, with the establishment of an Assembly of Parties, a new plenary body that addresses the types of issues normally concerning sovereign states. Since it is the organ of INTELSAT in which States, referred to as Parties, participate directly, it is described as the "principal" organ of INTELSAT. Each Party has one vote in the Assembly, which meets every two years unless extraordinary sessions are needed. Extraordinary sessions have been required for the coordination of other satellite systems with INTELSAT, in order to comply with the six-month review cycle established by Article XIV of the Agreement. The Assembly takes its decisions by the affirmative vote of two-thirds of the Parties present and voting on subjects that may come before it, such as: formulating and communicating views on general policy and long-term objectives to other organs; authorizing the use of the INTELSAT space segment or the provision of other satellites by INTELSAT for specialized telecommunications services; taking decisions on proposed amendments to the Agreement and proposed amendments to the Operating Agreement; considering complaints of governments; selecting a panel of legal experts; approving the permanent management arrangements; and confirming the appointment by the Board of Governors of the Director General. It is significant that, unlike other intergovernmental international organizations, virtually all INTELSAT decisions are taken by consensus. Formal roll call votes are rare.

In addition to the Assembly of Parties, there is another plenary body, open to all members, consisting of the "shareholders," and known as the Meeting of Signatories. This organ is also a one-representative, one-vote forum, which ordinarily meets annually. It considers and expresses to the Board of Governors it views on reports from the Board, and, in addition, can raise INTELSAT's capital ceiling [it has done so on two occasions, so that the current ceiling is $2,300,000,000 (U.S.)] beyond the 10% the Board can raise of its own volition; determines, within the parameters spelled out in the Agreement, the minimum investment share required to obtain a seat on the Board; approves as international traffic certain types of domestic traffic as permitted under Article III of the Agreement; establishes general rules on the approval of Earth stations to use the space segment, allotments of satellite capacity, and the establishment of charges for use of the space segment; proposed amendments to the Operating Agreement; and other general matters that are of concern to Signatories. As is the case with the Assembly, decisions are taken by two-thirds of the Signatories present and voting.

The successor to the ICSC, as has been noted, is the Board of Governors. Membership has been liberalized so that as many as 27 Governors can sit on the Board. Membership is open to any Signatory or group of Signatories that meets the minimum investment share requirement established annually by the Meeting of Signatories, plus up to five seats, within the permissible total of 27, on the basis of a group of at least five countries in an ITU-defined region agreeing to be jointly represented--said representation being without regard to whether the group meets the minimum investment share requirement.

The powers and responsibilities of the Board, set forth in 27 subparagraphs of Article X of the Agreement, make it clear that it is this organ that is the successor to the ICSC for most of the functions involved in the design, development, construction, establishment, operation, and maintenance of the INTELSAT space segment. The Board normally convenes four times a year in order to meet its wide-ranging decision-making responsibilities. On most commercial and operating matters, and in terms of expertise among organs other than the day-to-day management unit, the Board possesses the ultimate authority among the organs of INTELSAT. The Board appoints the Director General, subject to confirmation by the Assembly of Parties, and is the organ to which the Director General is most responsive as the legal representative and chief executive officer of the organization. The Board establishes general policies and grants broad authorizations during its regular meetings. It has the right to raise the capital ceiling established by the Meeting of Signatories an additional 10% when the financial need arises. Approval of plans, programmatic decisions, budgets, operating procedures, and the whole array of policy and program decisions are thus set in a manner similar to any large-scale commercial operating organization.

Voting in the Board on substantive matters is quite different from that of the Assembly and the Meeting of Signatories. In Article IX, a 40% limit is set on the voting power of any Signatory represented on the Board, although currently the largest investment share (and, hence, voting share) is less than 25%. Voting entitlement is determined each March 1st on the basis of investment shares which, in turn, depend upon the degree of utilization of the system by Signatories. Thus decisions on substantive matters require agreement by at least four Governors, provided that they possess at least two-thirds of the voting/investment shares represented on the Board, <u>or</u> the support of all but three Governors without regard to the total investment share this majority may represent. Procedural matters may be decided by a simple majority of Governors, with each Governor having one vote. However, the Board is enjoined to try to act unanimously, and in over 95% of the decisions, consensus is achieved. The highly technical nature of INTELSAT, its commercial objective, and the clear principles that underlie the decision-making process, have all served to allow this pattern of consensus to evolve and be successfully sustained for over a decade.

Anyone cognizant of the issues surrounding the negotiating of the interim arrangements in the early 1960's would have anticipated that the nature of the management arrangements for the "permanent" INTELSAT organization would be of major significance for many INTELSAT members. It was not only a question of "philosophical" importance to those favoring a "traditional" form of international organization, staffed by an international secretariat or directorate, but that the period of the interim arrangements had revealed difficulties in the use of the "loose" joint venture approach in which the largest venturer provides the management control and staff expertise to implement the decisions of the venture and to run the day-to-day operations.

Among the difficulties were: (1) The fact that the largest "shareholder" in the venture was a corporation subject to the laws and regulations of its own government and that the determinations of the venture could be and were subject to the review of a United States regulatory body. (2) Many felt that the interests of the largest shareholder were already overstated and overprotected by reason of its having an absolute veto over decisions of the governing body and then, in addition, having the "managerial" role as well. This American control was thus generally viewed as excessive. (3) The largest venturer (COMSAT) had ambitions in other aspects of satellite communications, including offerings to the United States military. (4) The largest venturer was a newly established entity, and there was concern that COMSAT might or did use its role as INTELSAT's Manager in its efforts to deal effectively with more established telecommunications entities in its own country, which could use the system only by going through the American company serving as Manager. (5) Closely related was the concern that the staff of the Manager was not adequately sequestered so as to ensure that the costs of the managerial activity were properly allocated and that the management and decision-making processes were always oriented in INTELSAT's favor.

In short, the ICSC members were frequently concerned that the staff of the Manager was not delineated and trained so as to provide "sensitive" individuals to serve the enterprise as a whole. There was also a sense that decisions taken by the ICSC could subsequently be interpreted in different ways, thus leaving the individuals at COMSAT, in which the ICSC members had little or no voice or control, to determine the actions of the organization.

In the minds of others, however, these difficulties were offset by the success of the organization in deploying and operating a global satellite system in a highly successful manner, in a period of just five years, which attracted the usage and participation of scores of countries. This was the very pragmatic view of: "Who cares about organizational and management theory? We want an organization that works well and efficiently."

Whatever the viewpoint, the question of management arrangements was prominent in the early days of the INTELSAT definitive arrangement negotiations and was the focus of attention and controversy. Perhaps surprisingly, the wide-ranging philosophies were ultimately reconciled. As a result, a fourth "tier" of organization was agreed upon at the 1970 continuation of the Plenipotentiary Conference and adopted in 1971, as set forth in Articles XI and XII of the Agreement and Annexes A and B thereto. An "Executive Organ" (instead of a "Secretariat" or "Directorate") was created for the purpose of conducting in-house the management duties attendant on the organization and doing so on an ever-increasing basis as it assumed functions and responsibilities over a period of years.

The Executive Organ, headed by a Secretary General, was made directly responsible to the Board of Governors. Its responsibilities were scheduled to increase in scope over a period of six years from the entry into force of the Definitive Agreements and its expansion to be accomplished in a two-phase process. The first phase, during which the Executive Organ was headed by a Secretary General, lasted until Dec. 31, 1976, at which time a Director General assumed the position of "chief executive" of the organization.

The Secretary General was responsible to the Board for the financial, administrative, and certain limited technical and operational functions of the organization, while the Manager under the interim regime, COMSAT, continued to perform the other largely technical managerial responsibilities. The Board of Governors would and did determine, in 1975, that the Secretary General was properly staffed to perform all management functions and he then became responsible for the conduct of all of the affairs of the organization. Under this arrangement, COMSAT continued until Feb. 12, 1979 (six years after the entry into force of the definitive arrangements) as a technical and operations manager, called the "Management Services Contractor" (MSC). Thus, the contracting out to the interim arrangements Manager spanned both phases of the evolution of management arrangements.

The second phase began on Dec. 31, 1976, when the first Director General of INTELSAT, Santiago Astrain, who had previously served as Secretary General of the organization, assumed office. He had been appointed by the Board in July of 1976 and confirmed by the Assembly of Parties in September of that year. Also, in September 1976, the Assembly of Parties approved the permanent management arrangements recommended to it by the Board of Governors following several years of study and discussion.

Upon the expiration, in 1979, of the Management Services Contract with COMSAT, the permanent management arrangements permitted the continuation of COMSAT in the role of contractor, but for precisely defined, highly technical functions and under two separate contracts--one for a four-year term and another for six years. In 1977, the Board approved the Director General's recommendation to negotiate such "service" contracts with

COMSAT, and subsequently approved them. In addition, the Director General concluded, with Board approval, Laboratory Services Contracts with COMSAT. Moreover, many managerial functions, such as tracking, telemetry, command, and monitoring (TTC&M) of the satellites and plans for introduction of new modulation techniques on a system-wide basis (such as the new digital communications system, TDMA) are performed by a number of Signatories under contract with INTELSAT (Australia, Brazil, Cameroon, Japan, Indonesia, Italy, France, the United Kingdom, and the United States). All of the functions and responsibilities for running the system and the organization itself are the responsibility of the Director General, who carries the trifold label of "Legal Representative," "Chief Executive," and, of course, "Director General."

There is also a provision in the Agreement designed to ensure that each of the organs performs its functions on an efficient and effective basis, while interacting as appropriate to the functions to be performed, in the overall interests of the organization and the Parties and Signatories. Thus Article VI of the Agreement ensures a "noninfringement" principle, while assuring a regular flow of reports and information to assist the different organs to perform effectively. The Director General and the staff of the Executive Organ are the key to smooth interaction in terms of support and information flow.

The Executive Organ is now staffed by some 600 people, of whom almost two-thirds are in professional positions. In addition, INTELSAT has an assignee program for training professionals from various countries around the world, and a young professional training program for "hands on" training of nominees from Signatory organizations. Over 50 nationalities are represented on the staff of the Executive Organ.

Financing and Ownership

The determination of ownership and use charges under the Interim Agreements was necessarily based upon forecasting of how the membership and use of the system would evolve over a period following the demonstration of the practicality of geosynchronous satellites. By the time the definitive arrangements were negotiated, there was ample experience to guide the negotiators in formulating provisions for inclusion in the final texts of the two Agreements. However, these issues were highly complex and depended greatly on the underlying principles motivating the negotiators.

As was the case with regard to the Interim Agreements, the principle agreed to for establishing investment shares and ownership rights was to base it on an objective standard pertinent to the activities of the organization. Basing ownership on use of the system was seen as being the best objective standard, but this was not easily implemented due to the need to correct for four factors, namely: (1) the desire to permit membership in the organization, even if there was little or no initial usage; (2) the "obligation" to permit nonmembers to use the system; (3) the desire to accommodate members who might not be in a position to make their full financial contributions based upon usage in periods of heavy capital requirement; and (4) other minor deviations from the "purity" of a concept of relating investment in the organization directly to actual usage of the system. Nevertheless, despite these difficulties, the Definitive Agreements came very close to realizing the ideal relationship of making the requirements for capital investment in the system closely tied to actual usage of the system by Signatories. The relationship was to be reestablished on an annual basis to permit accommodation to changes and to overcome the deficiencies found in the "fixed quota" system of the interim accords.

Each year, each Signatory's payment of space segment utilization charges for a six-month period is calculated as a percentage of total space segment charges collected. Each Signatory thus becomes entitled (and obligated) to own a financial interest in INTELSAT equivalent to this share. INTELSAT, by virtue of its juridical personality, has ownership rights in all INTELSAT property, including the space segment. It thus directly owns the satellites, related TTC&M equipment and other properties, such as the INTELSAT headquarters building. Flexibility has been provided for Signatories to take a minimum investment share if they are not yet users of the system, and for Signatories to take a lesser share than that which they are obligated to take on the basis of usage if there are other Signatories available to assume larger investment shares. This method seems to have worked, in that a sufficient number of Signatories have been willing to subscribe to this greater share, perhaps attracted by the 14% return on capital investment and the consequently larger voice in the Board of Governors. (It should be noted that the rate of return payment is in no way a profit. This is because if ownership and use are the same, which is normally the case, then the money is being paid to oneself. The rate of return is thus an equitable arrangement to compensate partners for the cost of money if their usage of the satellite exceeds their ownership share.)

This system of allocating investment shares is reflected in Article V(b) of the Agreement and in Article 6 of the Operating Agreement, and permits investment to be related to use in general, while also accommodating the variations to this principle deemed appropriate by the negotiators. A Signatory's financial interest is equal to its investment share as calculated by the Board pursuant to Article X of the Agreement and in accordance with the provisions of Articles 3, 4, and 6 of the Operating Agreement. Because not all users need be members of INTELSAT, and due to the variations noted above from a "pure" investment-related-to-use system, it was necessary to establish utilization charges for use of the system, and this was accomplished in a manner similar to that of the Interim Agreements.

Article 8 of the Operating Agreement sets the purposes that govern the Board of Governors in setting the utilization charge, and in obtaining the consequential revenue, such as to "have the objective of covering the operating, maintenance and administrative costs of INTELSAT, the provision of such operating funds as the Board of Governors may determine to be necessary, the amortization of investment made by the Signatories in INTELSAT and compensation for use of the capital of Signatories." Revenues earned by INTELSAT are thus to be applied first to meet the operating, maintenance, and administrative costs; second, to provide the operating funds referred to above, as determined by the Board; third, to repay to Signatories in proportion to their respective investment shares the capital in the amounts of the provisions of amortization decided by the Board; fourth, to repay a withdrawing Signatory; and, finally, to pay to Signatories in proportion to investment shares a compensation for use of capital from the balance available from the revenues obtained. This compensation level has consistently been set at 14% and consistently achieved.

Charges for use of the system are set in accordance with these noted clauses and Article V of the Agreement. Article V, in effect, embodies a "public utility" concept or "regulated monopoly" or quasimonopoly (natural monopoly) concept familiar to regulatory economists, i.e., average cost pricing. This approach is consistent with the concept of INTELSAT's being run as a single global system without other systems in competition (leaving aside competition from other means of carriage or transmissions such as high-frequency radio, troposcatter systems, standard submarine cable, microwave links across international boundaries, and newer fiber optic and light wave transmission facilities).

From the first, INTELSAT as an organizational entity, was seen as a "cooperative" designed to obtain benefits and economies for the entire world. It was not designed to subsidize developing countries, but rather to allow all countries, developed and developing alike, to share the benefits of space communications. Despite the extraordinary success of INTELSAT, it remains to be seen whether changes in this concept will be required in the future.

Authority is given to the Board of Governors to raise loans under certain circumstances (Article 10 of the Operating Agreement). This is an authority that the ICSC did not possess. A capital ceiling limitation is contained in Article 5 of the Operating Agreement. This was initially set at $500 million (U.S.). The

Board may and indeed has recommended increases to the Meeting of Signatories from time to time, which, as noted previously, has the authority to increase the ceiling. The ceiling was raised in 1976 to $900 million and again, in 1981, to $2.3 billion. It is difficult to predict whether this will sustain INTELSAT into the late 1980's and early 1990's, since the resulting sum of cumulative net capital contributions and contractual commitments is just over $2 billion in 1983, making INTELSAT the largest and most heavily capitalized satellite system in the world.

Procurement and Patent Policies

A major concern of many of the industrialized countries, in Europe in particular, during the early 1960's negotiations and again during those for the Definitive Agreements in the late 1960's, involved the adoption of procurement policies and clauses. The development of the European satellite and launch vehicle industry meant that significant national industrial interests were at stake. On the other hand, by the time the negotiations for the definitive arrangements were under way, a large number of countries in the developing world had become INTELSAT members, and they wished to see the most effective and economical procurement and contracting policies and practices possible for the organization.

The developing countries were united on the principle that they did not want to see any policies of "juste retour," set-aside programs or other artificial means of dispersing contracts and work if the INTELSAT organization itself were to bear the financial burden for such approaches. The basic principle many of these countries espoused was to have and encourage "international participation" only if it did not involve higher costs than would be incurred under a "best product at best price" philosophy. All negotiators recognized, however, that it was in the overall interests of the organization to have access to as many sources of supply as possible and to be in a position to encourage multiple bidders for major procurements of spacecraft and related equipment.

The result of a balancing and tradeoff of viewpoints was to adopt as INTELSAT's primary policy the principle of awarding contracts on the basis of open international tender and to select bidders offering the best combination of quality, price, and the most favorable delivery time. It was established as a secondary principle that "if there is more than one bid offering such a combination, the contract shall be awarded so as to stimulate, in the interests of INTELSAT, worldwide competition" (Article XIII of the Agreement). In practice, bidders have formed "international teams" in order to respond to INTELSAT procurement requests for proposals and a great deal of effort and consideration goes on within this context to allocate the work among the team members from different countries. Thus far, up through the INTELSAT VI contract, the prime bidders have been American companies, although it may be anticipated that with the very significant activities and development in Europe, Japan, and elsewhere, there may at a future date be a prime contractor bidding on behalf of a team from outside the United States' family of aerospace companies.

Similar debates and consideration went into formulating provisions for inclusion in the Operating Agreement with respect to procurement procedures and practices. One such issue, late in the deliberations, was the extent of the authority and flexibility to be given the Board of Governors in terms of granting exceptions to competition and open international tender in procurements. The final provisions in Article 16 of the Operating Agreement set forth strict standards under which the Board may deviate from full-blown competitive tendering and thus permit sole source or limited source procurements.

Much ado was also made with respect to patent, data, and licensing of information issues during the period of the interim arrangements, and, not surprisingly, it became an important issue in the negotiation of the definitive accords. The fundamental difference was over the rights INTELSAT should acquire and thus be able to limit others in its use, versus a fairly liberal approach pursuant to which INTELSAT would not acquire exclusive rights and would license the usage of rights acquired in patents, data, and information to Signatories for use in connection with both INTELSAT and non-INTELSAT purposes on fairly modest terms. Ultimately the latter approach was codified in Article 17 of the Operating Agreement and has been a little-disputed point since the negotiations.

Settlement of Disputes

Due to the short period within which the interim arrangements were determined, the arbitration provisions agreed upon for the formal settlement of disputes had to be very simple. In negotiating the Definitive Agreements more time was available and more elaborate arrangements were considered than those contained in the Supplementary Agreement on Arbitration, which entered into force in 1966. While many of the provisions of the Supplementary Agreement were carried forward into the definitive period in Article XVIII of the Agreement and Annex C as well as Article 20 of the Operating Agreement, the subject matter, parties, and procedures for the settlement of disputes were spelled out in greater detail and precision. The final provisions appear to be fairly comprehensive, although untested, since INTELSAT has yet to have the settlement of disputes portions of the Definitive Agreements invoked. Many feel that one very real hallmark of INTELSAT's success is the fact that arbitration proceedings have never been necessary.

Other Provisions

The Agreements address in more or less routine fashion other questions to be faced by an international organization, such as duration of the Agreements, withdrawal from the organization, amendments to the Agreements, rights and obligations of members, INTELSAT headquarters, privileges, exemptions and immunities, official languages, relations with other international bodies and similar matters--none of which require further elaboration here.

Overview of the INTELSAT Agreements

The INTELSAT constitutional documents, past and present, have been discussed at some length for several reasons. For one, they represent the thought process in the creation of a highly successful international entity to exploit technology in outer space for practical benefits on Earth. The definitive arrangements, like their predecessor documents, are remarkably thorough and clear in expressing the objectives set for the organization and in detailing the powers, authority, and responsibilities necessary to their achievement. They provide purpose and create flexibility to accommodate the dual nature of the organization: on one hand, an operating commercial venture; on the other, a unique international organization--one that reflects many signal achievements over its brief lifetime. In reviewing the Agreements, one is struck by the fact that they contain a detailed and direct statement of "business plans" that entities, public and private, often strive to express but do not always succeed in clearly articulating.

In short, the INTELSAT Agreements seem unequaled in the way that they allocate powers and responsibilities; anticipate needs, problems, and difficulties; provide flexibility for change; and give clear indications of future directions and opportunities. This has been proven not only by the ability of the organization to meet changing circumstances (as, for example, the need to raise the capital ceiling) but also because of the absence of any serious attempt--let alone any expressed need--to amend the Agreements, in the more than ten years since they entered into force. It is thus critical to the appreciation and understanding of INTELSAT, and the related growth of the organization and its facilities, to have a thorough knowledge of the INTELSAT Agreement and the Operating Agreement. The Agreements, in fact, set forth the approaches to the evolution of the design and development of the system; the broad approach to system economics; and even the problems and challenges that could face the organization.

System Design, Control and Operation

The Interim Agreement basically delineated the design of the system in the various phases it addressed: first, an experimental phase for the purpose of testing the proposition that satellites in geosynchronous orbit would be able to provide acceptable voice, data, and television services; second, that, after this experimental phase, the organization would proceed to establish an appropriately designed and developed space segment of satellites and related terrestrial facilities so as to provide global coverage at the earliest practicable date; and, third, to grow and expand from this global network, once established.

Both sets of Agreements set forth the requirement that the "global satellite system" of INTELSAT be available for use by all nations and areas of the world regardless of whether users chose to become investors in the INTELSAT system by means of ownership shares. The Agreements defined the scope of the organization, in terms of ownership and operational responsibilities, to include only the satellites and related facilities and not the Earth stations accessing the satellites. Responsibilities for the design and development and maintenance and operation of the satellites are spelled out in greater detail in the Definitive Agreements than in the Interim Agreements.

In keeping with the provisions and directions provided by the Agreements, the system has evolved in the incredibly short span of 18 years since Early Bird (INTELSAT I) was launched in April 1965. During this period, INTELSAT has developed the space segment to meet ever-wider service requirements and an increasing volume of traffic, while introducing innovative technology. These innovations include, for example, the introduction of mechanically and electrically despun antennas; spot beam technology; frequency reuse by both geographic separation and polarization discrimination; cross-strapping of transponders operating at both C-band (16/14 GHz) and Ku-band (14/11 GHz) frequencies; demand assignment techniques; varied Earth station facilities ranging in size from large "Standard A" [30-m (98-ft)] stations to small customer-premise antennas [3.3 m (10.8 ft)]; TDMA digital transmissions; high-speed data transmissions; and digital voice.

Satellite switching and signaling, demand access, onboard processing and other advanced modulation and encoding technologies are marked by introduction and use on the INTELSAT system. These and other design and operational techniques have made possible in the INTELSAT VI, currently under construction, a circuit capacity 170 times that of the INTELSAT I. If, for instance, one were to extrapolate a linear continuation of capacity expansion into the next century, an INTELSAT satellite with a 2007 launch date (most likely a space platform) would be capable of providing thousands of television channels or more than five million equivalent voice circuits.

Placing these astounding figures in context, one returns to the interim arrangements and the first phase of the INTELSAT program, the experimental/operational satellite known as Early Bird. This satellite tested operational use of geosynchronous satellites for the global system and was able to provide links between North America and Europe. It was a spin-stabilized satellite with a fixed squinted beam omnidirectional antenna through which a signal would be transmitted over a very broad arc, with the result that much of the power of the satellite was transmitted and lost in outer space. The satellite power, by virtue of its squinted beam antenna, was directed at the Earth's surface in the North Atlantic region only. The capacity of the Early Bird satellite was 240 simultaneous telephone calls, or one black-and-white television channel in lieu of telephone transmissions. Despite the absence of multiple-access capability, the limitation of nonsimultaneous use of either television or telephone communication, and a design wherein much power is lost in outer space, Early Bird, the world's first commercial communications satellite, was an unmitigated success. It proved the acceptability of geosynchronous orbit satellites for the carriage of telephony traffic; it successfully forged INTELSAT into an operating organization and attracted the attention of many developing nations outside the Atlantic region with regard to the benefits that satellite communications could bring them. Although designed to last only 18 months, Early Bird remained on active duty for five years after its deployment. Its performance led to the conclusion that satellites could provide reliable service over long operational lifetimes.

Building on the Early Bird experience, INTELSAT II satellites had multiple-access capability, permitting users within the zone of coverage to access the satellites simultaneously through many Earth stations by means of a new design of the transponder. These satellites possessed an antenna design that permitted energy and power to be directed into covering larger areas than the INTELSAT I did, and they were used to provide the first satellite coverage of the Pacific basin, including both Northern and Southern Hemisphere coverage. The capacity of the INTELSAT II satellite was the same as the INTELSAT I, but the design life was three years and this was achieved by all operational INTELSAT II satellites. The first INTELSAT II was successfully deployed in October 1966 and placed in operation in the Pacific Ocean region in January 1967. In addition, INTELSAT II satellites were positioned over the Atlantic in March 1967 and again in the Pacific in September 1967.

As noted earlier, the follow-on phase of the Interim Agreements called for the extension of the satellite system to provide global coverage, and the INTELSAT III satellite series was the one chosen to achieve this. It did so in July 1969, when an INTELSAT III satellite was placed in orbit over the Indian Ocean. It also represented an increase in design capability from the 240 circuits of INTELSAT I and II to 1200 voice circuits plus simultaneous television transmission. The satellite derived increased power and energy transmission efficiency by introducing a design innovation in the form of a mechanically despun antenna. It enabled the antenna in the spinning satellite to be driven by a motor to rotate at the same speed as the satellite but in the opposite direction. The result was to have the capability to focus the radiated power of the satellite fully onto designated areas of the Earth's surface and thus lose little energy in outer space. These improvements made it possible for the INTELSAT III satellites operating over the three ocean regions to provide 1200 voice circuits, four television channels, or a combination of telephone and television. This global network, established only a week before transmitting, live, man's first lunar landing to virtually every corner of the globe, set a new record global television audience of 500 million people.

By the time the global system was entering the 1970's, changes and improvements in satellite technology were even greater with the more powerful, larger, and flexible INTELSAT IV satellites. These were much greater in mass/weight than the previous series and required the use of the larger and more powerful Atlas/Centaur launch vehicle rather than the Thor-DELTA, on which all prior INTELSAT satellites were launched. Depending on the operational configuration in which the satellites were used, they could provide many more telephone and television channels--an equivalent of 4000 simultaneous telephone circuits and two television channels on average. The INTELSAT IV satellites had 12 transponders and achieved the great increase in capacity by innovative design, including two spot beam antennas (steerable from Earth so as to direct the power in the spot beams to given areas of the Earth) and had four global horns to provide for total regional interconnectivity within the Atlantic, the Pacific, or Indian Ocean regions.

These satellites represented a major gain in technological capability, since they were designed to achieve a seven-year lifetime. With the launch of this new generation of satellites in the early 1970's, INTELSAT was able to respond to the burst of overseas traffic demand, with traffic increasing from 1835 voice channels at the start of 1970 to 11,500 at year-end 1974--a remarkable sixfold increase in only five years.

With the demand for the use of the INTELSAT system growing at a highly accelerated rate, the requirements for greater operational flexibility also grew. More countries and different regions came on the system. Different traffic streams emerged. New Earth stations were constantly coming on the air. The frequency spectrum was not sufficient to accommodate all of these requirements, lacking an ability for reuse so as to maximize the power output at different portions of the C-band (6/4 GHz) then being exclusively used in INTELSAT satellites.

Thus the INTELSAT IV-A satellites, while a derivative of the INTELSAT IV satellites, were most significant in their ability to meet these frequency reuse requirements. By introducing a complex antenna array that spatially isolated the Eastern and Western Hemispheres of the Earth area covered, it was possible for the INTELSAT IV-A satellite to derive 6000 voice equivalent circuits plus two television channels in 20 transponders. The INTELSAT IV-A satellites, with a seven-year lifetime, although launched in the mid-1970's, are still operational today as part of the global network.

In 1976, in response to ever-expanding requirements for space segment capacity, INTELSAT embarked upon the INTELSAT V program, which produced satellites that are three-axis stabilized, in contrast with all of its earlier spin-stabilized spacecraft. The cylindrical spin satellite has solar cells on the outside of the drum to provide the power, but the INTELSAT V uses wings that extend from the boxlike body of the spacecraft to present the solar cells with maximum solar exposure. Like the INTELSAT IV-A, the satellite has hemispheric beams that are spatially separated from each other to allow frequency reuse. The INTELSAT V also has global, zonal, and spot beams, with the zonal beams being cross-polarized so as to attain a four-times frequency reuse capability in the C-band (6/4 GHz) frequencies. In addition, the INTELSAT V uses the Ku-band (14/11 GHz) frequencies in its spot beams so as to alleviate the congestion problems at C-band, primarily in Western Europe and North America. The INTELSAT V has approximately double the capacity of the IV-A, with a capability of 12,000 simultaneous telephone circuits and two television channels. Starting with the fifth spacecraft in the INTELSAT V series, each satellite is equipped with a Maritime Communications System (MCS). INTELSAT has leased this capability to the International Maritime Satellite Organization (INMARSAT) to enable it, together with other spacecraft capabilities obtained by INMARSAT, to meet its objective of a first-generation global maritime system.

In addition to the INTELSAT V satellites, there are under construction six INTELSAT V-A satellites, which have about a 25% greater capacity than the V satellites. These six V-A satellites use more extensive cross-polarization techniques through six additional transponders and also have fully steerable 4-GHz spot beams.

They are expected to be launched in the 1984-85 period. The last two or three satellites in the INTELSAT V/V-A series will also be equipped to operate in the 14/12 GHz range to provide customer-premise INTELSAT Business Services (IBS) through a class of new Earth stations ranging from 3 to 8 m (9.8 to 26 ft) in size.

In 1982, INTELSAT contracted for the provision of a new generation of satellites to succeed the V/V-A series--the INTELSAT VI. Initially, five satellites will be provided, and there are options for some 11 additional spacecraft. These satellites represent another quantum jump in capacity by providing as many as 35,000 telephone circuits and several television channels. The INTELSAT VI reuses the available frequency spectrum up to six times by spatial separation, cross-polarization, and by employing satellite-switched time division multiple-access techniques, yet another innovation for INTELSAT. It will also use frequency division multiple-access techniques in order to derive the maximum usage from the satellites and to provide the greatest operational flexibility possible. The INTELSAT VI satellites will contain both C-band and K-band transponders (38 and 10, respectively), with two sets of hemispheric beams, four sets of spot beams, and two sets of global beams for coverage. Each INTELSAT VI satellite will have six communication antenna, two global coverage horns, two large hemi-zone reflectors, and two spot beam reflectors. The satellite is a cylindrical shaped satellite with solar cells covering the drum in a manner similar to spacecraft prior to the INTELSAT V/V-A family. The lifetime of the satellites is anticipated to be ten years. (Each INTELSAT IV, IV-A, V, and V-A has had a seven-year lifetime.)

The INTELSAT satellites have been designed to accommodate ever-changing and increasing requirements--telephony, data, television, facsimile, high-speed services, and digital requirements of an international nature--while also being responsive to domestic needs of members wishing to lease capacity from INTELSAT and to the special needs of such organizations as INMARSAT. In this respect, digital voice processing techniques and other compression techniques could ultimately boost the capacity of an INTELSAT VI to the neighborhood of 100,000 voice circuits. This should be reassuring to those who are worried that fiber/optics may be winning out over satellite technology.

This brief description of the design of the system to meet these needs, as contemplated in the charter of INTELSAT, is primarily designed to demonstrate INTELSAT's responsiveness and innovation. More detailed technical descriptions of each series of satellites are to be found in Chap. 6.

Before addressing the growth in usage of the INTELSAT system, let us briefly examine the control and operation of the satellites.

At the time of the Early Bird program, and, indeed, in the INTELSAT II and III series, the control and operation of the space segment were relatively simple. The Manager performed all of the necessary functions and did so in conjunction with the operation of the United States Earth stations. The first few satellites, however, evolved into the global system, and the requirements for control of the satellite became more complex. This was true at all stages--as the satellite was placed into orbit, as it was tested, maneuvered, operated, monitored, and controlled over its lifetime, and even in its removal from geosynchronous orbit by using the remaining fuel to lift the satellite above the geosynchronous orbit.

As more and more Earth stations had access to the satellites, the need for up-to-date information on the health of the satellites, transponder by transponder, grew as did the need to monitor the use of the satellites by an ever-increasing number of users. Use of a variety of beams and frequency separation techniques, coupled with diverse uses of the satellites, added to the need to have more and more elaborate means of controlling and monitoring the space segment of INTELSAT. This, in turn, meant that entities located at different geographic points in each of the three ocean regions would be called upon to provide crucial assistance to the INTELSAT management teams responsible for the system, first the Manager and then the Executive Organ, under the direction of the Secretary General and now the Director General.

The result is that, today, INTELSAT has one of the most extensive and sophisticated networks in existence for the control and operation of its satellites. The INTELSAT IV, IV-A, and V series satellites are all in use, and are accessed by a number of different standards of Earth stations. There is the Standard A, 30-m (98-ft) station--the largest antenna operating to INTELSAT at C-band (6/4 GHz). There is the Standard B station, 10 to 11 m (32.8 to 36 ft) in size, also operating at C-band (6/4 GHz) for international services. There is also the Standard C Earth station, operating at Ku-band (14/11 GHz) and having antennas 13 to 15 m (42.6 to 49 ft) in size. Further, there are E-1, E-2, and E-3 Earth stations for INTELSAT Business Services (14/12 GHz); a Standard Z for domestic services; a proposed new Standard D [4.5 m (14.7 ft)] for rural services; and a proposed Standard F for C-band (6/4 GHz) business services. In addition, there are nonstandard Earth stations with authorized access to the space segment for international purposes on an individual basis, and numerous small (including receive-only) Earth stations using the INTELSAT system for domestic services on a dedicated leased capacity basis.

The complexities of operating this system require specialized capabilities worldwide. INTELSAT, by means of Signatories performing under contract, operates a number of specialized tracking, telemetry, command, and monitoring (TTC&M) stations which are collocated with Standard A Earth stations at Carnarvon, Australia; Tangua, Brazil; Zamengoe, Cameroon; Pleumeur-Bodou,

France; Fucino, Italy; Yamaguchi, Japan; and Paumalu, Hawaii, and Andover, Me., in the United States. There are also in-orbit test facilities (IOT) in Italy and Japan and communications system monitoring (CSM) facilities in France and at Etam, W. Va., in the United States. Additional TDMA reference and monitor station services (up to 16 in number) are to be established with the introduction of TDMA services on the INTELSAT system. The INTELSAT Operations Center (IOC) in Washington, D.C., is the hub for these and related worldwide facilities required to ensure the effective operation of the system. Recently, INTELSAT commissioned a new Communications System Monitoring (CSM) system to provide operational support for communications links in the system, which permits measurement at the IOC by specifying and selecting a particular communications link. The IOC receives the measurements from the ten Earth stations and 26 antennas described above as comprising the monitoring system. The measurements are stored, displayed, and analyzed at the IOC in order to assist the worldwide community of users to maintain the high quality of service of the network. By these means, INTELSAT is able to ensure close connection and cooperation between the operators of the space segment and the Earth stations and the users that access the satellites.

System Growth

The growth in utilization of the INTELSAT system has already been partially demonstrated in the preceding description of the INTELSAT satellites and their increase in capacity to meet traffic demand. Likewise, mention has been made of the growth in numbers of Earth stations and the diversification of antenna types having access to the INTELSAT system and other similar indications of system growth.

Yet all of this does not capture the truly dramatic nature of INTELSAT's expansion from the INTELSAT I era through the 1980's.[9] At year-end 1965, Early Bird had a total of 150 half-circuits in use; 80 hours of television were provided during the year; and 15 countries were served, through only one Earth station-to-Earth station pathway. By 1969, with the completion of global coverage, these numbers had exploded by a factor of 15 times. There were 1835 half-circuits in use and 1826 hours of television transmission and reception; there were also 52 countries or territories being served and 82 Earth station-to-Earth station pathways. But this was just the start! By the time the definitive arrangements entered into force in 1973, these figures had grown to 9814 half-circuits, 6817 hours of television transmission and reception, 100 countries or territories served, and 293 Earth station-to-Earth station pathways. A scant three years later, when the IV/IV-A programs were under way and INTELSAT was concluding arrangements for the procurement of INTELSAT V satellites, the comparable figures were 16,520 half-circuits, 12,952 hours of television, 115 countries or territories, and 491 Earth station-to-Earth station pathways. At the end of 1981, the growth had continued so that there were 50,250 half-circuits, 35,658 hours of television, 160 countries and areas, and 976 Earth station-to-Earth station pathways. In addition, from the time INTELSAT began to lease capacity for domestic applications in 1975, the number of countries availing themselves of this opportunity grew from 2 to 23 (in 1983).

By 1983, there were some 170 different political entities using the INTELSAT system, some 1200 Earth station-to-Earth station pathways, and some 60,000 half-circuits in service. All in all, a traffic growth of some 400 times the original number of circuits placed on the system in 1965! Basically, the INTELSAT Global Traffic Meeting (GTM), which convenes yearly, endeavors to forecast for five-year periods. It most recently has predicted that the system will realize roughly a doubling of traffic every five years. Figure 1 provides a graphic sense of the growth in the INTELSAT system's use, which has been paralleled by growth in membership.

Chapter 9 provides the details of INTELSAT's "conventional" services, and Chap. 10 provides an indication of the new services that are being planned for the INTELSAT system.

System Economics

As INTELSAT has extended and improved the satellites and related facilities and equipment comprising the space segment of the global system, and as usage and services have burgeoned, so have the economics of the system improved remarkably. The reduction in the charge for utilization of the space segment, at the same time that large capital investments are being made, is not only impressive but contrasts formidably with experiences in other high-technology fields. INTELSAT has achieved these results while serving not only heavy and lucrative traffic routes but, as noted previously in discussing the financial provisions of the Definitive Agreements, by averaging the costs of service to the members and other users alike over thin and heavy routes in all three ocean regions on a global coverage basis.

As Fig. 2 indicates, the INTELSAT system has produced meaningful cost reductions for the telecommunications satellite services customer.

Taking note of the cost of a unit of satellite utilization, which provides an indication of the trends in the cost of using the INTELSAT space segment, it can be seen in Fig. 3 that the charge

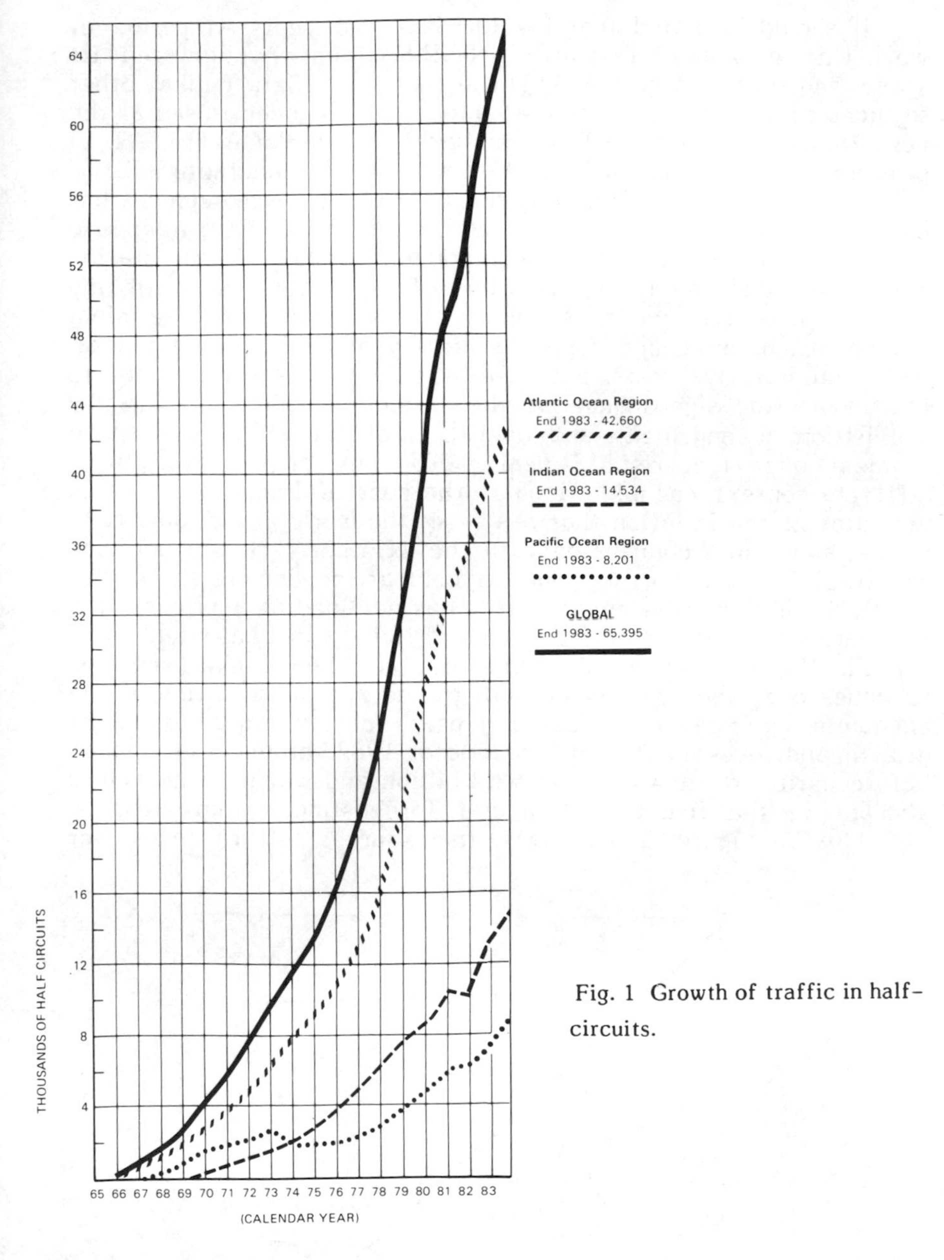

Fig. 1 Growth of traffic in half-circuits.

per unit of satellite utilization (a half-voice-circuit) has declined precipitously since the initial charge was established for a unit on Early Bird. Perhaps the significance of INTELSAT's lowering the cost of using the space segment can best be appreciated by observing that the charge by INTELSAT to those accessing the system represents less than 10% of the cost to the end-user and, in some cases, less than 5%.

It should be noted that for the last two years--a period of worldwide economic difficulty--INTELSAT has not lowered its space segment charge. Most interesting is that, unlike other services of a similar nature, the charge has not been raised at the very time when INTELSAT has embarked upon the INTELSAT VI program and is still having V and V-A spacecraft manufactured.

Of course, as noted earlier, the INTELSAT space segment has become a very small element of the overall charge for a telephone circuit to end users. Other components include Earth station equipment and facilities, landline charges between switching systems and the Earth stations, transmission and terminal equipment, maintenance costs, marketing and billing, and general and administrative costs, etc. Nevertheless, this serves only to underscore the significance of the economic efficiencies in the acquisition, management, and operation of the INTELSAT space segment over the past 18 years. To consider this facet in a different context, one might look at the cost-of-living index, as an indicator of the inflation that has beset the world in the same time frame, and gain a comparison with the INTELSAT charge per unit of utilization (see Fig. 4). Such a comparison demonstrates that the INTELSAT space segment charge has declined at approximately two-thirds the rate by which the cost of living index has risen!

Finally, a brief examination of INTELSAT's expenses and revenues over the past decade will provide an appreciation of the economic significance of declining usage charges in a period of growth and investment in the decade of 1982 through 1991, which is forecast to cost an estimated $2.2 billion and which is taken into account in the five-year financial forecasting system used by INTELSAT. Figure 5 indicates the sources of revenues for

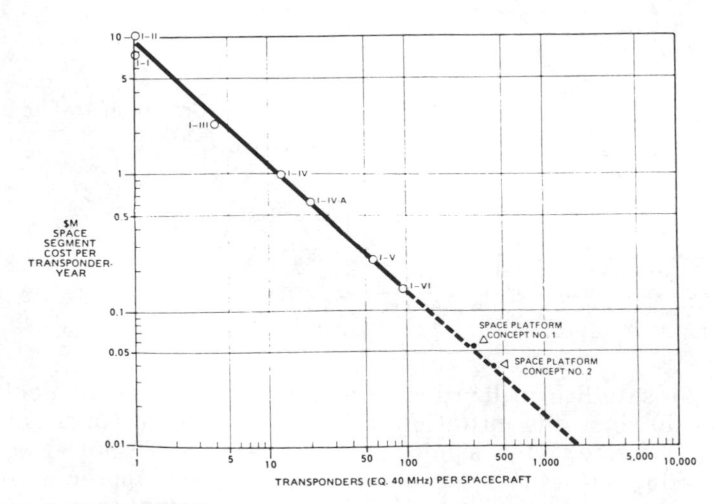

Fig. 2 INTELSAT series demonstrates economy of scale in communications spacecraft (past and present).

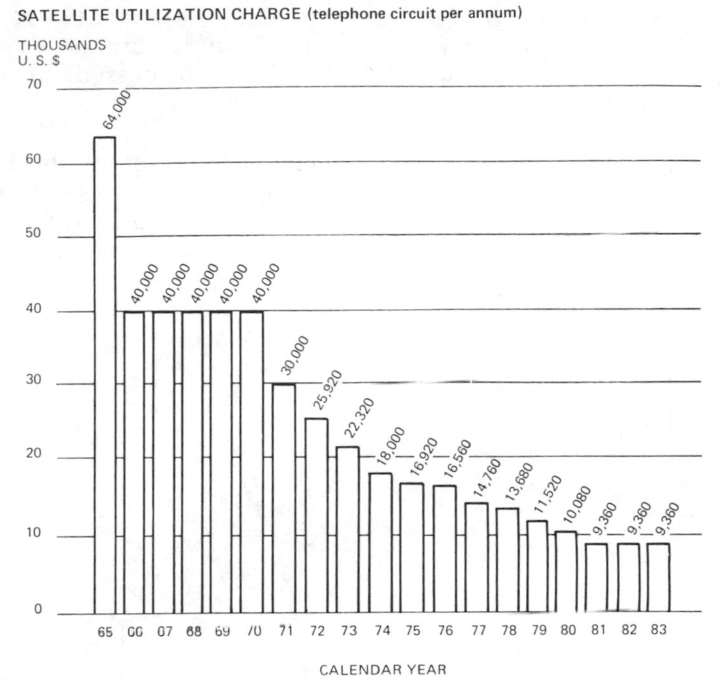

Fig. 3 Satellite utilization charge (telephone circuit per annum).

INTELSAT on a percentage basis, and Fig. 6 the disposition of total revenues. While the examination of the space segment utilization charge above discussed only the full-time half-voice-circuit charge ("unit of satellite utilization"), other services have had their charges based upon derivatives of the basic "unit of satellite utilization" charge, except perhaps where newer services have recently been introduced, where an element of "promotional" charging may be implied.

Some Policy Issues and Challenges for INTELSAT

The preceding sections of this chapter have addressed the history of INTELSAT, the Agreements constituting the organization and guiding its actions, the design of the system, and the growth in the system and organization, as well as the economic and financial results of the organization over a period of years. The establishment of the INTELSAT Organization and system has not, however, been free of difficulty and problems. Some of these have been discussed in a previous section and are well known. Some have come and passed. Others have emerged fully and persist in the form of either problems or challenges that the Organization must meet; while still others are just in the process of emerging and, while not fully defined, may be expected to be around for some time to come. Some of the most crucial policy issues and challenges facing INTELSAT are discussed in the following sections.

International Telecommunication Union Allocation of the Use of the Frequency Spectrum and Orbital Arc for Geostationary Satellite Systems. As noted in discussing the formation and philosophy of the INTELSAT Organization, as early as 1961-62,

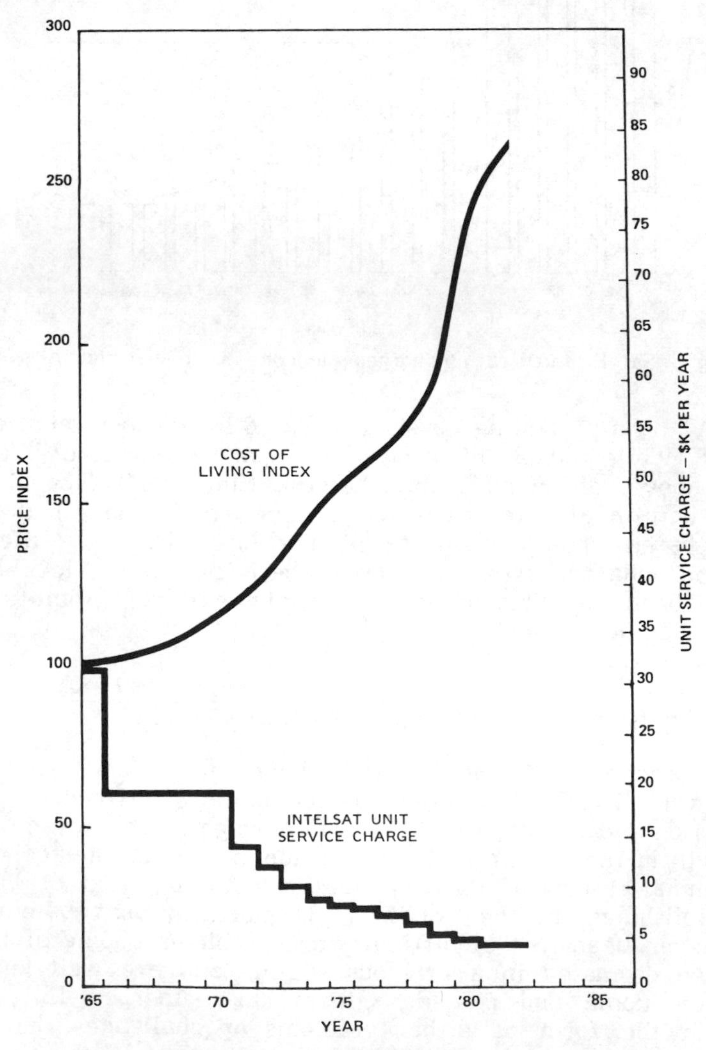

Fig. 4 Cost-of-living index vs INTELSAT unit service charge.

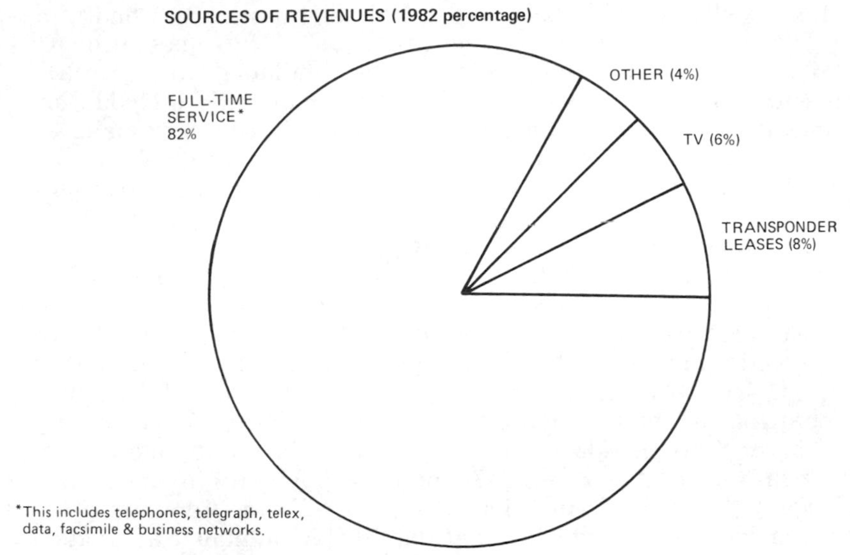

Fig. 5 Sources of revenues (1982) percentages).

there was widespread concern about the lack of adequate capacity in the frequency spectrum and orbital arc of outer space. Indeed, one of the reasons for establishing INTELSAT as a unique global commercial communications satellite system was to achieve efficient, economic, and equitable use of these vital international resources. The global interconnectivity that has been accomplished by INTELSAT is ample testimony to the wisdom of such a "sharing" approach.

Despite significant technical advances and the effort on the part of INTELSAT and other satellite system operators and regulators, the problem of inadequate capacity in the frequency spectrum and orbital arc continues. There can be no challenge to the assertion that the INTELSAT system has proved to be not only a cost-efficient system serving the entire world, but an extremely efficient system for preservation of the efficient use of the orbital arc and in deriving the maximum usage of available frequency spectrum. The INTELSAT VI, for instance, will derive 170 times more usable capability from a single orbital location than did the INTELSAT I.

The emergence of "domestic" systems serving the needs of various countries, and the extension of satellite systems apart from INTELSAT to serve the special needs of a "region" in addition to the requirements of international organizations such as INTELSAT and INMARSAT, test the ability to continue to achieve these results. Moreover, suggestions that there should be additional international systems, over and above those such as INTELSAT and

INTERSPUTNIK, which would serve their different areas and user groups, can only exacerbate the problems. The geostationary orbital arc and the radio frequencies available for satellite communications are limited natural resources and the introduction of duplicative and unnecessary systems can only serve to make an already difficult situation worse.

INTELSAT has displayed leadership in introducing a variety of techniques in the design of its satellites, and transmission techniques in the use of them. It has pioneered the use of "hybrid" satellites operating at both C- and K-bands and in frequency reuse techniques. Its requirements continue to increase, both for additional frequencies and use of orbital arc "slots," to accommodate the more than 170 user areas of the world that it serves. It would appear that its "premier" role should be given consideration as the international telecommunications community contemplates actions relative to frequency assignments and orbital arc allocations in the future. A major opportunity to meet this challenge will be in connection with the World Administrative Radio Conferences of the International Telecommunication Union in the 1985-88 time frame. It may be assumed that INTELSAT will undertake serious preparation for these WARCs and it can only be hoped that narrow industrial and short-term interests will be put

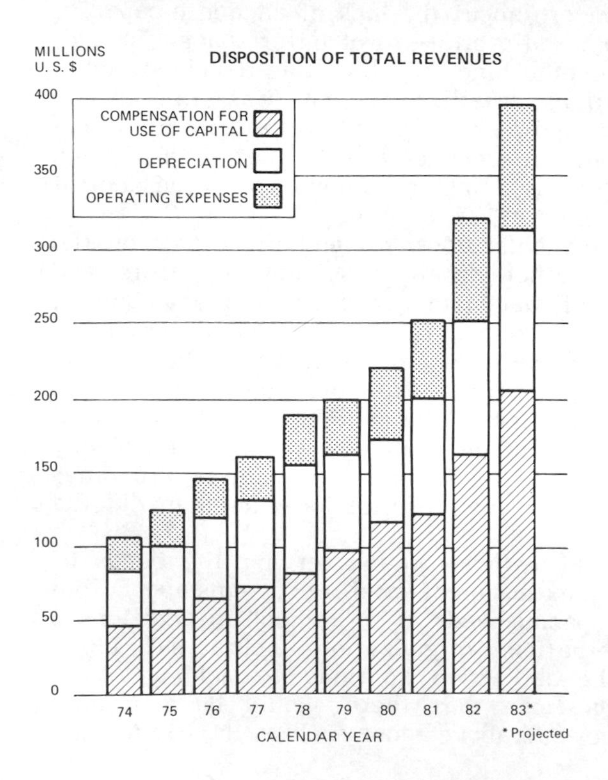

Fig. 6 Disposition of total revenues.

aside in recognition of the need to provide reasonable and equitable spectrum and arc capacity for use by INTELSAT to serve the entire world.

Other Satellite Systems and INTELSAT. As noted briefly in the foregoing discussion of the INTELSAT interim and definitive arrangements, the question of the "single global system" of INTELSAT and whether it was to exist to the exclusion of other systems has arisen with some consistency over the years, in different forms and for different reasons. That INTELSAT is the "unique" system serving the entire globe can hardly be disputed from the facts presented in the previous sections of this chapter. And, whether it should exist to the exclusion of other systems is not an issue, since other systems are contemplated to exist under the provisions of the INTELSAT Agreement (Article XIV in particular) and do, in fact, exist and operate. Many domestic satellite systems, beginning with Canadian and Soviet domestic systems, have been established and have achieved technical coordination to the satisfaction of all, even though the proliferation of satellites has made this more and more difficult.

There have been difficult and complicated coordinations of satellite frequencies (power outputs as well as orbital locations of satellites have been adjusted between INTELSAT and its members under this Article), but, in general, they have been resolved expeditiously and satisfactorily. In addition, of course, INTELSAT itself continues to provide satellite capacity to some two dozen countries for their respective domestic usage--a number expected to double by the end of the decade.

In sum, as under the interim arrangements, the era of the INTELSAT definitive regime--the last decade--treats satellite systems apart from INTELSAT for strictly domestic usage as being noncompetitive and nonharmful to INTELSAT. Thus only "technical" coordination that is of mutual interest need take place with the Board of Governors and, in some cases, the Assembly of Parties, in accordance with paragraph (f) of Article XI. The rights and obligations of INTELSAT members with respect to "international systems" (excluding both international and domestic "national security systems"), however, are a more complicated matter. As noted above, there were interests as early as during the negotiation of the Interim Agreements in considering the possibility of INTELSAT members establishing and using separate systems to carry international traffic--particularly on the part of highly industrialized countries with developing aerospace industries.

The issue occupied a great deal of attention during the negotiation of the definitive arrangements and the result was that members agreed that, in addition to the technical coordination required for domestic systems, they would also coordinate any such "international" systems with INTELSAT, through the Board to the Assembly of Parties, in order to determine if they would cause

INTELSAT "significant economic harm" (paragraph d of Article XIV). The Assembly would then make recommendations, which would contain its "findings" with respect to both the impact on INTELSAT in terms of technical compatibility in the use of the radio frequency spectrum and orbital space, as to whether there would be significant economic harm to INTELSAT.

Among the objectives was that of not prejudicing the establishment of direct telecommunications links through the INTELSAT system among users. While a thorny and continuing problem for INTELSAT to address, the establishment, use, and acquisition by INTELSAT members of other systems has not been widespread up until 1982, and those presented for coordination were not seen as being "competitive" with INTELSAT. Thus unanimous action by the Assembly of Parties has, to date, been possible. The so-called "regional" systems that have been successfully coordinated pursuant to Article XIV(d) include: ARABSAT, EUTELSAT (covering portions of Western Europe), and certain United States/Canadian services. In each of these cases, no harmful impact upon INTELSAT in economic and technical terms has been found--in the main because the amount of traffic that has been diverted from the INTELSAT system or lost to the INTELSAT system as potential traffic has been minor.

Basically, there have yet to be efforts to divert heavy amounts of traffic from transoceanic streams from the INTELSAT system that would clearly impact directly and negatively on the fundamental purposes of the organization to provide global tariffs and coverage through its system; with the user countries and peoples most severely affected being those from developing areas.

However, INTELSAT must be diligent in anticipating and defending itself against such possible actions and has demonstrated the ability to do so in responding to recent applications in the United States to provide what must be clearly perceived as directly harmful satellite service over the heavy North Atlantic stream. A letter from the Director General of INTELSAT to the United States Party (the Department of State), voicing INTELSAT's objection, was unanimously sustained by a strong resolution of the Meeting of Signatories.[10] On the other hand, there may continue to be limited international systems of a regional nature that would not cause INTELSAT fundamental harm. In sum, INTELSAT will have to continue to seek practical and "balanced" solutions to such circumstances while reserving combative responses for situations that challenge its very viability and existence.

Other Competition and Challenges Facing INTELSAT. While satellite technology under the aegis of the developing INTELSAT system has been outstanding, and efforts are under way to continue this innovation and creative movement, other means of providing international transmission and service capabilities have also seen major technical development. In effect, such developments challenge INTELSAT to make even greater progress that can quickly accrue to the benefit of consumers around the world. In particular, advanced fiber/optic systems suitable for use in transoceanic submarine cables will provide tardy competition for the INTELSAT system--in a healthy manner--in the near future.

Such systems now planned for implementation in the Atlantic basin in the 1980's will provide major increases in capacity at much lower cost than all of the previous submarine cable systems in various ocean regions. Should such systems, which are capable of providing tens of thousands of telephone circuits and multiple-landing (destination) points at relatively low cost per channel year, emerge as expected, they will force INTELSAT to strive to make even greater progress in technical innovation and to bring down even further the cost of using the INTELSAT system while offering more and more diverse and responsive services via the satellite system. This competitive technological situation, which will take place primarily on transoceanic and heavy-stream routes, is sure to provide maximum benefits to users. It can be construed as being "healthy competition" in contrast to proposals to divert traffic from the INTELSAT system to other satellite systems intending to serve only lucrative routes. Such traffic has already been planned in a highly technical and economically efficient manner and in a way that will benefit many countries and users, rather than enrich a few.

INTELSAT is geared to take competition from cables as a result of its long-term exploratory research and study programs addressing such concepts as deployable offset antennas, 23/32 GHz and optical intersatellite link subsystems, low bit rate signal processing, all solid-state amplifiers, and the potential use of the 30/20 GHz bands in the Earth-to-space and space-to-Earth modes. INTELSAT has system-related technologies bearing on future systems (beyond the INTELSAT VI), including optical links, new digital systems, and a variety of spacecraft-related improvements affecting switching, interconnectivity, lifetime and Earth coverage, and transmission capabilities.[11]

Conclusion

It may be concluded that INTELSAT has met and continues to meet in exemplary fashion its mission requirements to serve the entire world with the most effective and efficient satellite system and services possible. There seems to be no doubt that, despite challenges and changes in the telecommunications environment since the creation of this unique organization and system, it can and will continue to introduce technical innovation and needed services of great variety to all users and to benefit telecommunications consumers throughout the world, particularly in developing areas.

References

[1]Naisbitt, J., Megatrends, Warner Books, New York, 1982, p. 12.

[2]U.N. Resolution 1721, 1961 U.N. GAOR, Supp. 17, U.N. Doc. A/5100, Dec. 21, 1961. U.N. Resolution 1962, 1963, 18th U.N. GAOR, Supp. 15, U.N. Doc. A/5515, Dec. 13, 1963.

[3]Treaties and Other International Acts (TIAS) 4892.

[4]White House press releases of Jan. 1, 1964 and July 24, 1961.

[5]Proposal of President Kennedy of Feb. 7, 1962, introduced by Senators Magnuson and Kerr as S.2814, Hearings Before the Senate Committee on Aeronautical and Space Sciences, 87th Congress, Second Session (1962).

[6]White House press release of Aug. 31, 1962.

[7]Treaties and Other International Acts (TIAS) 7532.

[8]INTELSAT Interim Agreement, Article IX(b)(iv).

[9]INTELSAT Annual Report, 1982, p. 32.

[10]Letter dated Apr. 5, 1983, from INTELSAT Director General, S. Astrain, to the United States Deputy Secretary of State, K. Dam.

[11]INTELSAT Annual Report, 1982, p. 22.

6. INTELSAT Satellites

Emeric Podraczky* and Joseph N. Pelton†
INTELSAT, Washington, D.C.

Introduction

The story of INTELSAT satellites is complex and interesting. In fact, it is not one story, but a series of stories: stories that involve people as well as technological breakthroughs and set-backs. These stories bespeak an important transition, from the valiant efforts by a small and dedicated group of scientists and engineers who helped build Early Bird, to complex management structures that employ virtual armies of scientists, engineers, and technicians throughout the world to design and manufacture today's highly sophisticated INTELSAT V and INTELSAT VI satellites. They also involve the story of how these satellites are launched into geosynchronous orbit and, once launched, how the tracking, telemetry, command, and monitoring (TTC&M) functions maintain and operate the satellites in orbit. This supporting TTC&M process also has become much more sophisticated, demanding--and costly--as the satellite lifetimes have been stretched from a few months to a decade, thanks to the advent of the current superreliable satellites.

The story of INTELSAT satellites must, of course, begin with Early Bird.

I. Early Bird

In November 1963, the Communications Satellite Corporation (COMSAT) placed a contract with the Hughes Aircraft Company to build a satellite, referred to at that time as HS-303. That satellite, eventually to be known as Early Bird, was to become the world's first global communications satellite. The successful

---

Invited paper received October 11, 1983.

*Director, Engineering.
†Director, Strategic Policy Formulation.

launch and operation of the SYNCOM II satellite (also built by Hughes Aircraft Company) had previously demonstrated that a geosynchronous orbit could be achieved with a lightweight satellite and that such a satellite could provide reliable microwave service between space and Earth. In many respects, therefore, the design for Early Bird was based upon the design of the SYNCOM satellite. Harold Rosen and Donald Williams, both of Hughes Aircraft, were the primary design engineers for both.

The HS-303 spacecraft was a spin-stabilized solar-powered unit, with an orbit transfer mass of 68 kg (150 lb) and an orbital mass of 38.5 kg (85 lb). Its solar cells were designed to produce, at the end of an 18-month lifetime, a power output of 40 W. The total bandwidth was a mere 50 MHz. The satellite employed an intermediate-frequency-type translation transponder, which received in the 6000-MHz band and retransmitted from space to Earth in the 4000-MHz band. The quite small, cylindrical-shaped satellite had a diameter of only 72 cm (28 in.), and contained two receivers that operated continuously, with one carrying signals in the west-to-east direction (from the United States to Europe), while the other one carried signals east to west (from Europe to the United States). These transponders thus had a bandwidth of 25 MHz each, and the output of the two receivers was combined to feed a single 6-W traveling wave tube (TWT). The output of the traveling wave tube fed the transmit antenna, which had a toroidal, or donut-shaped pattern, with an 11-deg beam width squinted at an angle so that the point of the maximum radiated power was 7 deg above the equatorial plane of the satellite. The purpose of this angle was to maximize the signal at the location of the North American and European Earth stations. This seemed a perfectly reasonable thing to do at the time. It substantially increased power levels at all five Earth station locations equipped to operate to this satellite--namely: the United States Earth station at Andover, Me., as well as the four European stations, located in Goonhilly Downs (England); Pleumeur-Bodou (France); Raisting (West Germany); and Fucino (Italy). At the time there were no other Earth stations either in existence or planned. Thus, coverage of Africa, South America, or the Middle East was not available. This lack of universal coverage was to change quickly, but for the time being Early Bird's squinted toroidal beam optimized operational capacity.

Early Bird was, in effect, designed almost as if it were a submarine cable in the sky and, as such, was able to serve only one transmit and reception point on either side of the Atlantic. Thus only one European and one North American station could handle the traffic at a time. All European traffic had to terminate at one of these Earth stations, and a terrestrial routing carried the traffic to the other countries. In that early configuration, there was one European station designated as the operating station, another as the immediate standby, while the third and fourth stations were considered to be on a maintenance status. During the initial period of operation, the traffic was rotated from station to station. The Italian station at Fucino, which had a smaller antenna, was used to carry traffic on Saturdays and Sundays. This station was thus able to share in the commercial operations and to provide useful information concerning operation between Earth stations with dissimilar transmit power and reception capabilities.

Early Bird was launched from Cape Canaveral, aboard a thrust-augmented DELTA vehicle, under contract with the United States National Aeronautics and Space Administration (NASA). The vehicle (DELTA No. 30) was the second thrust-augmented DELTA launched from Cape Canaveral. The previous such launch was DELTA No. 25, which successfully placed the SYNCOM III in geosynchronous orbit in August of 1964. Although a geosynchronous orbit inclined as much as 15 deg from the equatorial plane could still be visible between Andover, Me., and all of the European stations, such a large inclination angle would have constituted a problem because of the sophisticated tracking, data, and antenna pointing system that would be needed to maintain continuous operation. A noninclined orbit had clear-cut advantages for a geostationary satellite. There was, however, some doubt as to whether the thrust-augmented DELTA could place Early Bird, with its fully loaded hydrogen peroxide fuel tanks and apogee motor weighing a total of 57.6 kg (148.6 lb), into the proper transfer orbit needed to keep a low inclination angle.

It was for this reason that two plans were prepared for the launch. If the nominal performance was achieved from the thrust-augmented DELTA, then the spacecraft would be reoriented in the transfer orbit to permit the apogee motor, when it was fired, to remove all inclination and circularize the orbit. The second contingency plan was to place the spacecraft in an orbit with an inclination of 7 deg. Fortunately, the thrust-augmented DELTA did perform extremely well, and a perfect circular geostationary orbit was achieved. Early Bird's apogee motor, developed by the Jet Propulsion Laboratory, weighed some 32 kg (70 lb) and contained a total of 17 kg (60 lb) of high-energy propellant. It was capable of developing a thrust of 509 kg (1120 lb), for a sustained burn of some 20 s and, thus, imparted a velocity of 1482 m/s (4862 ft/s) to the spacecraft.

The liftoff on April 6, 1965, occurred at 2347 Greenwich time, and the first computation of the transfer orbit was made using radar data from Cape Canaveral. This computation provided predictions of the first Andover visibility time, which occurred as the spacecraft approached second apogee. Solar aspect pulses that appeared on the spacecraft telemetry from Carnarvon, Australia, five hours after launch, indicated that spacecraft attitude was 8-14 deg off the predicted value. Very high-frequency (VHF) telemetry was received at the proper time at Andover, and commands were sent to turn on the microwave transponder so that microwave polarization measurements could be used to provide a second reference reading on spacecraft attitude. Precision tracking on the microwave signal was also used to compute refinements to the transfer orbit. By second apogee, computations showed that the spacecraft attitude was unsatisfactory, being 8 deg off the nominal. The solution was to use the spacecraft hydrogen peroxide control system in the brief fire pulse mode to reorient the spacecraft so that after the apogee motor firing the velocity vector would lie in the equatorial plane. After reorientation and additional computation to improve the accuracy of the computed orbit, it was estimated that after the apogee motor firing the spacecraft velocity would be about 30 m/s (100 f/s) less than the required velocity for stationary orbit.

At fourth apogee, the hydrogen peroxide system was operated in the continuous mode to add the needed velocity. This raised the perigee altitude from 1454 km (901 mi) to 1747 km (1083 mi).

The accumulation of spacecraft attitude data during the second and fourth transfer orbits provided a more precise determination, and an attitude correction of a few degrees was made just prior to sixth apogee. At sixth apogee, two and a half days after launch, the apogee motor was fired, and the spacecraft was in a nearly perfect synchronous orbit, with an inclination of only 0.087 deg. This confirmed "perfect" spacecraft orientation and proper operation of the apogee motor. Eleven hours later the hydrogen peroxide system was used to reorient the spacecraft and make the spin axis perpendicular to the plane of the orbit. This was accomplished on April 10th, at 0100 hours Greenwich time, just three days and two hours after launch. For the years that followed, the spacecraft attitude and position were satisfactorily maintained to allow continuous communications between North America and Europe.

After two days of precise tracking and orbit computation, it was possible to plan the velocity correction maneuvers required to circularize the orbit. Using the radial jets in the pulse mode at apogee, on April 13th, the velocity was increased to raise perigee and change the satellite drift rate. About 12 hours later, at perigee, the radial jet was used to decrease velocity and bring apogee nearer the synchronous altitude. On April 14th, a velocity correction was made to raise perigee to the circular synchronous altitude, and almost completely "stopped" the satellite's movement with respect to the Earth by reducing the drift rate from 1.5 to 0.22 deg of longitude per day.

On April 22nd, the peroxide system was used to change the spacecraft velocity by 2 m/s (6.6 ft/s), resulting in a drift rate of 0.05 deg east per day. This precise value was expected to provide a drift that would be decreased to zero in a few months by the triaxiality forces of the Earth. Subsequent tracking data indicate that the satellite drifted eastward from April 22nd to July 1st, when, at 28.0 deg west longitude, the drift reached zero.

The next velocity maneuver was not required until December 1965, more than seven months after the April 22nd operation. The quantity of hydrogen peroxide remaining in the spacecraft by December 1965 was about 1 kg (2.2 lb) in each system, while the more or less biannual velocity corrections were estimated to require only about one-twentieth of a kilogram (0.44 lb). Thus Early Bird clearly demonstrated that, with a nominal launch and apogee motor fire, the quantity of fuel would not be an important limit on the system lifetime. The redundant communications equipment that had been installed on Early Bird therefore became significant. A minimum stationkeeping lifetime of three years became possible, even though the originally expected lifetime had been conservatively estimated to be 18 months.

With the spacecraft synchronized, attention was turned to the spacecraft communications performance. Two groups of tests were conducted. The performance test plan was designed to determine how well the spacecraft performance in orbit would meet the contract specifications. The second group of tests, the detailed experimental plan, was designed to determine system parameters in optimizing the spacecraft performance with respect to the operating ground stations.

The performance test plan was completed in the four days following spacecraft reorientation. The contract specification for effective isotropically radiated power (e.i.r.p.) was 10 dBW. Measurements at Andover, Me., indicated that the power was greater than 10.25 dBW for the lowest of the four combinations of two carriers and two output tubes. Spin ripple measurements (i.e., variations in transmit performance caused by the spinning motion of the satellite), as measured by the microwave signal, showed less than one-quarter of a decibel variation in level during one revolution of the spacecraft. Measurements of intelligible cross talk indicated that the specification of -45 dB could be met only with transponder No. 2. For this reason, that traveling wave tube was used for all commercial operations throughout the summer.

The experimental plan tests were conducted for 18 h/day for a very exhaustive week. System characteristics with a variety of loading conditions were measured for all stations. It was determined that the 240-circuit capacity could be realized by the four high-capacity stations and that a 6-dB margin existed in fair weather. It was also determined that the antenna gain and receiving sensitivity for the four large stations were similar to within 1 dB.

Tests on system characteristics for monochrome television were continued until all stations reached agreement on the optimum operating parameters.

Thus, only three weeks after launch, the spacecraft and the Earth stations had been properly tested, and the space segment of the system was ready for use. The terrestrial network operators then began circuit lineup. For the 120-channel multiplex system to be used for the first commercial service, it was necessary to allow four weeks for adjustments for a complete end-to-end circuit level test. These tests were completed on schedule and the system was ready for operation in June. On June 28, 1965, commercial service was inaugurated, with 65 telephone circuits utilized for 18 h each day, Monday through Friday, with some weekend traffic through the Earth stations in France and Italy.

The communications quality of Early Bird was good and, indeed, operation exceeded existing standards--being some 6-dB lower in noise level than the submarine cables then in operation. The question of the effect of time delay was more complicated. The applicable recommendations of the Consultative Committee for International Radio (CCIR) of the time follow.

1) Acceptable delay without reservation: 0-150 ms.

2) Provisionally acceptable delay: 150-400 ms. In this range, connections may be permitted--in particular, when compensating advantages are obtained.

3) Provisionally unacceptable: 400 ms and higher.

The Early Bird system used modern echo suppressors and had a time delay of 260 ms from Earth station to the satellite and back down to the other Earth station. In addition to this, an allowance of about 30 ms for the terrestrial networks at each end must be added. This totaled 320 ms for the one-way delay, and thus it was only provisionally acceptable by CCIR standards.

Since little experience was available on commercial satellite circuits, there was a natural concern about the delay factor.

On Sept. 1, 1965, a program of call-backs was initiated in the United States by the American Telephone & Telegraph Company to question satellite users and to collect statistical data on their reactions to the quality of service. Tests of this type were also conducted by the United States, England, France, and West Germany. After thousands of call-backs, users indicated a very high level of acceptance of the service, with favorable comparisons being made vis-à-vis submarine cables.

The Early Bird satellite project proved many things but, most fundamentally, it proved that a working network of a geosynchronous spacecraft plus Earth stations could provide a high-capacity, reliable communications link between North America and Europe.

II. INTELSAT II

"The INTELSAT II satellite, the largest communications satellite ever launched into a synchronous orbit, is the first in a new generation of bigger and more powerful spacecraft being built by Hughes Aircraft Company ... for the International Telecommunications Satellite Consortium (INTELSAT). The new satellite, twice as large as Early Bird and with more than two times the power, is another step toward the day when even larger multi-purpose satellites will form a global system of satellites around the equator." Thus was the now "primitive" INTELSAT II satellite described in the February 1967 issue of the Telecommunication Journal.[1]

By today's standards, of course, the INTELSAT II was a very small and uncomplicated spacecraft; yet, at the time, it represented a number of major new developments that were firsts in the field. The INTELSAT II, in fact, represented the evolution of communications satellites as a unique technology--as opposed to Early Bird, which was, in effect, designed to emulate a submarine cable in the sky. A major limitation of Early Bird was that it could link only two points, rather than a complete Earth station network with multidestination access. What was considered an advanced satellite antenna design at the time was developed to permit direct contact with several ground stations simultaneously and thus provided for greater commercial use without any degradation for loss of power and the transmitting signal. Development of this new capability was no accident. In fact, in many ways, the INTELSAT II was tailor-made to meet the urgent need of NASA's communications network (NASCOM) in support of their manned space programs.

NASA had already established an extensive worldwide tracking, telemetry, and command system to support those manned space programs. Yet there were still major gaps in their communications system over the oceans. Through COMSAT, the United States Signatory in INTELSAT, NASA inquired whether a special INTELSAT satellite program could be quickly implemented to support its communications needs through ground terminals as well as ship-based terminals that could provide tracking, telemetry, and command services to support the Mercury and Gemini spacecraft.

After intense discussions, it was decided that, in order to produce an INTELSAT II, NASA would finance enough of the cost of the INTELSAT II program to defray the incremental costs of upgrading the Early Bird design. Thus, in a matter of weeks negotiations were completed and the Interim Communications Satellite Committee (ICSC) agreed to squeeze the INTELSAT II program between Early Bird and the next planned system, now known as the INTELSAT III satellite program.

The Telecommunication Journal[1] further described the INTELSAT II satellite:

"In design concept the satellite follows the same basic principles developed by Hughes for Early Bird. This includes spin stabilization, a toroidal antenna beam which continually encompasses the Earth and a simple gas jet system for attitude control and stationkeeping.

"The satellite is 142 cm (56 in.) in diameter and 67 cm (26.5 in.) in height, exclusive of antennas and apogee motor nozzle. It provides communications capability which will consist of mutiple-access high-quality voice and digital data, television or a combination of both."

Thus, for the first time, an interactive satellite network, connecting a large number of Earth stations simultaneously, was possible, and the inherent flexibility of a space-connected multipoint-to-multipoint network was introduced. No longer would a communications satellite be an imitation of a point-to-point submarine cable, but a new force in global communications.

The Telecommunication Journal[1] article continued:

"Weight of the new spacecraft was 162 kg (356 lb) at launch. The outer surface of the satellite is covered with 12,756 N-on-P silicon solar cells which deliver 85 watts electrical power under normal operating conditions. The additional power is used to add increased effective radiated power to provide greater geographical coverage. The extra weight is used in providing the additional electrical power and a long life control system.

"The spacecraft structure consists basically of a central-stiffened tube which directly supports the apogee motor and communications antennas. An aft radial bulkhead and rib assembly supports the majority of the payload electronics, while a forward bulkhead provides support for lateral and radial loads.

"Both ends of the structure are closed by thermal shields, and the shield at the antenna end also serves as the antenna ground plane.

"Four traveling wave transmitter tubes are provided which may be used in any combination consistent with the available power. The normal communications mode of operation throughout the satellite lifetime will be two or three tubes in operation, even during the eclipse period.

"With the increased weight of the satellite, 87 kg (191.5 lb) in orbit, as compared to the 38.5 kg (85 lb) Early Bird, the hydrogen peroxide fuel supply has been doubled.

"The basic communications system is comprised of two redundant linear repeaters, with a 125-MHz bandwidth, a 6-dB noise figure, and four 6-W traveling wave tubes of which one, two, or three may be turned on in parallel. The transmitting antenna is a four-element biconical horn array that has a relatively flat gain across the passband.

"The telemetry subsystem is similar to that of Early Bird and consists of two VHF encoders, two transmitters, and eight whip antennas. The encoders modulate the two VHF transmitters, and the 4-GHz beacon signals.

"The spacecraft's self-contained propulsion system (the Aerojet General SVM-1 apogee motor) supplies the velocity boost to inject the satellite into synchronous altitude. The apogee motor imparts a velocity of approximately 1830 m/s (6000 ft/s) to the 162-kg (356-lb) spacecraft for injection into the synchronous orbit."

The first INTELSAT II (F-1), launched on Oct. 26, 1966, was intended for a "stationary" synchronous orbit position over the Pacific. Because this satellite failed to achieve stationary orbit due to the launch failure of a DELTA rocket, a second INTELSAT II (F-2) was launched on January 11, 1967, and was successfully positioned in a stationary orbit to provide multiple-access communications in the Pacific area.

Voice and teletypewriter communications were thus provided between one of the United States gateway stations in Washington and Hawaii and two NASA Apollo tracking stations: one in Australia and one onboard a ship in the Pacific. The balance of the satellite capacity was used to support communications to stations in Japan, Thailand, and the Philippines.

The stationkeeping constraints imposed by INTELSAT's first three satellites were remarkably different. Early Bird was maintained between longitude limits of 25 and 40 deg west, sufficient to ensure higher-than-minimum antenna elevation angles for the participating Earth stations. As a result, stationkeeping maneuvers for Early Bird were necessary only at eight- to nine-month intervals.

Stations communicating with the Pacific INTELSAT II were spread over much greater distances from the subsatellite point, the two farthermost being Brewster Flat, Wash., in the United States, and Bangkok, Thailand. The small mutual region of Earth station coverage, coupled with the satellite's orbital inclination of 2.0 deg, required that the satellite be maintained between longitude limits of 174 and 176 deg east to ensure that antenna elevation angles for these two stations did not become less than 5 deg. Stationkeeping maneuvers for this satellite thus were required approximately every two months.

The allowable longitude drift of the Atlantic INTELSAT II was less than that allowed for Early Bird, because coverage required a satellite link between Andover, Me., and a ship in the Indian Ocean. Also, since the orbital inclination achieved with the Atlantic INTELSAT II was 1.5 deg off the equatorial plane, velocity corrections were required to compensate for this as well. The new effort was a requirement for longitudinal corrections at about three-month intervals.

III. INTELSAT III

The INTELSAT III satellite series represented a number of important operational, technical, and institutional firsts for INTELSAT, and, in many ways, can perhaps be considered a major turning point in INTELSAT's development. Key elements of the INTELSAT III program follow:

1) It was the first satellite wherein INTELSAT engaged in a full international Request for Proposal (RFP) process. (The

INTELSAT I had been awarded to Hughes Aircraft Company by COMSAT prior to the formation of INTELSAT, and the INTELSAT II was basically a modification and expansion of the INTELSAT I satellite design in response to the urgent multidestination access needs of NASA.)

2) INTELSAT III, with five times the capacity of the INTELSAT I and II satellites, represented a major technological innovation. The INTELSAT III program was, in effect, a statement of policy that said: "We believe that, with the availability of a new, lower-cost, high-quality transmission facility, a new market for our services will quickly develop."

3) The INTELSAT III program introduced INTELSAT to the many political, technical, and operational problems inherent in the operation of a global satellite system. Specific questions that arose included: a) How should INTELSAT respond to the issue posed by the increased cost of international subcontractor participation? b) To what extent should patents and data developed under INTELSAT contracts be released? To whom? And under what constraints, in terms of competitive use in regional systems? c) How should INTELSAT respond to in-orbit satellite difficulties, and what type of in-orbit or on-the-ground "sparing" philosophy should INTELSAT embrace?

4) Finally, the INTELSAT III series marked the definitive commitment by INTELSAT to the operation of a global three-region geosynchronous satellite system.

A. INTELSAT Undergoes Its First International Procurement of Satellites

The procurement of INTELSAT III probably could be said to have been initiated in June 1965, when a detailed technical review and amendment of an international Request For Proposals (RFP) was undertaken by the Technical Subcommittee of the Interim Communications Satellite Committee (ICSC). This was followed by a review of the specifications by the ICSC itself, in August 1965, where further modifications to the proposed technical specifications were made and the RFP approved for issue. One of the significant modifications to this RFP was the ICSC's decision to allow additional time for translation and international mailing of the RFP. When the responses were received, however, only three American contractors--namely, Thompson, Ramo, Wooldridge, Inc. (TRW), RCA, and Hughes Aircraft--had responded. Following a detailed review of the proposals by the Technical Subcommittee, the Manager (COMSAT), and the ICSC, TRW was selected as the preferred contractor for final negotiations, even though Canada and the Nordic Group (Sweden, Norway, and Denmark) reserved their positions because they believed that RCA's lower cost proposal merited parallel negotiations. But, perhaps more significantly, the ICSC, in selecting TRW, proposed that the negotiations should seek to increase the level of international participation.

In April 1966, the ICSC approved the award of the contract to TRW and, in addition, awarded the option for increased costs to broaden the level of international participation in the contract. An important aspect of the INTELSAT III program was the abandonment of the idea that nonsynchronous satellites could provide a more cost-effective service. The concept had been studied under several research contracts that examined whether a number of simple, low-orbit satellites could be launched simultaneously at a fairly low cost per satellite. The relatively low orbit of those satellites would have required much lower transmission power from the satellite, which would have increased their relative technical performance vis-à-vis the geosynchronous satellite system, and have eliminated much of the "delay" problem.

On the opposite side of the coin, however, were two big negatives. The rapid velocity of low- or medium-altitude satellites with respect to the Earth, and the need for frequent pointovers as satellites appear and disappear, indicated that very sophisticated tracking and telemetry equipment and costly Earth terminals would be required to maintain operations. This aspect, in particular, would have inhibited the use of the global satellite system by smaller or developing countries. A second factor, of perhaps greater importance, was that, even with elaborate tracking networks picking up the various satellites as they came over the horizon, nonsynchronous satellites would still result in periods of brief outages. In the technical studies performed, it was shown that those outages would be of very brief duration in the Atlantic Ocean region; but over the much larger Pacific Ocean region, they would be for more extended periods--perhaps several minutes at a time. The Australian delegation to INTELSAT, in fact, was among the first to oppose serious consideration of this system configuration, and blocked further planning of subsynchronous satellites. At any rate, with the decision to proceed with the INTELSAT III, all consideration of nonsynchronous satellite operation and noncontinuous service was abandoned.

The INTELSAT III satellite had the same diameter as the INTELSAT II satellite, namely, 1.42 m (some 56 in.), but its height was 1.04 m (41 in.), making it almost twice the volumetric size of the INTELSAT II. The INTELSAT III also had a mass in orbit of 152 kg (268 lb) or, again, about twice the mass of the INTELSAT II.

The expanded bandwidth used in this satellite (some 450 MHz for the uplink and the same for the downlink); the significant increase in radiated power as a result of the increased solar cell capability of the satellite; and, even more significantly, the mechanically despun antenna that allowed a high-gain antenna to always be pointed toward the Earth resulted in a fivefold increase of capacity. This allowed some 1200 voice circuits, or up to four television channels. The mechanically despun antenna was mounted on the spin axis at one end of the spacecraft, thereby allowing the antenna to direct conical transmit antenna beams toward the Earth, and also allowed a much higher gain reception than an omnidirectional antenna. A combination of Earth and sun sensors, as well as TTC&M signals from the Earth, allowed the spacecraft to be continuously positioned to maximize the performance derived from the pointed beams. The antenna drive system included a solenoid motor, which operated functionally as a rotary solenoid, with 128 equally spaced positions--an arrangement that allowed the motor to be rotated from one position to the next each time a pulse was applied to the motor, thus achieving a very high level of rotational control. An encoder on the motor shaft provided a measurement of angular position of the antenna with respect to a reference point, establishing exactly where the antenna was and pointing at all times with reference to the spacecraft and thus, by reference, to the Earth as well.

The motor was increased in rotational speed by a train of pulses from an oscillator in the motor starting circuit. When the motor achieved the optimal speed, the drive mechanism was automatically switched from the starting circuit to the phase-lock loop circuit. After completing this sequence, the antenna was despun with respect to the spacecraft, but not necessarily properly positioned with regard to the Earth. The repositioning of the spacecraft and the antenna was accomplished by a digital positioning circuit that modified the phase-lock loop output by adding or subtracting, as required, one pulse per spacecraft revolution. This change in the motor speed resulted in the gradual reorienting of the antenna around to the optimal position, pointing directly to the subsatellite point representing the center of the Earth.

The spacecraft attitude determination system consisted of a pair of redundant Earth sensors operating in the carbon dioxide infrared spectral region, plus a sun sensor. The Earth sensors were designed to sweep across the Earth once per spacecraft revolution and produce pulses coincident with the leading and trailing edges of the Earth crossing. These data were then transmitted to the ground via the telemetry system. The spin axis attitude was determined on the ground by measuring either of two parameters--namely, timing the interval between the horizontal pulses produced by each sensor or comparing the leading- and trailing-edge pulse times of one sensor with a known Earth cord pulse length. The sun sensor was provided as an adjunct to the Earth sensors to accelerate the attitude determination process, particularly in the initial transfer orbit.

The antenna despin control was provided by ground commands based upon analysis of the Earth and sun sensors information. The ground control of the antenna was actually provided by substituting artificial Earth pulses transmitted from the ground station, while the timing of the artificial Earth pulses was controlled by comparing and adjusting the time interval between the reception of the antenna reference pulse and the Earth sensor pulse.

The spacecraft's primary electric power was provided by a fixed array of 3.6 kg (8 lb) mounted on the outside draw of the spacecraft, which was covered with 2 x 2 cm (0.78 x 0.78 in.) N-on-P silicon solar cells. The solar cell array was designed to provide a minimum of 131 W of power at the end of five years--a minimum standard that was well exceeded in actual operation. Nickel-cadmium batteries, similar in design to those used on the INTELSAT I and II, were used to support spacecraft operations during the eclipse season. The spacecraft power control unit both regulated the main spacecraft bus voltage and controlled the battery charge and discharge operations. A diagram of the spacecraft power subsystem is provided in Fig. 1.

The INTELSAT III, rather than using hydrogen peroxide fuel, as did previous spacecraft, utilized a hydrazine propellant system that was used in both the radial and axial thrusters, activated by ground command. The positioning and orientation system consisted of two independent assemblies, with each equipped to provide half the total impulse requirements. A capability to cross-connect the assemblies was also provided in the design. Each assembly had two spherical hydrazine tanks, a radial and axial thruster, and interconnect plumbing to provide redundancy. Each system had the capability to provide 350-m/s (1100-ft/s) velocity increments to the spacecraft either in the steady-state or pulsed-mode operation.

The apogee motor was a solid propellant that employed a glass filament-wound encasing. Electromechanical safe and arm devices were used in the ignition circuit to ensure the proper firing of the motor. The spacecraft frame consisted of a central magnesium

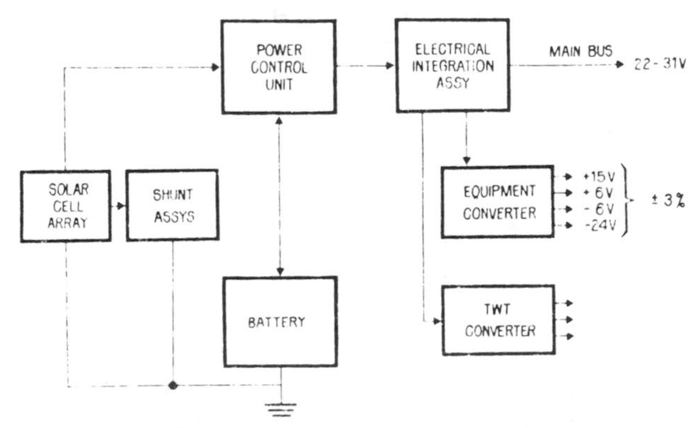

Fig. 1 INTELSAT III electrical power subsystem.

cylinder and an aluminum honeycomb sandwich platform structure that was used to support the spacecraft antenna and communications subsystem equipment. The thermal control for the spacecraft was achieved by passive means, using selected surface coatings and insulations as required. The INTELSAT III communications subsystem was actually quite straightforward--consisting of a despun receive and transmit antenna, a linear microwave repeater, and a command filter. A block diagram of this system is provided in Fig. 2. Both of the "global beam" antennas produced a 19.3-deg-wide cone that, in effect, was shaped by the need to provide Earth coverage at 17.3 deg with an allowance of ± 1 deg for pointing error.

The satellite repeater had two radio frequency channels, each with a 225-MHz transmission bandwidth, with 40 MHz of guard band between the two channels. The frequency plan for the INTELSAT III was as follows.

1) Band A: receive, 5930-6155 MHz; transmit, 3705-3930 MHz.

2) Band B: receive, 6195-6420 MHz; transmit, 3970-4195 MHz.

The end-to-end gain of the communications system was approximately 130 dB. This gain included a 13.5-dB receive antenna gain; a 104-dB repeater gain; and a 130-dB transmit antenna gain. The total gain was 13.5 dB.

The antenna system for the INTELSAT III consisted of an orthomode transducer, a linear-to-circular polarizer to feed the conical horn antenna, and a flat-plate reflector over the horn aperture to reflect the antenna beams into the equatorial plane of the satellite. The orthomode transducer and polarizer served as an integral assembly mounted directly on the antenna base. The orthomode transducer isolated the transmit and receive inputs into the antenna. The outputs of the orthomode transducer were

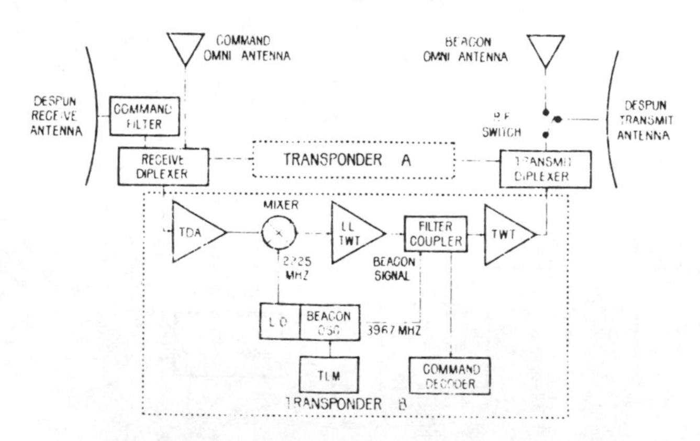

Fig. 2 INTELSAT III communications subsystem.

linearly polarized. The isolation between the transmit and receive antenna beams was accomplished by the orthogonal polarizer and frequency separation.

IV. INTELSATs IV and IV-A

With the INTELSAT IV satellite series, in the early 1970's, came yet another major watershed in the development of satellite communications technology. By the time the INTELSAT IV satellite was conceived, a fundamental transition in philosophy had already occurred. This was the shift from thinking of satellites as "experimental," to the recognition that satellites were now the dominant force in the communications revolution and that their applications virtually knew no bounds. For domestic, mobile, overseas, or any type of future communications application, satellites had to be seriously considered because of their reliabilty and cost efficiency.

Thus, for the INTELSAT IV generation, bold thoughts were emerging as to the size, scope, and service capabilities of the communications satellites of the 1970's. These thoughts were reflected in many ways: in extending the lifetime of the satellites; in increasing their physical size and mass to achieve higher capacities; and in designing and implementing highly sophisticated and complex communications devices. Although some possible new technologies, such as multibeam antennas, were not included in the INTELSAT IV spacecraft design, the prospects for the future were recognized and the R&D activities for those areas were re-emphasized for future spacecraft designs. It was thus safe to say that, with the INTELSAT IV, big and bold concepts for the future were the order of the day, and planning became a much longer-term process, stretching over a 15-year cycle. Alternatively, tentative, experimental, and small-scale steps were now passé.

Closely paralleling the development of INTELSAT IV was the planning for the Canadian geostationary ANIK satellite system. Further, planning and regulatory proceedings were underway in the United States that resulted in a virtual explosion of domestic satellite systems. Not surprisingly, many of the technological developments associated with the INTELSAT III and INTELSAT IV satellites were applied, directly or indirectly, to many of these first domestic satellite systems. Of course, the process also worked in reverse: INTELSAT clearly benefited technologically from the many innovations developed under the NASA Applications Technology Satellite Program, which ran from 1968 to 1974.

The INTELSAT IV decision, however, was not as clear-cut and easy as this retrospective history might make it appear. There were many who thought that a less-bold design would have been preferable. There were also numerous technical disputes on a number of issues related to the INTELSAT IV design, including whether it should be gravity-gradient or spin-stabilized and, in particular, whether such a large structure might develop a sustained problem and thus tend to topple over into a "flat spin." Certainly, one aspect of the INTELSAT IV program that distinguished it from its predecessors was the fact that it involved an extremely long lead time from the beginning of the analysis to the final decision to award the construction contract.

The story of INTELSAT IV actually began as early as December 1965, when COMSAT, acting as Manager of the INTELSAT system, presented a proposal for the design study of a multipurpose geosynchronous communications satellite that would include an aeronautical and maritime communications capability. Although this contract was ultimately awarded to Philco Ford, it was done over the strenuous objection of the French Signatory; in fact, there was a serious threat that the French Signatory would go to arbitration over whether it was obligated to pay for this study. The extent and nature of the objections (primarily from Europe, and especially France) ultimately served to kill the idea of making the INTELSAT IV a multipurpose satellite that incorporated a VHF or UHF capability to provide mobile communications services.

In January 1966, INTELSAT decided to issue a separate RFP for INTELSAT IV design studies. By September of that year, after lengthy discussions and review within the ICSC and the Technical Subcommittee, it was finally agreed to negotiate with both Hughes and RCA for such a study. There was also discussion at that time as to whether a contract on advanced stabilization techniques, particularly gravity-gradient techniques, was required. By March of 1967, however, the situation was greatly eased by the decision not to study advanced stabilization techniques and to remain with a spin-stabilized design. It was further decided to award contracts to Hughes Aircraft and to Lockheed for design studies. The contract was not awarded to RCA, which refused to accept INTELSAT's standard contractual terms and conditions.

As those design studies were going forward, however, several INTELSAT members (particularly the French Signatory) again proposed that serious study be given to modifying the INTELSAT III satellite to expand its capacity. As a result of this request, the Manager and the Technical Subcommittee of INTELSAT developed parallel draft specifications for both an INTELSAT III-1/2 and an INTELSAT IV spacecraft, this work being completed in the first and second quarters of 1968. Ultimately, however, the ICSC did decide to proceed with the INTELSAT IV design, as modified by the Technical Subcommittee, and reviewed the final technical proposals of Hughes and Lockheed, which led to the award of the INTELSAT IV contract, in October 1968, to Hughes Aircraft Company. The procurement award was for four INTELSAT IV spacecraft, but, by October 1969, a decision was made to purchase another four spacecraft of essentially the same design as the initial four. This second award decision was made, however, only after the United States Signatory had strongly pressed the idea of procurement of a modified INTELSAT IV design to include a greater design capability. This proposal, however, was strongly resisted by European Signatories, with one of the concerns expressed being that modifications possibly were being accommodated to the design needs associated with the United States COMSTAR domestic satellite system, on which COMSAT was then embarked.

Thus the ICSC and the Technical Subcommittee were very deeply involved in the decision-making process of the INTELSAT IV procurement. During the period that led up to the procurement of the eight INTELSAT IV spacecraft, some eight different spacecraft options and some nine different system configurations were considered for the provision of services for the 1971-85 time period, with capital investments considered ranging from $700 million to over $1.2 billion. During the latter part of those analyses and discussions, several modified versions of INTELSAT IV were considered. There was also heated discussion and ultimate rejection of COMSAT Laboratories' proposal for an experimental satellite program to demonstrate satellite-switched time-division multiple-access (SS-TDMA) techniques in space (possibly in conjunction with the European Space Research Organization). There was also a proposal for an early INTELSAT V satellite program, with a 1975 implementation date. Finally, a study was made concerning the idea to design a special satellite of smaller capacity for the Indian Ocean region, as well as a much higher-capacity INTELSAT V program for the 1978-80 time period.

The end result of all of these studies for the early to mid-1970 time frame was a decision, in January 1972, to procure additional INTELSAT IV satellites of a modified design--with about a 50% capacity increase--rather than proceed to an early procurement of an INTELSAT V. This decision was the result of weighing the somewhat greater capacity of an early INTELSAT V and its higher cost against the shorter delivery schedule and higher confidence in the technology of a modified INTELSAT IV. This decision was understandably influenced by the difficulties with the bearing assembly on the INTELSAT III despun antenna, which developed in July 1968. The ultimate "freezing" of the antenna bearing, and the failure of the INTELSAT III (F-2) satellite, served to convince INTELSAT's decision-makers that a more conservative next step was warranted. During this same time frame, greater flexibility was given to COMSAT, as Manager, to maneuver and restore services under contingency situations, and increased attention was given to system reliability concepts, in terms of spare satellites, multiple satellite paths, and Earth station pointover techniques to maximize the speed and efficiency of satellite restoration procedures.

Ultimately, eight INTELSAT IV and six INTELSAT IV-A satellites were procured at a capital cost of $188 million, with associated launch vehicle costs of $287 million for some 14 Atlas/Centaur launches. These were the spacecraft that sustained INTELSAT satellite operations for more than a decade, with service being provided particularly on the IV and IV-A satellites from 1971 through the early 1980's.

The INTELSAT IV/IV-A spacecraft differed significantly from the INTELSAT III spacecraft in several regards. First of all, the power from the solar cells mounted on the outside surface of the satellite's very large drum produced a power of 569 W for the INTELSAT IV and 708 W for the IV-A. The INTELSAT IV satellite, with 3.2 times the power of the INTELSAT III and 12.4 times the power of INTELSAT I when operating to Standard A Earth stations, thus was frequency limited rather than power limited, as had been the case with all previous INTELSAT satellites. The antenna system--particularly the IV and IV-A system--increasingly dominated the spacecraft and claimed a higher percentage of its mass and volumetric characteristics. The cylinder diameter of the INTELSAT IV/IV-A spacecraft was exactly the same [2.4 m (7.87 ft)]. The INTELSAT IV-A, however, was almost 1.5 m (4.92 ft) taller than the INTELSAT IV. This difference in height [5.3 m (17.4 ft) vs 6.8 m (22.3 ft)] was mainly due to a larger and more complex antenna system. The IV and IV-A, however, were significantly bigger than the previous spacecraft and formed the basis of a new class of Atlas/Centaur-type spacecraft in the worldwide satellite market.

The INTELSAT IV and IV-A satellites both included small global beam antennas, as well as much larger antennas, to transmit both to the east and to the west. Those antennas [at 1.5 m (4.92 ft) and somewhat larger for the IV-A] were the precursors of much larger antenna structures that were to follow. The communications system for the satellites utilized almost all of the available 500-MHz band (in fact, 486 MHz were utilized), but, in the case of the INTELSAT IV, this frequency was subdivided into twelve 36-MHz transponders. Thus, with guard bands, a total usable bandwidth of some 432 MHz was realized. INTELSAT IV-A, by being able to create two geographically separated zonal and four hemispheric beams (one for the east and one for the west), was actually able to reuse frequencies in most of the available band and thus was able to provide service on 20 transponders of 36 MHz each, or a total of some 720 MHz of capacity.

It was this increase of frequency, as well as a 24% increase in spacecraft power, that allowed the INTELSAT IV-A to derive a total of some 6000 voice circuits plus two color television channels, in comparison to the 4000 circuits plus two color television channels for the INTELSAT IV spacecraft. The global beam transponders on the INTELSAT IV and IV-A satellites produced a downlink power of 22.5 dBW at beam edge and a 29/31 dBW spot beam at beam edge. The INTELSAT IV spot beam power at beam edge was 34 dBW; while the INTELSAT IV-A hemispheric beam power at beam edge was 26.5 dBW. Thus a wide range of service capabilities was made possible through the INTELSAT IV and IV-A spacecraft.

V. INTELSATs V and V-A

The powerful communications transmitters, sensitive communications receivers, and rf upconverters on the INTELSAT V spacecraft require 800 W of electrical power. Consequently, a

Table 1 INTELSAT V power summary.

Load	Synchronous average power		
	Autumnal equinox	Summer solstice	Eclipse
Communications	782.12	782.12	781.02
Telemetry, command, and ranging	43.50	43.50	43.50
Attitude determination and control	49.13	[illegible]	49.13
Propulsion	0.80	[illegible]	0.80
Electrical power	9.1[illegible]	[illegible]	9.10
Thermal control	109.59	[illegible]	32.69
I^2 Harness losses	10.00	10.[illegible]	9.30
Total load (buses 1 and 2)	1004.24	968.24	925.54
Battery charging at 7 years	100.72	29.99	...
Total solar array load	1104.96	998.23	...
Contract load contingency (10%)	123.09	117.09	...
Solar array power availability at 7 years	1354.00	1288.00	...
System power margin at 7 years	125.95	172.68	53.09
Battery DOD for 1.2-h eclipse, %	...	...	52.00

Table 2 INTELSAT V spacecraft mass summary.

Subsystem	Current baseline mass, kg	
	Centaur launch	STS launch
Structure/thermal	183.1	183.1
Propulsion	35.3	35.3
Electrical power	141.9	141.9
Communications transponder	174.6	174.6
Communications antenna	58.9	58.9
Telemetry, command, and ranging	28.0	28.0
Controls	72.5	72.5
Electrical integration	40.1	40.1
Mechanical integration	15.4	15.4
Total	749.8	749.8
Apogee motor	922.5	922.5
Propulsion fuel	172.6	185.1
Total spacecraft	...	...
Launch total	1869.3	1897.0
Mass margins	24.4	39.6

large solar array area of nearly 20 m^2 (65 ft^2) capable of constantly tracking the sun for maximum exposure, was designed to provide the needed electrical power for the communications and supporting subsystems. This type of solar array area was made possible through the use of body-stabilized spacecraft configuration with deployable, sun-oriented solar panels.

The spacecraft's three-axis-stabilized design is composed of a box-shaped main body measuring 1.65 x 2 x 1.75 m (5.4 x 6.56 x 5.74 ft), containing the electronics and propulsion subsystems, and a truss-type tower holding the antennas, with the tower extending from the Earth-facing surface of the body. The spacecraft is oriented in space, with the 2- by 1.75-m (6.56 by 5.74 ft) side facing north and south. The solar arrays extend from these surfaces approximately 7.8 m (25.6 ft) on each side of the spacecraft. The antennas are oriented with the large 4- and 6-GHz reflectors on the east and west sides. Overall, the fully extended spacecraft spans 17.5 m (57.4 ft).

The total spacecraft power requirements for synchronous orbit conditions are presented in Table 1. The resulting power margin is 125 W at end of life (EOL) for the autumnal equinox, when the satellite is experiencing maximum eclipse, and 170 W at EOL during the maximum eclipse at the summer solstice.

The total spacecraft mass is summarized in Table 2. The fuel budget (summarized in Table 3) illustrates the individual components that comprise the total fuel budget for the seven-year lifetime of each INTELSAT V satellite.

INTELSAT V was designed for launch by either the Atlas/Centaur or the ARIANE launch vehicle. The use of the Space Transportation System (STS) was considered, but several factors adversely affected its use: a delay in its development; the need to develop a new perigee engine; and the "safety requirements" that would have necessitated modification of the satellite "explosive" fuel systems. The perigee engine development proved to be the major impediment, however, since the performance of the new engine was such that it would have, through a so-called "pogo" effect, shaken the spacecraft apart. Furthermore, the modification required to dampen the effect would have added enough mass to have reduced the satellite's usable lifetime in orbit. Ultimately, the STS launch option was discarded in favor of the Atlas/Centaur. Thus, of the 15 launches scheduled for the INTELSAT V and V-A programs, ten were scheduled for the Atlas/Centaur and five for the ARIANE launch vehicle.

Insurance for these launches has been placed by INTELSAT at an average cost of about 8% of the insured value of the in-orbit satellite, with each being insured for approximately $75 million. The first six launches on the Atlas/Centaur vehicle have been successful, as was the first launch of an INTELSAT V satellite

Table 3 INTELSAT V fuel budget.

Maneuver	Centaur launch Magnitude	Centaur launch Fuel weight, kg	STS/SSUS launch Magnitude	STS/SSUS launch Weight kg
Transfer orbit				
Spinup	45 rpm	1.5	N/A	...
Active nutation damping	10-min time constant	4.3	10-min time constant	4.3
Reorientation	48 deg	1.2	140 deg	3.4
Drift orbit				
AM dispersion correction, including coverage reorientation	65.8 m/s	28.5	92.1 m/s	40.2
Spindown	45 rpm	1.4	45 rpm	1.4
Synchronous orbit				
N-S stationkeeping	347.5 m/s	106.0	347.5 m/s	106.0
E-W stationkeeping	29.0 m/s	11.7	29.0	11.7
Attitude maintenance	7 years	12.3	7 years	12.3
Residuals		2.0		2.0
Total fuel requirements without pressurant	...	172.6	...	185.1

aboard an ARIANE launch vehicle. The elaborate precautions that INTELSAT takes to monitor the design, construction, and launch operations for its satellites have been a critical factor in INTELSAT's obtaining perhaps the lowest launch insurance rates in the world market.

The INTELSAT V communications subsystem provides an rf bandwidth capability of 2137 MHz, which is three times that of its predecessor, INTELSAT IV-A. It accomplishes this by means of extensive frequency reuse of 4 and 6 GHz, which is achieved through both spatial and polarization isolation and by introducing 11/14 GHz operation into the INTELSAT frequency plan. The frequency reuse scheme is accomplished by multiplicity of antenna coverages, which allows the spacecraft to transmit right (hemi) and left (zone) circularly polarized signals at 4 GHz to many east and west locations at the same frequencies. This provides a 4:1 frequency reuse factor for these locations.

A switching network interconnects the various coverage areas and allocates channels between hemispheric (hemi), zonal, and global coverages. The hemi/zone transmit and receive antennas consist of large, offset-fed parabolic reflectors [2.44 and 1.5 m (8 and 5 ft) in diameter, respectively] illuminated by clusters of square feed horns (feed elements). The feed elements are electronically controlled or "excited" through power division/phasing networks to produce the shaped hemispherical and zone beams. Both the hemispherical coverage and the zone coverage beams are generated simultaneously employing opposite senses (right-hand and left-hand) of circular polarization with low ellipticity ratio. Each antenna geometry (reflector size, focal length) is chosen to produce appropriately shaped beams with a sharp edge roll-off, low sidelobes, and good isolation between the coverage regions.

The east and west hemi beams are shaped to accommodate the appropriate ground station locations in the east and west hemispheres, respectively, as seen from all specified satellite locations in all oceans. A single east hemi pattern and a single west hemi pattern satisfy all these requirements with no switching.

The required zone coverages are also specified by designated ground station locations. In this case, the difference between the Atlantic, Pacific, and Indian Ocean distributions is so great that two pairs of beams are required: one pair (east and west) for Atlantic and Pacific coverage, and a second pair for Indian Ocean coverage. Four coaxial switches (east and west transmit, east and west receive) electronically reconfigure the antenna feeds for the two locations.

The hemi/zonal antenna feed consists of a closely packed array of 88 square waveguide feedhorns attached to multiple-layer air-supported stripline power division/phasing networks. The east and west hemispherical beams are formed by a fixed number of feed elements, each element being excited through one of the two excitation ports. The antenna zone beams are similarly formed by clusters of elements; however, the zone clusters utilize oppositely polarized excitation ports. Furthermore, each zone beam utilizes several common feed elements and several other elements that are selected by ground command to provide differently shaped Atlantic/Pacific or Indian Ocean zone beams.

The feed array elements are made up of three basic parts: 1) a stepped aperture, which provides a matched transition between the square waveguide and the radiating aperture; 2) a septum polarizer, which converts linear signals from each excitation port into circularly polarized signals; and 3) a coaxial-to-waveguide adapter, which provides the transition between a rectangular waveguide and the power division/phasing network. The feed array elements are designed to survive the space environment, launch, and handling during manufacturing and assembly. The elements are constructed from graphite fibre reinforced plastic (GFRP) material in order to minimize antenna weight and maintain dimensional stability over the wide temperature range. The inside of each element is lined with a thin copper layer to achieve good electrical conductivity.

An extremely important part of the hemi/zone antenna design is the polarization purity required to achieve the necessary beam-to-zone isolation. The axial ratio required for achieving a carrier-to-interference ratio of 27 dB at the worst location in a beam is on the order of 0.5 dB. Achieving this level of polarization purity from a closely packed array of elements was one of the most demanding aspects of the entire development program for the INTELSAT V satellite.

The 11/14 GHz spot beam antennas are designed to provide communications coverage to high-traffic areas, using narrow beams that can be steered by command from the ground. The antennas provide essentially constant beamwidths at the downlink (transmit) frequencies of 10.95-11.70 GHz and at the uplink (receive) frequencies of 14.0-14.5 GHz. The transmit and receive beams are linearly orthogonally polarized.

Each spot beam antenna consists of a nominal 1-m-diam (3.28 ft) offset-fed reflector illuminated by a conical corrugated feedhorn. The reflector is mounted on a support structure through a two-axis gimbal to enable the reflector to pivot in two orthogonal planes. Two linear actuators are used to pivot the reflector and thus obtain beam scanning without moving the feed.

The antenna positioner equipment consists of east and west spot beam antenna positioner mechanisms (APM), an Earth coverage antenna positioner mechanism, and an antenna positioner electronics (APE).

Positioning of any axis is accomplished by selecting the desired axis, selecting the direction of stepping, and by generating a series of 250-ms pulses. The selected actuator will then advance one step per pulse in the desired direction until telemetry indicates proper orientation has been achieved.

Each spot beam reflector is a honeycomb sandwich using GFRP faceskins and an aluminum honeycomb core. The west spot reflector is parabolic in both the vertical and horizontal planes to produce circular (1.6-deg diam) antenna beams. The east spot reflector is parabolic in the vertical plane but distorted (shaped) in the horizontal plane to produce elliptical (1.8 deg by 3.2 deg) transmit and receive beams with the minor axis inclined 22.9-deg clockwise from true north, as seen from the satellite.

The two Earth coverage antennas (4-GHz transmit and 6-GHz receive) are circularly polarized conical horns. The basic design for the two antennas is identical except that the higher gain transmit antenna is designed to cover an 18-deg field of view, while the receiver antenna covers a 22-deg field of view. The transmit antenna is mounted on a single-axis gimbal, which enables the antenna beam to be steered up to ± 2.0 deg in pitch to reposition the beam toward the Earth's center whenever the spacecraft is pitched in the east-west direction. This provides a global equivalent isotropically radiated power (e.i.r.p.) that is more than 1.5 dB greater than that provided by previous INTELSAT spacecraft. The wider beam receive antenna is fixed on the spacecraft.

Receivers are implemented with all solid-state components and use microwave integrated circuit technology. The 6-GHz receiver begins with a four-state bipolar amplifier at 6 GHz, followed by a low-loss balanced mixer. The 24-GHz receiver employs a single-stage 14-GHz tunnel diode amplifier, followed by a low-loss balanced mixer. In both cases, the mixer is followed by a transistor amplifier. The number of stages in the transistor amplifier differs for each of the global, hemi, zone, and spot varieties of receivers. All 6-GHz receivers contain an interstage commandable attenuation that can provide either nominal or extra high gain.

The input multiplexer consists of even and odd channel filters that are circular coupled, with even and odd sets hybrid coupled. Each filter is followed by an isolator, a group delay equalizer, and a commandable switched attenuator. Some channels also include an amplitude (tilt) equalizer.

The heart of each multiplexer channel is an eight-pole, pseudoelliptic function filter made of graphite fibre reinforced plastic (GFRP). It provides the required out-of-band rejector and in-band amplitude response. Each filter is assembled from prefabricated pieces, each of which consists of halves of two cylindrical cavities and an iris plate.

The INTELSAT V 4-GHz TWTAs are nearly identical to the INTELSAT IV and IV-A TWT designs in terms of vacuum envelope design, cathode loading and gun geometry, degree of overvoltage, and general stress levels. One difference from earlier configurations is the inclusion of an amplitude (slope) equalizer on the input to each tube to insure meeting the allocated 0.2-dB peak-to-peak variation over each channel. This was required because, unlike previous INTELSAT spacecraft, a given tube may be driven from several different sources, and it is therefore not possible to perform a unique end-to-end equalization of a channel. Also, most INTELSAT V channels are at least twice as wide as those of previous spacecraft in the series, but the same 1-dB peak-to-peak passband flatness requirement has been imposed.

The electronic power conditioner (EPC) for both 4-GHz TWTAs differs from those of previous INTELSAT spacecraft in that it employs a switching regulator to accommodate the unregulated 26.5- to 42.5-V bus with uniformly high efficiency. The heater supply voltage is dc to minimize spurious modulation of the beam by EPC switching transients. Indicative of the fact that all aspects of the satellite design must be carefully monitored, regardless of whether they represent new or breakthrough technologies or not, is the fact that the dc/dc power converter on the INTELSAT V developed problems of "arcing" during testing and redesign of the EPC, resulting in a delay of several weeks in the overall program.

The 11-GHz TWTA represents several firsts for an INTELSAT spacecraft: the first 11-GHz TWTA, the first dual-collector TWT, and the first impregnated cathode. The tube is evolved from a 20-W design developed for another program. Modifications for INTELSAT V included a slightly smaller cathode, lower electrode voltages, and a lower cathode temperature to insure extremely long cathode life.

The 4-GHz output multiplexers are an advanced contiguous design; that is, filters for adjacent channels are collocated on a common waveguide manifold. Consequently, single transmit antennas are used rather than duplicate odd and even antennas. The design of the output multiplexer places the 3-dB rejection points of each of the contiguous band channel filters halfway between adjacent channels. The filters are each of the singly terminated design.

The filter in each of the multiplexer channels is a six-pole, dual-mode quasielliptic design with two extra couplings: one between the first and fourth resonators and the other between the third and sixth resonators. This design was developed first and then used as the basis for the element design values in the singly terminated version. The complete multiplexer design requires the use of additional reactive cavities on the manifold to provide appropriate adjacent-channel reactances for those filters that carry the highest and lowest channels being multiplexed.

The controls subsystem provides active stabilization for the spacecraft to keep the antenna beams fixed on Earth throughout the mission. Figure 3 is a simplified block diagram of the controls subsystem. In transfer orbit, the spacecraft is spin-stabilized by means of active nutation control electronics, which fire hydrazine thrusters. Attitude determination is derived from Earth sensor and sun sensor data, which are processed by the altitude determination and control electronics (ADCE).

For rate damping and acquisition, the spacecraft is despun, and rate damping is performed about all three axes to less than

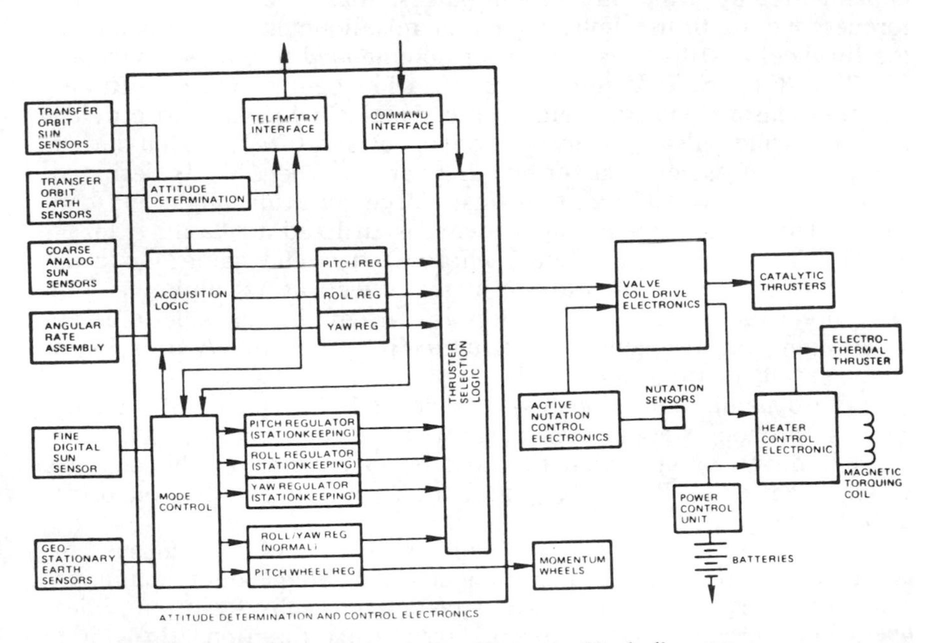

Fig. 3 INTELSAT V controls subsystem block diagram.

0.5 deg/s. Acquisition is commanded from the ground. During this procedure the spacecraft performs a series of automatic maneuvers to point the roll axis at the sun and to rotate the spacecraft at 0.5 deg/s about the roll axis. The solar arrays and reflectors are deployed, and the solar array drive is required to slew both arrays 90 deg to orientate the arrays normal to the sun line. Six hours after sun acquisition, the ADCS performs Earth and yaw acquisition, and the spacecraft is finally pointed at the Earth centroid. The solar array drive tracks the sun by being able to rotate up to 15 deg/h. The momentum wheel is then spun up in preparation for transition from stationkeeping to normal on-orbit control.

The normal mode control logis will automatically establish autonomous pitch control via the geostationary infrared sensor (GEO-IRS) and establish autonomous roll/yaw momentum bias control using GEO-IRS roll error signals to constrain the spacecraft motion to a small angle limit cycle. Small offsets are implemented by introducing bias commands into GEO-IRS output in roll and pitch axes. Due to external disturbance torques, the flywheel will accumulate angular momentum that finally results in increased or decreased wheel speed until a saturation limit is reached. At this point, a wheel unload pulse is automatically generated, removing the wheel from saturation.

The stationkeeping mode occurs during corrections for north-south or east-west stationkeeping. These corrections are implemented by firing thrusters in pulses, thus inducing disturbance torques due to thrust imbalance and misalignment. In this mode, the flywheel is either operational or commanded to a preset speed.

The INTELSAT V automatic nutation control (ANC) utilizes nutation sensor signals and electronic signal conditioning to provide thruster firing pulses for active control of satellite nutation during the spinning phase of transfer and drift orbit. The ANC is designed to operate in two different modes: large nutation angle and end game. The large nutation angle mode is utilized to capture large initial tipoff nutation angle, whereas the end-game mode is utilized to control small nutation angle as well as to minimize spin axis precision. Mode switching is automatically selected by determining the frequency of thruster firings of the ANC system during satellite nutation control.

The dynamic conditions experienced by the spacecraft during various mission phases, the transfer orbit, and operational orbit attitude dynamics environment must be considered in the control system design. Particular attention is given to on-orbit disturbance torques.

The telemetry, command, and ranging (TC&R) subsystem consists of two functionally redundant and independent command and telemetry channels. The command subsystem provides operational ground control for all spacecraft functions through a microwave link consisting of two ring-slot antennas, two command receivers, and two command units. A command transmission consists of a microwave carrier FM-modulated by a sequence of tone bursts at three discrete frequencies, resulting in 1, 0, and execute functions. The command receivers demodulate the microwave carrier and transmit the baseband tones via the spacecraft wire harness to the command unit bit detector. The command unit processes these discrete tones, and formats and executes specified discrete pulse, relay, and proportional command types. The command receivers and command units are fully cross-strapped, providing complete redundancy during all modes of operation.

The telemetry subsystem provides two independent and redundant data channels for transmission of diagnostic data received from sensors, transducers, and subsystem status. One of the two telemetry units processes and formats all incoming bilevel digital and analog telemetry for transmission via two phase-modulated telemetry beacon carriers. The telemetry unit has the capability of operating in three selected modes: PCM, PCM dwell, and FM real time. The telemetry digital bit stream is utilized to biphase modulate a 32-kHz subcarrier, which phase modulates the telemetry transmitter in the PCM and PCM dwell mode operation. The FM real-time mode is used for real-time attitude pulses (sun sensor, Earth sensor, and command execute) or nutation sensor signal. The occurrence of a sensor pulse or nutation signal switches the frequency of the interrange instrumentation group (IRIG) channel 13 subcarrier oscillator (SCO)

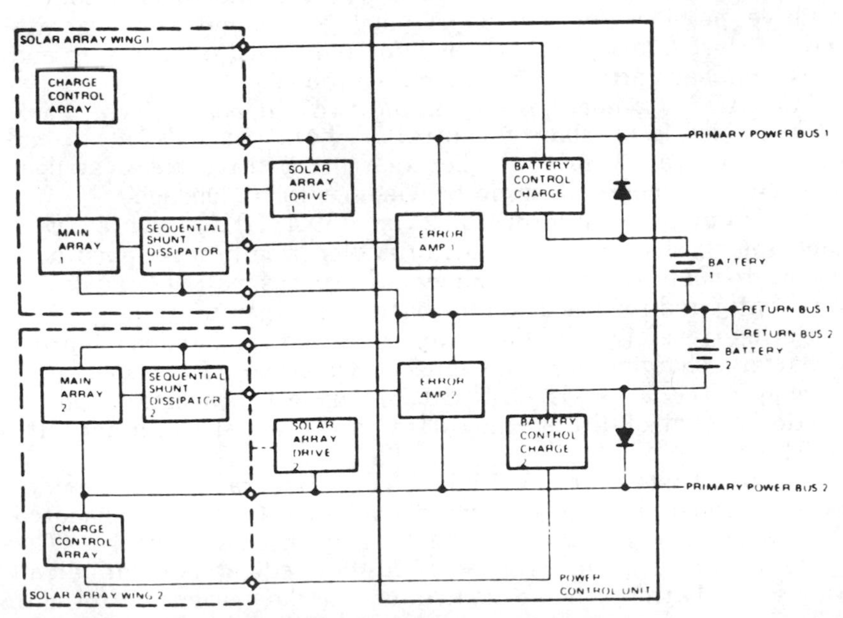

Fig. 4 INTELSAT V electrical power subsystem.

from its pilot tone to a frequency depending on the sensor pulse or nutation angle. The SCO output phase modulates the telemetry transmitter.

The telemetry transmitter has the capability of two transmission modes: 1) via the directional antenna for normal on-orbit Earth coverage; and 2) through one of three selectable TWTA's via a dual toroidal beam antenna for omnicoverage during transfer orbit. These outputs are both available whenever power is applied to the transmitter.

The electrical power subsystem (EPS) for the INTELSAT V spacecraft is a dual-bus, direct energy transfer system designed to accommodate a continuous spacecraft primary load of approximately 1.3 kW for a seven-year equinox synchronous orbit lifetime. Primary power is provided by two separate sun-oriented planar solar array wings. The power output of each solar array wing is regulated by a separate sequential linear partial shunt regulator. During periods of sufficient solar array power for support of spacecraft loads, power is supplied by two 28-cell nickel-cadmium batteries. The interrelationship of the major EPS elements is illustrated in Fig. 4.

The solar array consists of two single-axis sun-oriented wing assemblies. Each assembly consists of a deployment mechanism, three rigid panels, and an orientation mechanism connected to the solar array drive system. The solar array drive assembly (SADA) for the INTELSAT V spacecraft consists of a dual, two-channel solar array drive electronics (SADE) and two solar array drive mechanisms (one for the north solar array and one for the south). The drive provides for the support and positioning of the arrays about the spacecraft pitch axis and for the transfer of power and signals from each array to the spacecraft module.

The SADE is a dual box containing two redundant sides. Each of these sides is capable of controlling both channels (north and south) of solar array drives. The solar array drive has a stepper motor with two independent motor windings for redundancy.

The total panel area of 18.12 m^2 (59.4 ft) is covered with 17,568 solar cells. The solar array is electrically interfaced with sequential shunts to achieve the necessary bus voltage regulation and is configured to provide direct battery charge current.

During transfer orbit, the array is stowed so that load support and battery charging are accomplished with two outer panels (one per wing). The array is designed to support synchronous orbit operation at end-of-life equinox with an electrical power capacity of 1254 W.

In the first four INTELSAT V spacecraft, the battery configuration consisted of two nickel-cadmium batteries connected to the applicable bus through the battery discharge diodes. The battery charge current is controlled by dedicated solar array sections and battery charge controllers in the power control unit (PCU). The charge current is applied sequentially to each battery on a 50% duty cycle. The remaining INTELSAT V and V-A satellites, however, include the much higher-performance and longer-lived nickel-hydrogen batteries developed under the INTELSAT Research and Development Program (see Chap. 13).

Each nickel-cadmium battery assembly consists of 28 hermetically sealed prismatic cells connected in series. The nominal discharge voltage is 33.6 V, with a capacity of 34.0 A/h.

Open-circuit protection is provided for the batteries by diode bypass networks connected across each cell. Temperature sensors are utilized to provide temperature control inputs for battery heaters throughout the spacecraft's mission.

The power control electronics (PCE) consists of a PCU and two shunt dissipator assemblies. A key feature of the PCE is the provision of two independent primary buses. The outputs of one solar array wing and one battery are dedicated to each bus, with the capability provided to parallel connect or separate the two buses by command, as required. The output of each solar array wing is independently regulated to 42±0.5 V dc by use of a sequential linear partial shunt regulator.

The PCE provides sequential battery charge control and individual battery reconditioning capability by ground command. Single-part failure criticality is eliminated by use of circuit redundancy, and alternate modes of operation are selected by command. All spacecraft electroexplosive devices (EEDs) are controlled by the PCE, which employs redundant fail-safe circuitry for these important functions.

The propulsion subsystem for the INTELSAT V works in conjunction with the controls subsystem to maneuver the satellite. It consists of two screen-type surface-tension propellant/pressurant tanks that are manifolded to two redundant sets of thrusters. Two 22.2-newton (N) thrusters are used during transfer drift orbit for orientation, active nutation, and orbit

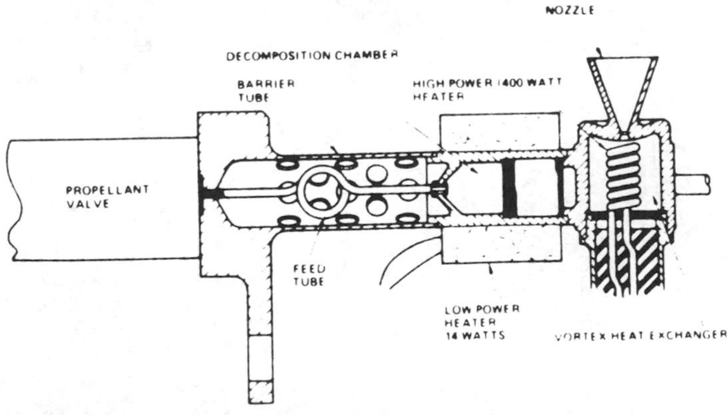

Fig. 5 INTELSAT V electrothermal thruster configuration.

velocity correction. Spacecraft spin/despin, east-west stationkeeping, and pitch and yaw maneuvers are performed by 2.67-N thrusters. These thrusters also serve as backup to the 0.3-N electrothermal thrusters, which are designed to perform the north-south stationkeeping function. Roll maneuvers are performed by 0.44-N thrusters. Latching isolation valves separate the tanks into half-systems. The plumbing is arranged so that, through use of these isolation valves, either tank can be used to feed one or both sets of thrusters.

For high reliability, previously space-qualified hardware has been used wherever possible. Propellant tanks and electrothermal thrusters (ETT's) were development items chosen for their particular benefits to the satellite design.

Two titanium propellant tanks are provided for storing the hydrazine propellant. The internal propellant management device feeds fuel to the thrusters under zero gravity conditions as well as 1g conditions in all tank positions.

ETT's were selected for INTELSAT V because potentially they can deliver a mission average specific impulse I_{sp} of 304 s by heating hydrazine propellant to 2206°C (4000°F) prior to ejection. The thruster assembly is described in Fig. 5. The development of the ETT, however, proved to be exceptionally difficult and was a significant factor in the delay of the INTELSAT V program. In-orbit experience with the early INTELSAT V's indicated that the ETT's were only a moderate success. An improved model was

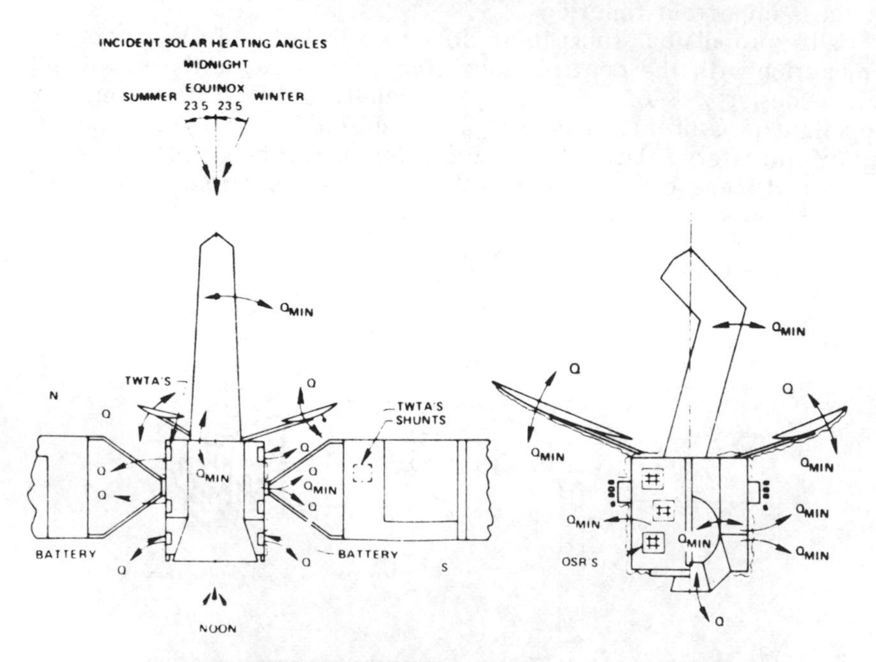

Fig. 6 INTELSAT V synchronous orbit thermal control.

flown with Flight 5, and early performance indicates some improvement over the earlier models.

Thermal control of the INTELSAT V spacecraft is accomplished using conventional passive techniques, including selective location of power-dissipating components, selective use of surface finishes, and regulation of conductive heat paths. The passive design is augmented with heater elements for components having relatively narrow allowable temperature ranges. The design approach provides simple and reliable temperature control. The thermal subsystem is configured to provide flexibility for variations in the spacecraft heat load, including payload growth, through easily accomplished modification of insulation blankets and radiators. Overall temperature control is achieved by: 1) the thermal energy dissipation of components in the communications and support subsystem modules; 2) the absorption of solar energy by the solar array, antenna module, and main body radiators; and 3) the reemission to space of infrared energy by the solar array, antenna modules, and main body radiators.

High thermal dissipators, such as the TWTA's, are located on the north and south panels of the main body so that they may efficiently radiate their energy to space via heat sinks and optical solar reflector radiators. The north and south panels were selected to contain the radiators because these panels are least affected by transient daily and eclipse variation in solar incidence.

The east and west panels, antenna deck, and aft surfaces are covered with multilayer insulation to minimize the effect of solar heating on equipment temperature control during a diurnal cycle. Thermal control of the tower supporting the various antenna reflectors and horns is achieved by the use of a three-layer thermal shield around the tower. Thermal control of the antenna reflectors, positioners, feed assemblies, and horns is achieved by the use of thermal coatings, insulation, and aperture covers, as necessary.

The thermal control concept indicating heat transfer paths is illustrated in Fig. 6.

The three main elements of the spacecraft are the antenna module, communications module, and support subsystem module. The latter two modules form the main body. Each of these modules is manufactured separately and then integrated in the final stages of the manufacturing process.

The INTELSAT V-A spacecraft, following the tradition established with the INTELSAT IV series, gives a significant increase in operational life, reliability, and communications capability for minor modifications to the spacecraft platform. C-band capacity in the global beams is increased with the use of cross-polarization, while C-band spot beams are introduced. Additional coverage in the east and west zones combine to give almost 15% additional C-band capacity. This results in an increase

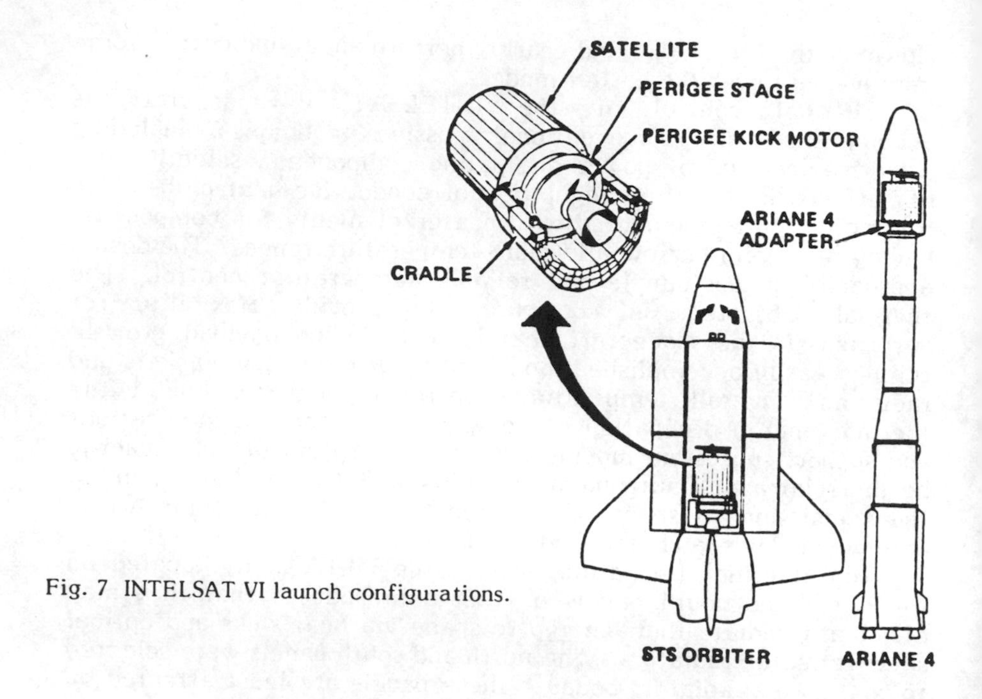

Fig. 7 INTELSAT VI launch configurations.

of 3000 equivalent voice circuits, thus bringing the total to 15,000 while still retaining the two color television channels.

VI. INTELSAT VI

The INTELSAT VI series continues the process of technical innovation represented by previous generations of INTELSAT satellites by introducing the following additional features:

1) Sixfold reuse of the 6/4-GHz spectrum.

2) Dynamic switching of the interconnections of six of the satellites' antenna beams, for use with satellite switched time division multiple access (SS-TDMA).

3) Limited introduction of newly allocated (WARC-79) frequency spectrum at 6/4 GHz.

4) Increased use of the 14/11-GHz spectrum and twofold frequency reuse.

5) Ten-year design lifetime.

6) Compatibility for launch with STS, ARIANE 4, and Titan 34D.

The INTELSAT VI satellite is the largest and most sophisticated commercial satellite ever placed under contract. Hughes Aircraft Company is building five of these satellites for INTELSAT, with options for up to 11 more. The first is to be

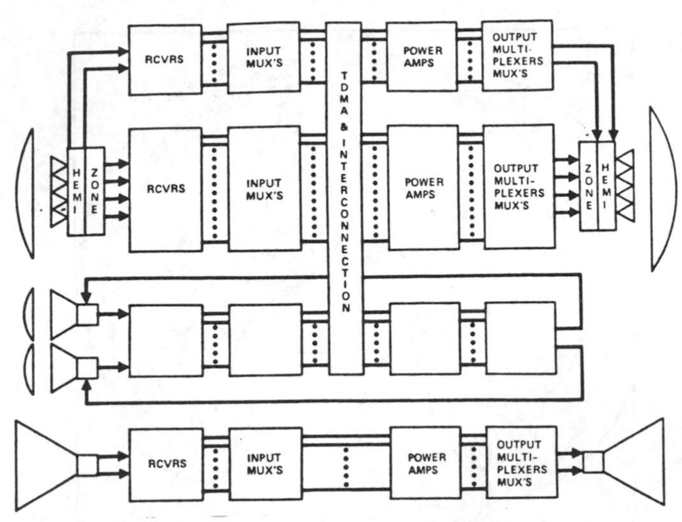

Fig. 8 INTELSAT VI communication repeater.

launched in 1986, and subsequent satellites are to be delivered at four-month intervals. The satellites are designed for ten years of operation in orbit.

A. Overall Configuration

The satellite is 11.8 m (38.7 ft) high when in orbit. The antenna configuration is dominated by the two large C-band reflector antennas that, along with the other antennas and communications equipment, are despun and Earth-pointed in orbit. The 4-GHz transmit antenna is 2 m (6.56 ft) in diameter and the 6-GHz receive antenna is 3.2 m (10.5 ft) in diameter. Six separate beams are generated by each of these antennas by means of a complex feed array. These six beams, consisting of two nonoverlapping copolarized hemispheric beams, and four nonoverlapping zone beams polarized orthogonally to the hemispheric beams, provide for sixfold reuse of the 6/4 GHz C-band frequency spectrum. A 6-GHz receive horn antenna and a 4-GHz transmit horn antenna each provide two cross-polarized Earth coverage beams. Two spot beam antennas of approximately 1 m (3.28 ft) in diameter provide twofold reuse of the 11/14-GHz K-band frequency spectrum.

The INTELSAT VI dual telescoping solar panel develops over 2 kW of power throughout the spacecraft's ten-year operational life. The inner fixed panel is approximately 2.2 m (7.2 ft) in length. This panel surrounds the heat-generating communications equipment and contains a 0.74-m (2.43 ft) cylindrical band of quartz mirrors, which are the primary path for heat rejection

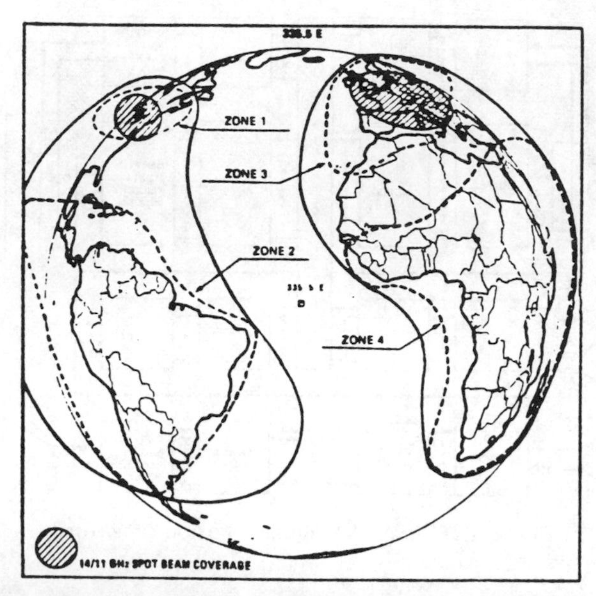

Fig. 9 INTELSAT VI coverage contours for the Atlantic Ocean region (from 335.5° east longitude).

from the communications equipment. Moderate-efficiency (K 4-3/4-type) solar cells are used on the panel because of their low solar absorption. The deployable outer panel is approximately 3.8 m (12.47 ft) in length. This panel contains high-efficiency (K 7) solar cells. During launch, the panel is placed over the inner panel to form two concentric cylinders much like two cylindrical barrels of a folded telescope. This folded configuration provides an efficient structural path for carrying the launch loads from the outer panel, and results in more than a twofold increase in solar panel power without the attendant mass increase normally associated with such an increase in power generation capability. Once in orbit the outer panel is extended, like opening a telescope, to expose this inner panel. When the operation is completed, full power is obtained. Eclipse power is provided by two 44-A/h nickel hydrogen batteries.

The satellite carries an integrated bipropellant propulsion subsystem used for apogee boost, orbit maintenance, and attitude control. Two 490-N thrusters are used for apogee boost and six 22-N thrusters are provided for stationkeeping. The bipropellant system provides fuel for ten years of operation.

The INTELSAT VI satellite is designed for launch by the ARIANE 4, the Space Shuttle, and the Titan 34D. Its launch configuration on each of these boosters is shown in Fig. 7. The diameter of the outer solar panel is 3.6 m (11.8 ft), to conform to the ARIANE shroud. In its launch configuration, with the antennas folded and the outer solar panel retracted over the inner panel, the satellite is 5.3-m (17.4 ft) high. The satellite's dry mass is approximately 1800 kg (3978 lb). The ARIANE 4 places the satellite in synchronous transfer orbit, and the satellite bipropellant reaction control system provides the necessary velocity increment for establishing the final synchronous orbit. Thus the only interface with the ARIANE is a simple conical booster adapter.

The STS, on the other hand, places the satellite in a low-altitude circular orbit. The large velocity increment necessary to establish the synchronous transfer orbit is provided by a solid perigee propulsion stage, carried with the satellite. The satellite and the perigee stage are held in the Shuttle bay by a cradle that provides all the mechanical and electrical interfaces between the satellite perigee stage and the Shuttle. A simple latch-and-spring mechanism is used to eject the satellite and perigee stage from the STS. This ejection provides linear velocity, which carries the satellite and perigee stage away from the orbiter, and a slow rotation that gives gyroscopic stability. This frisbee ejection concept was developed for Hughes's wide-body satellites. The perigee motor is fired by an onboard timer approximately 45 min after ejection from the orbiter. This propulsive maneuver establishes the synchronous transfer orbit. After the perigee motor is jettisoned, the remaining transfer orbit operations are similar to those of the ARIANE launch. For STS launch, the INTELSAT VI uses 44.6% of the length of the orbiter bay and 45.5% of the STS launch weight capability.

B. Communications Equipment

A simplified block diagram of the large and complex INTELSAT VI communications repeater is shown in Fig. 8.

At C-band, each of the large reflector antennas generates two hemispheric beams and four zone beams, which are polarized orthogonally to the hemispheric beams. The INTELSAT VI satellites will be located in the Atlantic, Pacific, and Indian Ocean basins, where the hemispheric coverage requirements are essentially identical, and where a single feed network is used to generate the two hemispheric beams for all three ocean basins. The zone beam coverage requirements are significantly different from ocean basin to ocean basin, and separate feed excitations are required for each basin. Examples of the beam coverage needed for the Atlantic and Indian regions are shown in Figs. 9 and 10. The large feed array for each of the C-band reflector antennas consists of 147 dual-polarized feed horns and four distribution networks. The orthogonally polarized hemispheric and zone beams are formed by simultaneously exciting the two polarizations for each feed horn in the array. A single distribution network for the

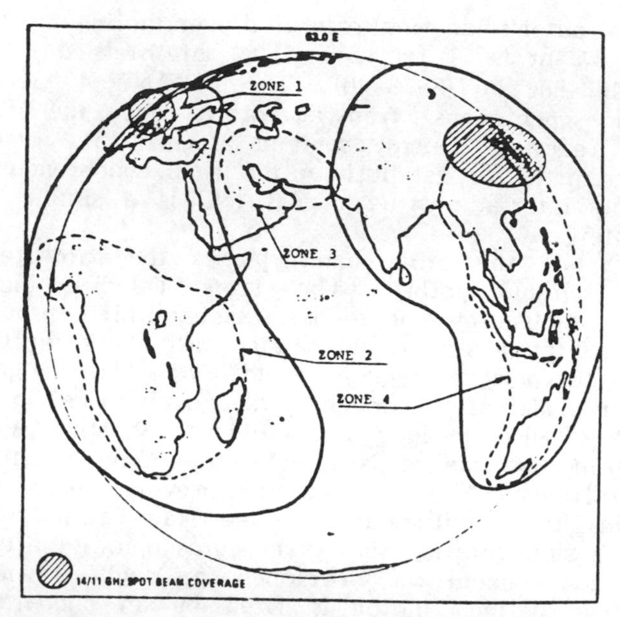

Fig. 10 INTELSAT VI coverage contours for the Indian Ocean region (from 63° east longitude).

hemispheric beams is used for all three ocean regions, while three separate distribution networks are used to form the Atlantic, Pacific, and Indian Ocean basin zone beams. A network of switches contained within the feed network connects the appropriate distribution network to the feed horn array. The antenna coverages are able to be reconfigured in orbit by means of commands transmitted from Earth.

In addition to the six C-band antenna beams formed by the hemi/zone antennas, global horns generate two C-band cross-polarized global beams, and the two K-band reflector antennas generate steerable spot beams. Thus a total of ten separate receive/transmit beam pairs are available on each satellite. Within these ten beams, a total of 50 transponders, or transmission channels, are provided in each satellite. Since the C-band channel 9 is shared by the global, the hemi, and the zone beams, a maximum of 48 transponders may be simultaneously active. Each transponder is provided with redundancy for the active components to assure a long and useful lifetime.

Satellite-switched TDMA operation is planned for transmission channels 1-2 and 3-4 within the six hemispheric and zone beams at C-band. The dynamic TDMA satellite switch in the communications repeater provides interconnection between these six receive and transmit beams so that Earth stations in one beam may simultaneously access, via TDMA, Earth stations in all six beams. Six input lines from the channel 1-2 uplinks in the six receive beams and six output lines to the channel 1-2 downlinks in the six transmit beams are cyclically interconnected through the dynamic TDMA satellite switch. The interconnection pattern for each switch state, and the duration of the state, are contained in the programmable memory in the switch control unit. This memory can store up to 64 separate switch states. The switching pattern repeats every 2 ms. Separate, identical dynamic TDMA switches connect the six receive and transmit beams for channels 1-2 and 3-4. Timing for both dynamic TDMA switches is provided by a commmon source so that separate synchronization of the two TDMA transmission channels is not required. In addition, a static interconnection switch is also provided for each of the C- and K-band channels 1-2 through 7-8 and C-band channel 9, in order to provide in-orbit flexibility in establishing static beam-to-beam interconnections. These switches can also be reconfigured by means of Earth-transmitted commands.

C. Management Approach

The management of a program with the scope of INTELSAT VI requires significant resources that extend well beyond those of the INTELSAT V and V-A satellite programs. Literally millions of man-hours will be consumed in building these spacecraft. The communications equipment alone encompasses the hardware listed below.

Type	Number
C-band receiver	16
K-band receiver	4
C-band driver amplifier	39
Input filters	50
4/11-GHz upconverters	20
C-band TWTAs	42
C-band SSPAs	15
K-band TWTAs	20
Output filters	50

The logistics of supplying, within a relatively short time, the large complement of communications and other equipment was recognized by Hughes as a central issue in the INTELSAT VI planning. Following the pattern successfully employed in prior INTELSAT satellite procurements, Hughes assembled an international team of leading companies, led by Hughes, for the design, manufacture, and testing of the INTELSAT VI satellites. As part of this team, SPAR Aerospace Ltd. of Canada will design and build C-band receivers and driver amplifiers. COMDEV Ltd. of

Canada will supply output filters. Thomson-CSF of France will design and build all of the K-band TWTs and will design and build C- and K-band receivers and output filters. The Nippon Electric Company (NEC) of Japan will design and build all of the solid-state power amplifiers, all of the 4- to 11-GHz upconverters and K-band receivers.

British Aerospace of England, Messerschmitt-Bölkow-Blohm (MBB) and AEG Telefunken of West Germany, and Selenia of Italy are also members of the INTELSAT VI team. British Aerospace will design and build the cradle and its associated electronic units that carry the satellite in the Space Transportation System (STS). They will also be responsible for the C- and K-band reflectors. MBB will design and build solar arrays, and AEG will provide all of the K 4-3/4-type solar cells. Selenia will design and build telemetry and command antennas, the remote and central telemetry units, the K-band Earth coverage horns, and the K-band spot beam antennas. In addition to these design activities, the companies will participate in the manufacture and testing of certain equipment designed by Hughes.

The design, manufacturing, and testing processes of the INTELSAT VI international team members will be closely monitored by Hughes to assure conformance and compatibility with the overall satellite design and its requirements. As the equipment is produced by the team members, it will be delivered to Hughes, who will integrate it into the satellite. This will then be followed by an extensive series of subsystem and system level tests, carried out by Hughes prior to the launching of the satellite.

VI. Conclusion

The INTELSAT VI satellites will introduce a number of new technologies to achieve a tripling of the communications capability relative to the previous generation of satellites. Some of these technologies, such as bipropellant propulsion systems, more efficient solar cells, and highly shaped antenna beams, achieve the mass reductions needed to carry the large quantities of communications equipment. Other technologies, such as sixfold spectrum reuse and SS-TDMA, achieve a greater efficiency in the use of the orbit-spectrum domain. The latter will also have an impact on the system operations and will require some modest modifications of the Earth segment. By these means, the traffic carried by the INTELSAT system will be able to grow in an economical and orderly manner.

Acknowledgment

The authors wish to thank Andrew F. Dunnet for his invaluable contribution to this chapter.

References

[1]Reproduced with the permission of the International Telecommunication Union (ITU), Geneva, Switzerland, from "INTELSAT-II Communications Satellite," ITU Telecommunication Journal, Vol. 34, No. 2, pp. 48, Feb. 1967.

The authors wish to acknowledge their use of material from the following sources, which have been excerpted with permission:

Feigen, M., Barter, N. J., and Slaughter, R. G., "The INTELSAT III Satellite," IEEE International Conference on Communications, ICC68, pp. 639-645, June 1968. © 1968 IEEE.

Martin, E. J., and McKee, W. S., "Commercial Satellite Communications Experience," IEEE Spectrum, Vol. 4, No. 7, pp. 63-69, July 1967. © 1967 IEEE.

Votaw, M. J., "The Early Bird Project," IEEE Transactions on Communications Technology, Vol. Com-14, No. 4, pp. 507-511, Aug. 1966. © 1966 IEEE.

Bennett, S. B., and Braverman D. J., "INTELSAT VI Technology, AIAA Paper A-82-44665, 33rd International Astronautical Congress, Paris, France, Sept.-Oct. 1982.

Rusch, R. J., Johnson, J. T., and Baer, W., "INTELSAT V Spacecraft Design Summary," AIAA Paper 78-528, pp. 8-20, AIAA 7th Communications Satellite Systems Conference, San Diego, Calif., April 1978.

7. The Earth Segment

Kunishi Nosaka*
Kokusai Denshin Denwa Company, Tokyo, Japan

I. General Considerations

A. Growth of Earth Stations

When the first INTELSAT satellite, Early Bird, was put into orbit in 1965 over the Atlantic Ocean region, only five countries maintained Earth stations (the United States, the United Kingdom, France, Germany, and Italy). These Earth stations had a gain and noise temperature antenna performance nearly equivalent to that for the Standard A Earth station (except for Fucino, Italy) and could be regarded as the prototype for the Standard A Earth station. Due to Early Bird's limited satellite power and bandwidth, operation by multiple-access mode was not possible, and the four Earth stations in Europe were operated on a rotation basis to communicate with the United States Earth station. After services were further expanded to the Pacific Ocean region by INTELSAT II in 1966 and to the Indian Ocean region by INTELSAT III in 1969, an international communications network on a global basis was established and many Earth stations were constructed in the three ocean regions in the INTELSAT system. The performance of the new satellite series in terms of power and bandwidth was improved from the Early Bird, and those Earth stations could set up direct circuits to many destinations by the multiple-access mode. As of September 1969, countries operating Earth stations in the INTELSAT system included:

1) Atlantic Ocean region: Brazil, Canada, Chile, Germany, France, Italy, Mexico, Panama, Peru, Spain, the United Kingdom (two), and the United States (three)--for a total of 16 stations.

2) Pacific Ocean region: Australia (two), Japan, the Philippines, Thailand, and the United States (three)--eight stations in all.

Invited paper received October 11, 1983.
*Vice Director, Research and Development Laboratories.

3) Indian Ocean region: Bahrain, Japan, and the United Kingdom--three stations in all.

Since then, the advantages of satellite communications for the economy and the capability of providing direct links among many countries with flexible expansion have been recognized. To meet the traffic growth of international telecommunication, the performance of satellite transponders has been improved continuously, with each generation adopting the newest technologies (frequency reuse, new frequency bands, and spot/shaped beam antenna). The technical characteristics of INTELSAT standard Earth stations have also been modified and expanded to be consistent with the improved performances and additional capabilities of the successive satellites series so that available capacities of the space segments are fully utilized while trying to minimize the burden of additional cost for modification of the Earth stations. The emergence in 1974 of nonstandard Earth stations accessing INTELSAT's leased transponders (later specified as Standard Z) and, in 1982, of coast/ship Earth stations accessing the maritime packages leased to INMARSAT have increased the variety of Earth station categories in the INTELSAT system.

Figure 1 illustrates the growth of Earth station antennas in the INTELSAT system. As of December 1982, the number of operational Earth station antennas in the INTELSAT system (composed of 94 INTELSAT members and 31 non-INTELSAT members) was 168 Standard A Earth stations, 53 Standard B stations, 4 Standard C stations, 223 Standard Z stations, and 2 nonstandard stations.

B. The Standardization of Earth Stations

Earth stations are an important segment, together with communication satellites, in the operation of the satellite communication system. INTELSAT's main objective is to provide a global public communication service to all areas of the world, and all Earth stations that access the INTELSAT system must be assured of high-quality and reliable service as well as compatibility with other Earth stations. The economics of construction, operation, and maintenance of Earth stations must also be considered carefully in the tradeoff with the cost of space segments as a whole.

To satisfy those requirements (although, in principle, it is the responsibility of Earth station owners to procure Earth station facilities and establish compatibility with all other stations with which they wish to operate), INTELSAT has specified technical performance characteristics for several types of standard Earth stations (A, B, C, and provisional E accessing the facilities of the INTELSAT global system; Z accessing the leased space segment capacity for domestic use; and coast/ship Earth stations accessing the INTELSAT-V maritime package).[1] Nonstandard Earth stations whose performances do not meet the standards may also be approved by INTELSAT for accessing the INTELSAT global system and the leased space segment capacity on a case-by-case basis.

All Earth stations in the INTELSAT system are owned and operated by the individual participating countries. When an Earth station owner wants to access the INTELSAT satellites, an application for approval of the Earth station must be submitted to INTELSAT, and the conformity of its performance characteristics with the standard performance characteristics will be verified by INTELSAT prior to commencing operation.

The Standard A Earth station was introduced at the outset of INTELSAT operation in 1965 in the 6/4 GHz frequency band and is the most widely used station in the system. Because of its large-sized antenna, about 30 m (98 ft) in diameter, the satellite capacity can be used most efficiently.

The Standard B Earth station, with an antenna about 11 m (36 ft) in diameter in the 6/4 GHz frequency band, was developed by INTELSAT in 1976. Since then, it has been used as a more economical alternative to the Standard A station and is particularly suitable for Earth stations with small traffic requirements.

The Standard C Earth station was introduced to access the 14/11 GHz band transponders of INTELSAT V. This Earth station is generally implemented by Earth station owners as a second or third station for the trunk connection of circuits, and its operation is limited within the coverage area of the 14/11 GHz spot beam.

The Standard E Earth station is used for new business services. This station has a small antenna and can be installed on the customer's premises with a frequency band of 14/11 and 14/12 GHz to provide the integrated digital networks among multinational offices.

The Standard Z Earth station is for access to INTELSAT leased space segments in the 6/4 and 14/11 GHz frequency bands for the provision of domestic services. The standard performance characteristics have specified only the minimum Earth station capabilities needed to achieve acceptable levels of interference to other systems.

The coast/ship Earth stations accessing the maritime package of INTELSAT V are operated by participating countries of INMARSAT. For those stations, INTELSAT has specified performance characteristics describing the technical parameters that must be controlled to assure the INTELSAT space segment of adequate interference protection.

The details of the technical performance characteristics of the above Standard Earth stations are described in Sec. II of this chapter.

C. Important Factors for Earth Station Design

1. Choice of Standard Earth Station. Countries that wish to join the INTELSAT system for the first time usually plan to construct their Earth stations mainly for the purpose of telephony and television transmissions, and they may wish to access the primary satellite to take advantage of full direct connections to all other stations with the same coverage. In this circumstance, they will be advised to first construct either a Standard A or B Earth station, depending on initial and future requirements of telephone channels. Even though the Standard B Earth station is a more economical choice when traffic demand is small, it is generally upgraded to the Standard A Earth station later when traffic growth warrants a larger investment. Countries that construct the Standard B Earth station should also take into account the space segment unit charge, which is 50% higher than that for Standard A stations, and the quality of television reception, which is acceptable but low compared to that for Standard A stations.

In any event, when countries construct their Earth stations, the Earth station facilities should be designed so that they are flexible and adaptable to the possible expansion of future traffic requirements, changes in carrier frequency, type, size allocations, and accessing satellites that are conceivable later.

To provide multipoint links among all stations in the same coverage, they will access a Primary satellite. But the increase in traffic demand has necessitated the deployment of additional satellites (Major Path satellites), and countries with heavy traffic have built their second and third antennas for Standard A and C Earth stations. These additional antennas for multiple satellite operation will also provide path diversity for major traffic streams for service protection and backup capability in the event of unforeseen station failures. INTELSAT has set up guidelines as to when additional antennas should be provided in the INTELSAT system and suggests that Earth station owners consider the provision of a second antenna when their forecasted traffic level is in the range of 300-400 circuits, and a third antenna when the traffic level is in the range of 800-1000 circuits.

By 1982, 12 countries had more than two antennas operating in the Atlantic Ocean region. For example, the United States has four Standard A and one Standard C Earth stations, and mainland France has three Standard A and one Standard C Earth stations. In the Indian Ocean region, five countries had more than one antenna in operation in 1982.

2. Operating Radio Frequencies and Rainfall Effect. Table 1 shows the frequency bands in which various Earth stations operate.

Since available satellite radiation power is limited, the increase in propagation loss, particularly through the atmosphere and the ionosphere, should be avoided as much as possible. The

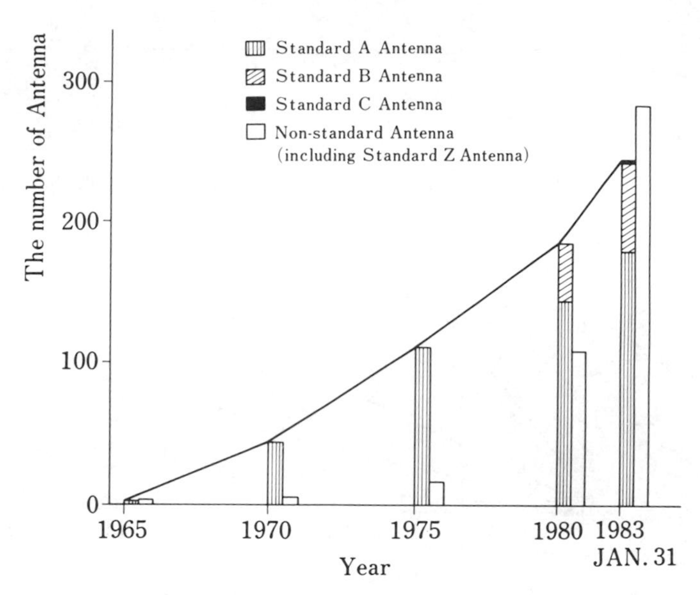

Fig. 1 Growth of Earth station antennas in the INTELSAT system.

Table 1 Operating frequency for standard Earth stations accessing INTELSAT IV, IV-A, and V

Standard	Uplink, GHz	Downlink, GHz
A	5.925-6.425	3.7-4.2
B	5.925-6.425	3.7-4.2
C	14.0-14.5	10.95-11.2
		11.45-11.7
E	14.0-14.25	10.95-11.20
Z	5.925-6.425	3.7-4.2
	14.0-14.5	10.95-11.2
		11.45-11.7
Coast station	6.417-6.425	4.1925-4.2005

frequency band between about 1 and 10 GHz, which is called the "radio window," is suitable for satellite communication, because this frequency range is less liable to suffer from the effects of atmosphere and ionosphere. INTELSAT selected 6 GHz for uplink and 4 GHz for downlink for the first operation of Standard A Earth stations. But even in the 6/4 GHz band, heavy precipitation will incur radio signal attenuation due to absorption by raindrops. Typical rainfall attenuation data in the 6/4 GHz region at the rain climate zone K (CCIR Rep. 563-2) are shown in Fig. 2. An Earth station operating at a site with a heavy rainfall rate accessing to a

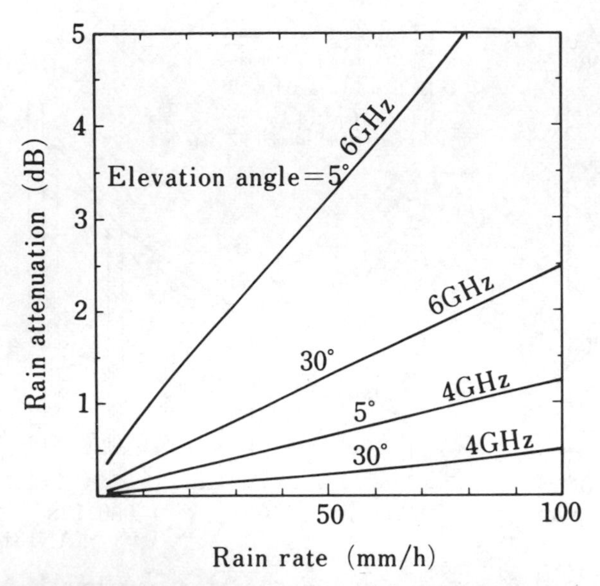

Fig. 2 Rainfall attenuation in 6/4 GHz at the rain climatic zone K (CCIR Rep. 563-2).

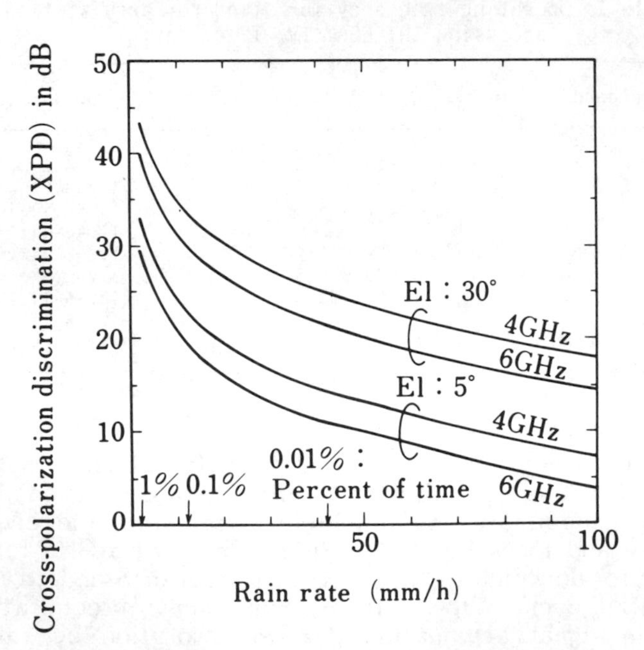

Fig. 3 Degradation of cross-polarization discrimination by rainfall (circular polarization, 6/4 GHz, rain climatic zone K).

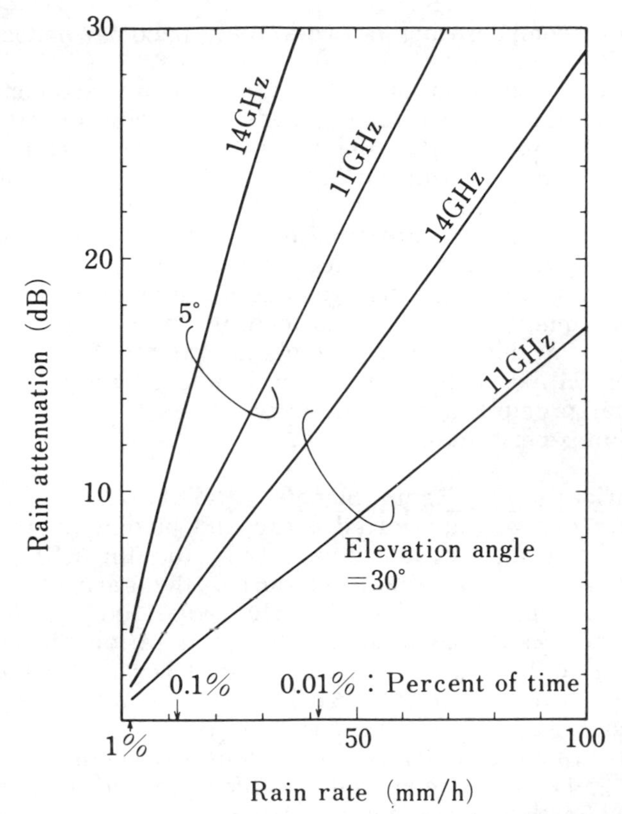

Fig. 4 Rainfall attenuation in 14/11 GHz at the rain climatic zone K (CCIR Rep. 563-2).

satellite with a low elevation angle should anticipate this degree of signal attenuation for operation in the 6/4 GHz band.

Ionosphere scintillation is another effect that causes strong short-term fluctuation of radio signals due to the spread F and the sporadic E layer in the ionosphere. This phenomenon is frequently observed at locations on the geomagnetic equator and at high latitude in the nighttime during springs and falls when there is relatively high solar activity.[2]

The level of cross-polarization discrimination (XPD) at Earth stations that access an INTELSAT V operating in the dual-polarization mode must be large enough to separate the copolar signals from the cross-polar signals. The axial ratios of satellite antennas and Earth station antennas are specified to meet this requirement. But heavy rainfall will degrade the XPD of radio signals received at Earth stations, particularly at a site with a low elevation angle (see Fig. 3).[3] In this case, the installation of a depolarization compensating network may become necessary to improve XPD.

At a frequency band of 14/11 GHz, signal attenuation by scattering and absorption through clouds and precipitation is more significant (see Fig. 4). But the resulting narrow beamwidth of satellites and Earth stations at the higher frequency will compensate partially for the above adverse effect, depending on the climate of the Earth station site and the elevation angle. At sites where a substantially heavy rainfall rate in a year is anticipated and the elevation angle is unfavorably low, the transmission quality may occasionally be very poor or unacceptable. In this case, the implementation of site diversity configuration will be effective. A preliminary study shows that correlation of precipitation between two sites separated by more than 10–30 km is very small.[4]

3. Polarization of Radio Signals. INTELSAT I and II had satellite antennas of the slotted array and biconical array type, respectively, and they could radiate only the linearly polarized radio signal at the 6/4 GHz band during the early days of the INTELSAT system. When a linearly polarized radio signal propagates through the ionosphere, the polarization plane will be affected by up to 9 deg at 4 GHz and up to 4 deg at 6 GHz due to the Faraday rotation effect.[2] Even if the antenna tracking device is implemented at Earth stations, there is a mismatch of polarization tracking between the transmitting and receiving frequencies, and some reduction of received signal powers at satellites and Earth stations is unavoidable.

INTELSAT III was equipped with a mechanical despun antenna, and, since then, INTELSAT has adopted circular polarization in the system at 6/4 GHz, eliminating the above disadvantage of linear polarization.

The axial ratio of satellite antennas for INTELSAT III, IV, and IV-A was about 3 dB, or the cross-polarization discrimination (XPD) was about 15 dB, and the required performance of the antenna axial ratio at Earth stations was also determined at the same level. For INTELSAT V with an offset parabolic antenna for hemi/zone coverages, the axial ratio was improved to 0.75 dB (XPD = 27 dB), and the frequency reuse by dual-polarization has become possible with the improved axial ratio of Earth station antennas, which was specified to be 0.5 dB (XPD = 31 dB). The rainfall effect on the degradation of cross-polarization discrimination was described in Sec. IC2.

INTELSAT V has transponders in the 14/11 GHz band. The effect of Faraday rotation in this frequency region is very small, and INTELSAT adopted linear polarization without the requirement of a polarization tracking device. The transmission and reception polarization at Earth stations for a given coverage is orthogonal, and the transmission polarization at Earth stations in the west coverage and east coverage is also orthogonal in order to achieve the necessary isolation between two spot beam coverages.

4. Gain-to-Noise Temperature Ratio of Earth Stations. The ratio of antenna gain G (in decibels) to system noise temperature T (in degrees Kelvin) at the receiving Earth station determines the basic configuration of Earth stations.

G is proportional to the square of the antenna diameter, and T is a sum of the thermal noise from the sky, the ground, and the equivalent noise contribution from the receiver. A larger gain-to-noise temperature ratio (G/T) (dB/K) means a higher receiving antenna gain with low sidelobe characteristics and a lower receiver noise figure. When the Earth station's elevation angle is low, the antenna will pick up the thermal noise from the ground and T is a function of the elevation angle, which varies according to the skyline of the Earth station sites (see Fig. 5). It should be noted that heavy rainfall also contributes to the increase of thermal noise from the sky, and Fig. 6 shows its effects in the 4 and 11 GHz frequency bands. The carrier-to-noise (C/N) power ratio in a downlink for a given receiving signal power from the satellite is improved in proportion to the G/T figure.

Thus INTELSAT has specified the G/T for the Standard A Earth station as 40.7 dB/K and for the Standard B Earth station as 31.7 dB/K under clear skies for the 4 GHz band in the direction of the satellite with an elevation angle of not less than 5 deg. This G/T value for Standard A can be achieved with a typical configuration of a 30-m antenna and an uncooled parametric

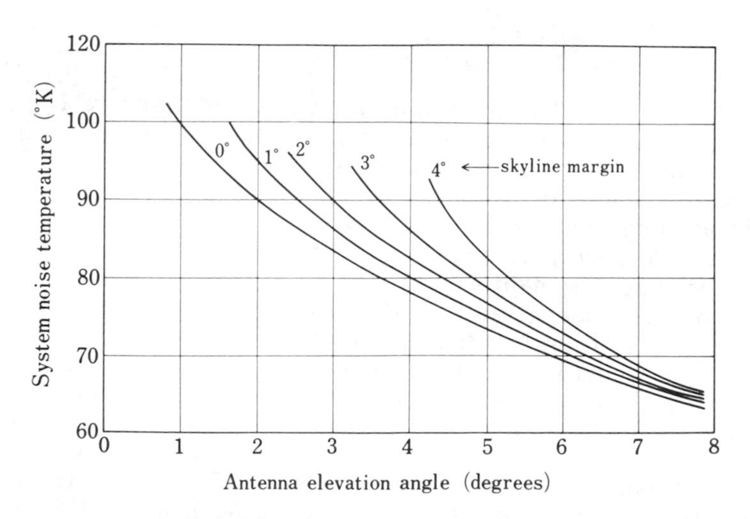

Fig. 5 Increase in system noise temperature (in 4 GHz).

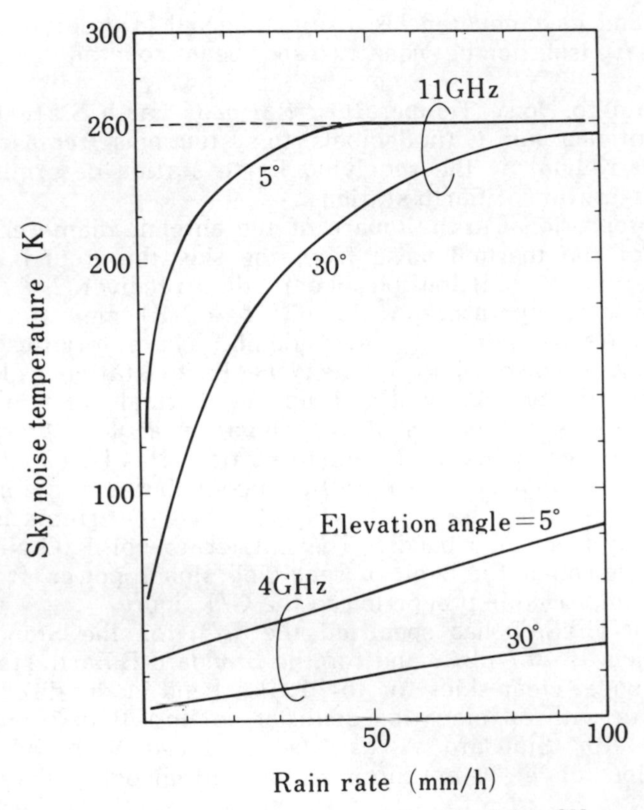

Fig. 6 Increase in sky noise temperature in 4 and 11 GHz at the rain climatic zone K.

amplifier of 50 K. The specified G/T value for Standard B can be achieved with a typical configuration of an 11-m antenna and an uncooled parametric amplifier of 50 K.

For the Standard C Earth station, which operates at the 14/11 GHz band, long-term rainfall data should be taken into account to ensure an adequate performance, and the G/T is given in terms of performance requirements associated with 10% and 0.017% of the time in a year for operation in the direction of a satellite with an elevation angle of not less than 10 deg.

For a Standard C Earth station at a site where the rainfall rate is not so heavy, the specified G/T values can be achieved with a typical configuration of a 12-20-m antenna and an uncooled parametric amplifier of 100 K or a thermoelectric-cooled field effect transistor (FET) amplifier of 125 K. But for a Standard C Earth station with a low elevation angle at a heavy rainfall site, the G/T requirements for 0.017% of the time are stringent due to a large rainfall attenuation of about 20 dB or more and a substantial increase of system noise temperature of about 200 K, and the site diversity configuration may become unavoidable.

5. Reliability of Earth Stations. The failures or malfunctions of one Earth station in the network affect not only a large number of circuits associated with the station but also can affect those of many other Earth stations. Therefore Earth stations should be designed and operated to maintain a high standard of reliability and stability for their station equipment. For instance, it is necessary to maintain carrier levels and frequencies within the tolerable ranges specified by INTELSAT, and it is advisable that the reliability of Earth station operation should be at least 99.8%.

The average reliability of the total INTELSAT Earth station system for the years 1977-1982 is shown below:

1977	1978	1979	1980	1981	1982
99.931%	99.889%	99.819%	99.893%	99.920%	99.930%

To achieve high reliability, it is particularly desirable to have redundant facilities for the subsystems whose failures would result in total station outage. Typical subsystems that have been implemented with redundant configuration are the high-power amplifier (HPA) and the low-noise receiver (LNR).

The Earth station subsystems reliability averages for all INTELSAT Earth stations in 1982 are shown below:

Antenna	99.9785%
HPA	99.9896%
LNR	99.9961%
Power	99.9812%
Total	99.9299%

Introduction of time-division multiple-access (TDMA) terminals will increase interdependency among participating Earth stations for the successful operation of an entire TDMA system, and more stringent reliability of TDMA terminals is recommended by INTELSAT with the objective of achieving 99.99% reliability.

6. Cost Estimate of Earth Stations. The operational cost of Earth stations is related to the initial investment cost, operation/maintenance cost, and space segment charge.

The initial investment cost of Earth stations depends on the selection of antenna size, the complexity of communications equipment for different modulation methods, the number of circuits, and many other requirements. For example, the antenna cost is the most expensive part, but still it will vary with diverse operational requirements due to local geographical and meteorological conditions at the site as well as to the supporting facilities that control the antenna. The HPA configuration will be affected by the number and size of transmitting carriers, the requirements of redundancy, and the supporting facilities operated by the station staff. The initial investment costs of Earth stations are assumed to be $13 million, $7.5 million, and $2.5-3 million (U.S.) for the following three cases, respectively.

Case 1: A Standard A Earth station with large traffic that implements 10-kW TWT, FM transmission equipment with two transmit carriers and ten receive carriers, and redundant facilities for HPA, LNR, and FM equipment. Assumed investment cost is $13 million excluding the building costs.

Case 2: A Standard A Earth station with medium traffic that implements 3-kW Klystron, FM transmission equipment with one transmit carrier and five receive carriers, and redundant facilities for HPA, LNR, and FM equipment. Assumed investment cost is $7.5 million excluding the building costs.

Case 3: A Standard B Earth station with small traffic in the 5-50 channel range that includes 3-kW TWT and SCPC terminals. Assumed investment cost is $2.5-3 million excluding the building costs.

The maintenance/operation cost includes management and labor salaries, materials, and electrical power. The space segment charge per half circuit (unit) is determined by INTELSAT. The annual charge as of 1983 is $4680 for an FDM/FM channel operated by a Standard A station and $7020 for SCPC and CFDM/FM channels operated by a Standard B station.

The total operational cost of Earth stations per year is derived by the following method.

The total cost of initial investment (Ci) is converted to the annual depreciation cost (Cd) and the annual return of investment (R), which are assumed to be about 10% each of Ci. The annual

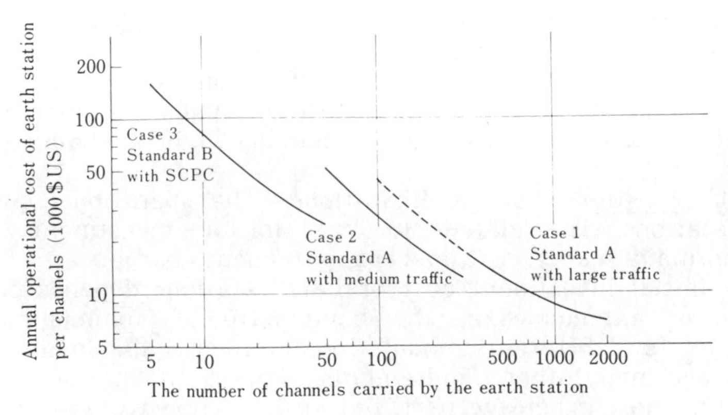

Fig. 7 Annual operational cost of Earth station.

cost of maintenance and operation (Cmo) is also assumed to be 10% of Ci.

Then, the total annual operational cost (Ca) is derived from the following formula:

Ca= Cd + R + Cmo + N (space segment charge per unit)
= 0.3 x Ci + N x $4680 (in U.S. dollars for Standard A)
or
= 0.3 x Ci + N x $7020 (in U.S. dollars for Standard B)

Figure 7 shows the annual operational cost of typical Earth stations with the given assumptions.

II. Performance Characteristics of Standard Earth Stations

A. Standard A Earth Stations (BG-28-72 Rev. 1)[1]

Standard A Earth stations, which are typically known by large antennas about 30 m (98 ft) in diameter with a G/T of 40.7 dB/K for 6/4 GHz operation have been specified for the efficient use of satellite powers and frequency bands and widely used in the INTELSAT system since 1965. (Through INTELSAT V, the interconnection of 6/4 GHz and 14/11 GHz bands is possible.)

Although the cost of initial investment for the Standard A station is higher than that for other types of standard Earth stations, it has the advantage of accommodating sizable traffic to many destinations with flexible expansion capability. It can also provide a color television transmission with CCIR quality of 54 dB. INTELSAT encourages Earth station owners to construct the Standard A Earth station when their traffic requirements exceed more than 25 to 30 circuits and future growth is anticipated.

Main performance characteristics of the Standard A Earth station accessing INTELSAT IV, IV-A, and V are described below.

1. Antenna. The gain-to-noise temperature ratio (G/T) in the direction of the satellites and for the polarization chosen for each satellite series under clear sky conditions and for any frequency in the 3.7-4.2 GHz band must satisfy the following condition:

$$G/T \geqq 40.7 + 20\log f/4 \text{ dB/K}$$

where G is the antenna gain at the input of receiver in dB; T is the receiving system noise temperature at the input of the receiver in dB Kelvin; and f is the receiving frequency in GHz.

Sidelobes pattern: To avoid interference with adjacent satellite systems and terrestrial systems operating in the same frequency band, no more than 10% of the transmit sidelobe peaks (at angles greater than 1 deg away from the main beam) should exceed the limits defined in the following formula:

$$G = 32 - 25\log\Theta \quad 1 \text{ deg} < \Theta \leqq 48 \text{ deg}$$

$$G = -10 \text{ dB} \quad \Theta > 48 \text{ deg}$$

where G is the gain of sidelobe envelope in dB at any frequency between 5.925 and 6.425 GHz, and Θ is the angle between the main beam and the direction considered. The same pattern is recommended for the receive sidelobes for the 3.7-4.2 GHz frequency band.

Polarization and axial ratio: For access to INTELSAT IV and IV-A, the Earth station transmits left-hand circular polarization and receives right-hand circular polarization. The voltage axial transmission ratio in the direction of the satellite must not exceed 1.4, and the same requirement is recommended for reception.

For access to INTELSAT V, which is operated in the dual-polarization mode, the Earth station transmits and receives both left-hand and right-hand polarizations or either of them, depending on their access to global, hemispheric, and zone coverage transponders. The voltage axial ratio of no more than 1.06 is mandatory for the transmission in the direction of the satellites and recommended for reception. (For old antennas, the axial ratio of 1.09 is accepted.) But for those Earth stations that will be placed only in the global beam transponders due to their geographical location, the voltage axial ratio is relaxed to 1.4.

At a few geographical areas, rain depolarization may degrade the isolation between polarizations to less than 12 dB for 0.01% of the time. In such a circumstance, the Earth station is asked to consider the use of a compensating network.

2. Types of Transmission. Carrier types available as transmission modes among Standard A Earth Stations and to other Standard B and C stations are shown in Table 2.

Table 2 Carrier types to be used from Standard A station

	To Standard A	To Standard B	To Standard C
FDM/FM	Yes	...	Yes
Companded FM	Yes	Yes	...
TV (30 MHz)	Yes	Yes	...
TV (17.5 MHz)	Yes	Yes	...
TDMA	Yes	...	Yes
SCPC/PSK	Yes	Yes	...
SPADE	Yes	...	...

Table 3 FDM/FM carriers for INTELSAT IV, IV-A, and V

Bandwidth unit, MHz	No. of channels
1.25	12[A]
2.5	24, 36, 48, 60, 72
5.0	60, 72, 96, 132, 192
7.5	96, 132, 192, 252
10.0	132, 192, 252, 312
15.0	252, 312, 432, 372,[B] 492[B]
17.5	432
20.0	432, 612, 792, 492,[B] 552[B]
25.0	432, 792, 972, 492,[B] 552,[B] 612[B]
36.0	972, 792,[B] 1092,[C] 1332,[D] 1872[E]

[A]INTELSAT V only.
[B]INTELSAT IV-A and V only.
[C]INTELSAT IV and IV-A only.
[D]INTELSAT IV-A only.
[E]INTELSAT IV only.

Table 4 Transmission parameters of television

	Half transponder television	Full transponder television
Allocated bandwidth	17.5 MHz	30 MHz
Earth station e.i.r.p.	88 dBW for IV, IV-A 85.4 dBW for V	88 dBW for IV, IV-A 85.4 dBW for V
Operating C/N at receiver	17 dB	16.2 dB
Video S/N	49 dB	54 dB
FM subcarrier for sound		
Top baseband	15 kHz	
Subcarrier frequency	6.6 or 6.65 MHz	

FDM/FM: Available standard FDM/FM carriers in the INTELSAT system are determined in terms of capacity (12-1872 channels) and occupied bandwidth (1.125-36 MHz). Requirements for minimum carrier transmit power in the range 70.4-98.6 dBW and other transmission parameters at Earth stations are also defined depending on the carrier size, type of transponder (global, spot, hemisphere, and zone) for each accessing satellite series (INTELSAT IV, IV-A, and V) and cross-strapped interconnection links (6-11 GHz in INTELSAT V) to assure a satisfactory carrier-to-noise ratio at receiving Earth stations. Table 3 summarizes the FDM/FM carriers available.

Television: For the transmission of both 525/60 and 625/50 television (TV) carriers, two types of band allocation (30 and 17.5 MHz) are specified for access to the global transponder. At present, television is transmitted mostly by 17.5-MHz carriers to increase band efficiency. Associated sound programs (up to a single 15-kHz program) are combined with normal video signals by the FM subcarrier method. Table 4 summarizes the main parameters of television carriers.

Single channel per carrier (SCPC): There are four modes of SCPC operation. Mode 1 is the SPADE system and is used for the transmission of voice and voiceband data operated on a demand-assigned basis in the global transponders among Standard A Earth stations. Mode 2 is SCPC/PCM/PSK and is used for the transmission of voice and voiceband data at or below 4800 bit/s on a preassigned basis. Mode 3 is SCPC/PCM/PSK, which is used for the transmission of voiceband data above 4800 bit/s with 120/112 forward error correction (FEC) on a preassigned basis. Mode 4 is SCPC/PSK, which is used for the transmission of 56-kbit/s data with 7/8 convolution code or 48/50 kbit/s with 3/4 codes on a preassigned basis. Modes 2, 3, and 4 can be assigned to either global or frequency reuse transponders for links among Standard A or between Standard A and B stations. For the SPADE system, carriers are assigned in pairs to form a full circuit. For SCPC of modes 2, 3, and 4, carrier assignments are at random and, in some cases, they are placed in different transponders. SCPC equipment of all modes must prepare frequency synthesizers to transmit and receive the carriers at the assigned frequency. The main parameters of SCPC are shown in Table 5.

The main differences of the SPADE system from other modes of SCPC are the additional requirement of common signaling channels for demand assignment and frequency allocation of pilot for automatic frequency control (AFC).

Time-division multiple-access (TDMA): TDMA traffic terminals at the Earth stations transmit and receive multiple bursts at a rate of 120 mbit/s and four-phase PSK modulation

Table 5 SCPC transmission parameters[A]

Parameter	Value
Allocated bandwidth	45 kHz
Transmission rate	64 kbit/s
Voice	7 bits PCM/A-law (voice activation)
Modulation	Four-phase coherent PSK
E.i.r.p. to Standard A	63 dBW for IV, IV-A; 61.5 dBW for V (global) 60.5 dBW for V (hemi/zone)
E.i.r.p. to Standard B	69.8 dBW for IV, IV-A (global); 67.5 dBW for IV-A (hemi/zone), V (global); 66 dBW for V (hemi/zone)
Operating C/N	15.5 dB
Nominal bit error rate	10^{-6}
Nominal BER with FEC	$1\text{-}3 \times 10^{-9}$

[A]These parameters are common for all modes.

accessing the 72-MHz hemi and zone beam transponders of INTELSAT V and its successors. There are two types of TDMA channels: Type 1 is TDMA/digital speech interpolation (DSI) for the transmission of voice and voiceband data on an interpolated basis. Type 2 is TDMA/digital noninterpolation (DNI) for the transmission of voice and data (64 kbit/s-8.192 Mbit/s) on a noninterpolated basis.

The channel size of TDMA carriers is determined by the length of traffic data in the transmit burst in one TDMA frame of 2 ms and can be varied in steps of one channel of 64 symbols. One TDMA traffic terminal has the capability of transmitting up to 16 nonoverlapping traffic bursts per frame and receiving up to 32 nonoverlapping traffic bursts (which contain up to 32 subbursts of DSI and/or DNI) in total per frame in the transponder hopping condition. One DSI subburst (interpolated) contains up to 128 satellite channels (DSI gain = the number of terrestrial channels/the number of satellite channels), and one DNI subburst (preassigned noninterpolated) is expandable from one to a maximum capacity of 128 channels of 64 kbit/s unit. One subburst may be shared by DSI and DNI channels.

The main parameters of TDMA carriers are summarized in Table 6.

Companded FDM/FM (CFDM): CFDM may be implemented to replace SCPC links between Standard A and B Earth stations when the equivalent CFDM carriers capacity approaches 24 channels or to expand the capacity of regular FDM/FM channels between Standard A Earth stations on a selective basis to alleviate short-term capacity shortage.

Table 6 TDMA transmission parameters

Bit rate/modulation	120.832 Mbits/four-phase coherent PSK
Frame length	2 ms
Voice	8 bits PCM/A law
Earth station e.i.r.p.	89 dBW
Bit error rate performance	CCIR Rec. 522

Table 7 CFDM carriers

Bandwidth unit, MHz	Carrier size (channels)
1.25	24, 36[A]
2.5	48, 60, 72,[A] 96[A]
5.0	96, 132, 192,[A] 252[A]
7.5	192, 252[A]
10.0	252

[A]Available only between Standard A Earth stations.

Available standard CFDM carriers are determined in terms of capacity (24-252 channels), occupied bandwidth (1.25-10 MHz), and transmit e.i.r.p. (73.1-91.7 dBW), and these parameters vary depending on the operating Earth stations (A to A or A to/from B) and accessing satellite series of INTELSAT IV-A, V (global or hemi-zone). The technical characteristics of a compandor with a compression and expansion ratio of 2 are specified based on the INTELSAT study and field tests.

Available CFDM carriers are summarized in Table 7.

3. Radio-Frequency Out-of-Band Emission. Out of a total telephony noise budget of 10,000 pW in the FM system, 1000 pW are allocated to the noise source in the Earth station, within which 500 pW is the allowance for radio-frequency (RF) out-of-band emission caused by intermodulation (IM) noise and other spurious emissions from Earth stations. This IM product results mostly from the nonlinear characteristics of the transmitting high-power amplifier (HPA) operating in the multicarrier mode of FM, TDMA, and other transmission systems. Thus INTELSAT limits IM products not to exceed levels in the range 19.3-26 dBW/4 kHz, depending on the type of accessing transponders and satellite series, within the frequency band of 5925-6425 MHz at the transmitting Earth stations. E.i.r.p. of spurious emission outside the bandwidth unit for each carrier, excluding the above IM product at the transmitting Earth station, is also limited within 4 dBW in any 4-kHz band.

The limits of IM products caused by the mutual interaction between SCPC carriers are specified so that their levels do not exceed 31.5 dBW for $2 \leq N \leq 7$ and $(48.5 - 20 \log N)$ dBW for $N > 7$, where N is the number of configured SCPC carriers at the transmitting stations. IM products between SCPC carriers and any other non-SCPC carriers resulting from multicarrier operation must not exceed the limits in the range 21-26 dBW/4 kHz, depending on the type of accessing transponders and satellite series, within the 5925-6425 GHz band. The permissible level of spurious emission excluding multicarrier IM products, outside of the allocated SCPC bandwidth, is 4 dBW in any 4-kHz band within 5925-6925 MHz.

For the TDMA system, the spectrum spreading of the phase-shift keying (PSK) carrier due to Earth station HPA nonlinearity results in RF out-of-band emission, and its level is also limited not to exceed 23 dBW/4 kHz outside the TDMA bandwidth. The spurious tone and other undesirable signals, excluding the PSK spectrum spreading and multicarrier IM products by HPA nonlinearity, are also limited to 4 dBW in any 4-kHz band within 5925-6425 MHz.

4. Energy Dispersal. Because the frequency band allocated to the fixed-satellite service is shared in many cases with other terrestrial systems, mutual interference must be avoided, and a

Radio Regulation has specified the maximum allowable power flux density from the satellite at the Earth surfaces (CCIR Rec. 358-3). To satisfy this requirement, the energy of carriers transmitted from Earth stations is dispersed within the allocated bandwidth of carriers. This energy dispersal is also essential to reduce the mutual interference between carriers of the same frequency allocation in the INTELSAT and other satellite systems.

The energy dispersal technique for FM and television carriers in the INTELSAT system is based on the addition of a low-frequency symmetrical triangular waveform to the baseband signal prior to the FM modulator. For FM carriers, the frequency of the dispersal waveform is in the 30-150 Hz band that will be designated by INTELSAT for each carrier, and the level of dispersal waveform is determined so that at all times the maximum e.i.r.p. per 4 kHz of transmitted carriers does not exceed the maximum e.i.r.p. per 4 kHz of fully loaded carriers by more than 2 dB. For television carriers, the level of the dispersal waveform is determined so that the peak-to-peak deviation of television carriers is always 1 MHz. Furthermore, for a 17.5-MHz video carrier, this peak-to-peak amplitude of energy dispersal modulation is increased automatically from 1 to 2 MHz when no video signal is present. The frequency of dispersal waveform is 30 Hz for the 525/60 standard and 25 Hz for the 625/50 standard.

The energy dispersal of the TDMA system can be achieved by adding a pseudorandom sequence to the digital baseband signal at the transmitting Earth station and subtracting the same sequence at the receiving station. INTELSAT specified a common pseudonoise (PN) sequence produced by a 15-stage shift register.

5. Engineering Service Circuits (ESC). All of the INTELSAT Earth stations must prepare ESC for communication with the INTELSAT Technical and Operational Control Center/INTELSAT Operations Center (TOCC/IOC) and with other Earth stations operating in the same network. The voice and telegram engineering service circuit for FDM/FM carriers is implemented in each FDM multichannel baseband to provide inter-Earth station 4-kHz service channels. When a Standard A Earth station communicates with a Standard B Earth station, at least one preassigned SCPC must be prepared as an ESC with voice and telegram capability.

B. Standard B Earth Stations (BG-28-74 Rev. 1)[1]

The Standard B Earth station, with a medium-sized antenna and with a G/T of 31.7-dB/K, operating in the 6/4 GHz frequency band, was developed as a more economical Earth station and is particularly implemented by Earth station owners whose traffic requirements are less than several tens of channels. Standard B

Earth stations can establish direct communication links among multiple Standard B stations and to Standard A stations but cannot communicate with Standard C Earth stations.

In general, most of the mandatory technical performances specified for the Standard A station are applicable to the Standard B Earth station, and the main differences are antenna aperture size and a more limited range of transmission capabilities.

1. Antenna. The gain-to-noise temperature ratio (G/T) in the direction of the satellites and for the polarization chosen for each satellite series under clear sky conditions and for any frequency in the 3.7-4.2 GHz band must satisfy the following condition:

$$G/T \geqq 31.7 + 20\log f/4 \text{ dB/K} \quad (f \text{ in GHz})$$

The transmit antenna main beam gain is specified to be not less than 53.2 dB, which is not specified for the Standard A station. This requirement is necessary to avoid unnecessarily high-power radiation in the sidelobe directions.

Requirements for the sidelobes pattern, polarization, and axial ratio of transmit and receive antennas are the same as those for Standard A Earth stations.

2. Types of Transmission. Transmission modes available through Standard B stations are limited to SCPC and companded FDM for voice and data, television, and FDM/FM for television audio. (See Table 8.)

Television: For the transmission of 525/60 and 625/50 television standard, full transponder carriers of 30 MHz can provide video quality nearly equivalent to Standard A reception. The video quality achieved by 17.5-MHz television reception is low but acceptable. Television-associated audio is transmitted to Standard A/B stations and received from Standard A stations by SCPC/PCM/PSK or by a 2.5-MHz FDM/FM carrier. The main

Table 8 Carrier types to be used from Standard B stations

	To Standard A	To Standard B	To Standard C
FDM/FM	...	...	...
TV (30 MHz)	Yes	Yes	...
TV (17.5 MHz)	Yes	Yes[A]	...
SCPC/PCM/PSK	Yes	Yes	...
SCPC/PSK	Yes	Yes	...
SPADE	...	...	...
Companded FDM	Yes	Yes[B]	...

[A]With low but acceptable quality.
[B]Available only for stations with improved performance.

parameters for television transmission are summarized in Table 9.

Preassigned SCPC: There are three modes of SCPC operation on a preassigned basis. Mode 1 is SCPC/PCM/PSK for the transmission of voice or voiceband data at or below 4.8 kbit/s to Standard A and B stations. A frequency pair to implement a full circuit will be assigned by INTELSAT, and Earth stations must prepare the dual-frequency synthesizer to meet the assignment of the channel frequency plan. Carriers must be controlled by voice activation. Mode 2 is SCPC/PCM/PSK for the transmission of voice or voiceband data above 4.8 kbit/s with 120/112 FEC. Mode 3 is SCPC/PSK for the transmission of 48/50-kbit/s data with 3/4 FEC and 56-kbit/s data with 7/8 FEC. The main parameters are shown in Table 10.

Companded FDM/FM (CFDM): Due to the increasing cost of SCPC equipment as traffic requirements grow, CFDM carriers have been developed for the Standard B Earth station to replace SCPC links between Standard B and A/B stations. Available standard CFDM carriers between Standard A and B stations are described in Sec. IIA2. Implementation of CFDM between the Standard B Earth stations is permitted for access to INTELSAT V and its successors if the antenna has improved performance in either or both by an excess main beam gain of over 53.2 dB and a better sidelobes pattern below the envelope of 32-25logΘ such that the net improvement results in the determined values of E shown in Table 11. The carrier sizes are also limited compared with the regular sizes available for communication between Standard A and B stations. These particular constraints were necessary to meet the permissible emission levels of 23 dBW/4 kHz at 3 deg off beam, or CCIR Rec. 524-1. Available CFDM carrier sizes with specified transmit e.i.r.p. in the 76.6-84.4 dBW range between Standard B stations are shown in Table 11.

3. RF Out-of-Band Emission. The limits of intermodulation products caused by interaction between SCPC carriers and between SCPC carriers and any other carrier resulting from multicarrier

Table 9 Television parameters

Carrier bandwidth	E.i.r.p., dBW	Video quality, dB
30-MHz TV	85	47.6 (A/B→B)
	81.8	51.1 (B→A)
17.5-MHz TV	88	41 (A→B)[A]
	85	41 (B→B)[B]
	85	47.3 (B→A)

[A]At 10-deg elevation angle.
[B]At 50-deg elevation angle.

Table 10 Main parameters of SCPC/PCM/PSK and SCPC/PSK

	To Standard B		To Standard A	
E.i.r.p. for INTELSAT IV-A	67.5 dBW	for hemi/zone	63 dBW	
	69.8	for global	...	
E.i.r.p. for INTELSAT V	66	for hemi/zone	60.5	for hemi/zone
	67.5	for global	61.5	for global
Allocated bandwidth	45 kHz			
Transmission rate	64 kbit/s			
Modulation	Four-phase coherent PSK			
Voice	7 bits PCM/A-law (voice activation)			
Nominal BER	10^{-6}			
BER with FEC	$(1-3)\times10^{-9}$			

Table 11 CFDM carriers between Standard B stations (INTELSAT V)

Bandwidth unit, MHz	Carrier size (channels)	E, dB Global	E, dB Hemi/zone
1.25	24	2.6	1.1
2.5	48	3	1.5
5.0	96,[A] 132[A]	...	2.4
7.5	192[A]	...	2.5

[A]Hemi/zone beam only.

operation are specified in the same manner as for the Standard A Earth station. The permissible levels of spurious emission excluding IM products are the same as those described for the Standard A station (see Sec. IIA3).

4. Energy Dispersal. The energy dispersal required for television transmission is the same as that required for Standard A stations (see Sec. IIA4).

5. Engineering Service Circuit (ESC). Standard B Earth stations equipped only with SCPC terminals must prepare two voice mode channel units to communicate with TOCC/IOC and all other Earth stations operating within the same network. This ESC is provided by the terminal with the same transmission and functional capabilities as the SCPC channel unit.

C. Standard C Earth Stations (BG-23-73 Rev. 1)[1]

Standard C Earth stations started their first operation in 1982, accessing INTELSAT V in the 14/11 GHz frequency band. Use of this higher frequency band is available at Earth stations located in the area covered by the 14/11 GHz spot beam antennas of INTELSAT V and its successors. The main features of the Standard C station are the technical characteristics of its antennas and its RF facilities, which take into consideration rainfall attenuation effects and the limitation of transmission modes to FDM/FM and TDMA.

The general performance characteristics of the Standard C Earth station are described in the following sections.

1. Antenna. The gain-to-noise temperature ratio (G/T) in the direction of satellites for any 10.95-11.2 GHz and 11.45-11.7 GHz frequency band must not be less than the following values: G/T_1-L_1 = 39 +20log f/11.2 dB/K for all but 10% of the time and G/T_2-L_2 = B +20log f/11.2 dB/K for all but 0.017% of the time.

G is the antenna gain at the input of receiver in dB; Li is the predictive attenuation relative to a clean sky at the operating frequency f in GHz, exceeded for no more than 10% and 0.017% of the time for i = 1 and 2, respectively. Ti is the receiving system noise temperature at the input receiver in dB Kelvin at the operating frequency when the attenuation Li prevails. B is 29.5 and 32.5 dB for Earth stations located within the west and east spot beam coverage, respectively. When an Earth station is operated with site diversity configuration, the above requirement applies to the operating terminal at any given time.

Sidelobes pattern: Requirements for the sidelobes pattern of transmit and receive antennas are the same as those for the Standard A Earth station (see Sec. IIA1).

Polarizaton and axial ratio: The Standard C Earth station transmits and receives linear polarizations which are orthogonal to each other. Although the present specifications do not employ dual-polarization in the 14/11 GHz region, use of dual linear polarization is possible in the future. The voltage axial ratio of transmission in the direction of the satellite must not be less than 31.6 within the main beam axis, and the same performance is recommended for reception.

2. Types of Transmission. The available transmission modes of Standard C Earth stations are solely FDM/FM and TDMA carriers. Standard C stations can communicate with Standard A stations by the interconnections of 14/4 GHz and 6/11 GHz links.

FDM/FM carriers are defined in terms of capacity (12-972 channels) and occupied bandwidth (1.25-36 MHz). The requirement for clear sky e.i.r.p. in the 62.4-83.8 dBW range is defined for each carrier size in the 14/11 GHz direct link of east up and west up and cross-strapped links in the 14/4 and 6/11 GHz link. The carrier size and bandwidth unit for the 14/11 GHz direct link are summarized in Table 12.

Table 12 FDM/FM carrier in the 14/11 GHz link

Bandwidth unit, MHz	No. of channels
2.5	24, 36, 48, 60, 72
5.0	60, 72, 96, 132, 192
7.5	96, 132, 192, 252
10.0	132, 192, 252, 312
15.0	252, 312, 372, 432, 492
17.5	432
20.0	432, 492, 552, 612, 792
25.0	432, 492, 552, 612, 792, 972
36.0	792, 972

Earth stations must prepare means to compensate the uplink rainfall attenuation when the attenuation relative to clear sky conditions increases over 6 dB in the 14/4 GHz link for 0.01% of the year and 6.5–9 dB in the 14/11 GHz link for 0.017% of the year. The site diversity configuration or the transmitting power control will provide means to satisfy the above requirement (see Sec. IC).

The TDMA transmission parameters for the Standard C station are the same as that for the Standard A Earth station except for the requirement of transmitting e.i.r.p., which is specified to be 85 dBW for the west beam uplink and 88 dBW for the east beam uplink. These e.i.r.p. values include an uplink (14/4 and 14/11 GHz) margin of 11 dB. When uplink attenuation relative to clear sky is more than the uplink margin for more than 0.008% of the time in the 14/4 GHz link and for 0.005% of the time in the 14/11 GHz link, the transmitter power control or the site diversity configuration is necessary to satisfy the short-term performance criterion of CCIR Rec. 552.

3. RF Out-of-Band Emission. The e.i.r.p. of intermodulation products for multicarrier operation within the 14–14.5 GHz frequency band must not exceed 10 dBW/4 kHz. The spurious emission excluding the above IM product must be no more than 4 dBW in any 4-kHz band within the 14–14.5 GHz frequency band.

For the TDMA system, the spectrum spreading of the PSK carrier due to Earth station HPA nonlinearity must not exceed the values of 12.0 dBW/4 kHz in the 14.0–14.5 GHz band outside the TDMA bandwidth. The limited level of spurious tones and other undesirable signals is 4 dBW in any 4-kHz band within the 14–14.5 GHz band.

4. Energy Dispersal. The requirement for energy dispersal is the same as that for the Standard A station.

5. ESC. The technical characteristics of the ESC are the same as those for the Standard A Earth station. The Standard C Earth station should prepare to communicate to both Standard A and C Earth stations by an ESC.

D. Standard Z Earth Station (BG-52-75)[1]

Most Earth stations for domestic service accessing INTELSAT's leased space segment capacity have rather small aperture antennas that do not conform to the technical performance characteristics of any of the Standard A, B, and C Earth stations; they had to be approved by INTELSAT on a case-by-case basis.

In 1982, INTELSAT specified the performance of Standard Z Earth stations for domestic systems accessing INTELSAT's leased transponders in the 14/11 and 6/4 GHz frequency bands. It described the minimum Earth station capabilities and limited the maximum level of emissions to within an acceptable level of interference to other users of the INTELSAT space segment. These performance requirements do not include G/T, the maximum e.i.r.p. per carrier, modulation methods, and transmit antenna gain, which differ from those for the Standard A, B, and C Earth stations. The lessees can decide the best transmission designs for their particular requirements so long as their transmission plans are coordinated and agree with INTELSAT. The standardization of Standard Z has offered a more simple procedure for the approval of Earth stations for domestic service by INTELSAT.

General performance characteristics for the Standard Z Earth station are described in the following sections.

1. Antenna Subsystem. Requirements for the sidelobes pattern, polarization, and axial ratio of transmit and receive antennas in the 6/4 and 14/11 GHz frequency bands are the same as those for Standard B and C Earth stations, except for small-aperture antennas of $D/\lambda < 100$, which should satisfy the sidelobes pattern of $G = 32 - 25 \log \theta$ for $100(\lambda/D) < \theta \leq 48$ deg.

2. Types of Transmission. Transmission parameters are left to the lessees for their selection. INTELSAT has specified only the e.i.r.p. stability of carriers and the frequency tolerance for SCPC carriers and analog carriers such as FDM, CFDM, and television.

Earth stations must be able to control these parameters and their antenna tracking at all times to avoid interference with other services.

3. Off Beam Emission e.i.r.p. Density. This requirement is most important to achieve acceptable levels of interference with other systems. Off beam e.i.r.p. density D (dBW) in any 40-kHz

band in the direction of adjacent satellites must not exceed the criteria below:

$D = 42 - 25\log\Theta \quad 2.5 \text{ deg} < \Theta \leq 48 \text{ deg}$ (for the 6-GHz region)

$D = 0 \quad \Theta > 48 \text{ deg}$

where Θ is the angle in degrees between the main beam and the direction considered.

$D = 33 - 25\log\Theta \quad 2.5 \text{ deg} < \Theta \leq 48 \text{ deg}$ (for the 14-GHz region)

$D = -9 \quad \Theta > 48 \text{ deg}$

4. RF Out-of-Band Emission. The limits of RF out-of-band emission resulting from the multicarrier intermodulation product and spurious emission are the same as those for Standard B and C Earth stations in the 6- and 14-GHz frequency bands, respectively, except for the condition that these requirements apply only to the bandwidth outside the leased frequency band. The acceptable levels of intermodulation product and spurious emission within the leased bandwidth can be determined by the lessees.

5. Engineering Service Circuit. A dedicated link must be established between INTELSAT and the master station of the leased network. The lessee must also prepare the communication means between all Standard Z Earth stations in his particular network.

E. Standard E Earth Station (BG-53-86)[1]

The activities of many international corporations are expanding on a global basis and the provision of new international business services that can be supplied directly to users by Earth stations at or near the customer's premises through a satellite network has become attractive. The Standard E Earth station, with a small-size antenna of 3.5-8 m operating in the 14/11 or 14/12 GHz band, has been specified provisionally to provide the new business service of integrated digital transmission.

This specification dealt primarily with RF performances, and the establishment of business communication is left to INTELSAT members to accommodate a variety of services such as voice, high-speed data, and teleconferencing. There are three types of Standard E stations, and the selection of E-1, E-2, or E-3 will depend on user requirements with respect to site location, traffic requirements, and desired channel performances. Existing Standard A, B, and C Earth stations may be used to communicate with Standard E Earth stations for business services.

1. Operating Frequency. The Standard E Earth station will initially access the existing 14/11 GHz band of INTELSAT V and V-A and subsequently a modified INTELSAT V-A and its successors in the 14/12 GHz band. Operating frequencies are summarized in Table 13.

2. Antenna. Gain-to-noise temperature ratio (G/T): The three categories of the minimum clear sky G/T are specified for receiving frequency bands given in Table 13: G/T = 25 dB/K for Standard E-1; G/T = 29 dB/K for Standard E-2; and G/T = 34 dB/K for Standard E-3.

Sidelobes pattern: Requirements for transmit and receive sidelobes patterns, polarization, and axial ratio are the same as those for the Standard C station.

Tracking capability: The minimum requirements are manual tracking capabilities for satellite east-west movements by Standard E-1 and for east-west, north-south movements by Standard E-2, and auto-tracking capability by Standard E-3 when accessing INTELSAT V and V-A. When Standard E Earth stations access INTELSAT VI with better stationkeeping and inclination tolerances, E-1 and E-2 stations do not need the tracking capability, and E-3 stations will need the manual tracking capability for satellite east-west movement.

3. Types of Transmission. The only approved modulation method is four-phase PSK, and carriers can be transmitted by low bit rate QPSK/TDMA/FDMA, TDM/QPSK/FDMA, and QPSK/FDMA using preassigned frequencies. Available carrier sizes are determined provisionally by seven transmission rates, and the transmission rate includes the information rate and 10% overhead plus additional bits of 3/4 FEC. The maximum uplink e.i.r.p. (56.7-76.5 dBW for E-1, 54.5-74.3 dBW for E-2, and 49.3-69.1 dBW for E-3) and the associated bandwidth unit have also been specified. Using these parameters and 3/4 FEC, a performance of at least 10^{-6} for 99% of the year can be expected for most geographical locations where rainfall margins for the up- and downlinks are within the values in the range 2.5-5.8 dB assumed for

Table 13 Operating frequency for Standard E stations

Transmit, GHz	Receive, GHz	
14.0-14.25	10.95-11.20	(All regions; V, V-A, VI)
14.0-14.25	11.70-11.95	(Region 2, planned for modified V-A and VI)
14.0-14.25	12.50-12.75	(Regions 1 and 3, planned for modified V-A and VI)

the reference link. Scrambling is necessary to assure uniform spectral spreading. Table 14 summarizes the main parameters.

4. RF Out-of-Band Emission. The IM product resulting from multicarrier operation of the HPA must be limited to within 12 dBW/4 kHz, and the carrier spectrum spreading must be 26 dB less than the main carrier spectral power. The spurious emission, excluding the multicarrier IM product, must be no more than 4 dBW in any 4 kHz in the range 1636.5-1644.5 MHz, 5850-6425 MHz, and 14.0-14.5 GHz.

5. Communication Link with INTELSAT. The network's control station for business services must establish the communication link to INTELSAT and to all Standard E Earth stations in that particular network.

F. Possible Future Changes in Earth Station Requirements

All of the present technical performance specifications for standard Earth stations and transmission systems have been defined primarily for access to INTELSAT IV, IV-A, and V. With the advent of INTELSAT V-A after 1984 and INTELSAT VI after 1986, several additional changes to Earth station facilities may be required for the maximum use of capacities that will be provided by future deployments of improved satellite series, thus satisfying the increasing and diversified traffic demands. These possible changes in Earth station requirements are described below.

1. New Frequency. WARC 1979[5] allocated a new frequency band with an additional 725 MHz for the 6-GHz uplink (5.85-5.925 GHz and 6.425-7.075 GHz), 600 MHz for the 4-GHz downlink (3.4-3.7 GHz and 4.5-4.8 GHz), 500 MHz for the 14-GHz uplink

Table 14 Main parameters for digital carriers

Maximum information rate	Transmission rate	Allocated bandwidth	C/N for 10^{-6}
64 kbit/s	94 kbit/s	66 kHz	8.7 dB
128 kbit/s	187 kbit/s	131 kHz	
256 kbit/s	375 kbit/s	263 kHz	
1.544 Mbit/s	2.3 Mbit/s	1.61 MHz	
2.048 Mbit/s	3.0 Mbit/s	2.10 MHz	
4.096 Mbit/s	6.0 Mbit/s	4.20 MHz	
6.144 Mbit/s	9.0 Mbit/s	6.30 MHz	

(12.75–13.25 GHz), and 500 MHz for the 11-GHz downlink (10.70–10.95 GHz and 11.20–11.45 GHz). INTELSAT VI will incorporate an additional two 80-MHz hemitransponders operating in the new band adjacent to the present 6/4 GHz band. Earth stations that access transponders of the new frequency regions will require modification of or addition to their facilities to cope with the changes in satellite performance.

2. Polarization. INTELSAT V-A and VI incorporate the frequency reuse of global transponders, and Earth stations that access only global transponders of the present satellite series will require the improvement of the transmit antenna axial ratio from 1.4 to 1.09 or 1.06.

3. Antenna Sidelobes Pattern. Current requirements for the transmit sidelobes pattern for INTELSAT standard Earth stations are as described in the previous sections. CCIR is presently studying further improvement of transmit sidelobes levels in the copolar pattern to 29-25logΘ envelope and that in the cross-polar pattern to 23.6-20logΘ envelope. In 1981, INTELSAT surveyed the sidelobes pattern performances of the Standard A and B Earth stations commissioned since 1978 and it was demonstrated that 50% of the tested Standard A and B Earth stations satisfied the new proposed sidelobes gain. Thus, when CCIR adopts the new recommendation on sidelobes gain, INTELSAT may modify the requirement for standard Earth stations.

4. SS/TDMA. When the SS/TDMA system is implemented during the INTELSAT VI period, the TDMA traffic terminal may require a minor change that is currently under study by INTELSAT.

5. Frequency Reuse and G/T. In the early days of INTELSAT, use of the largest technically feasible Earth station antennas in the Standard A class was essential to obtain necessary traffic capacities with satellites having relatively poor performance. The large antennas were particularly effective, since no interference noise was present due to frequency reuse, and the increase in Earth

Table 15 Percentage of cochannel interference due to frequency reuse

INTELSAT satellites	Percentage of interference noise
IV	0%
IV-A	25%
V	55%
VI	78%

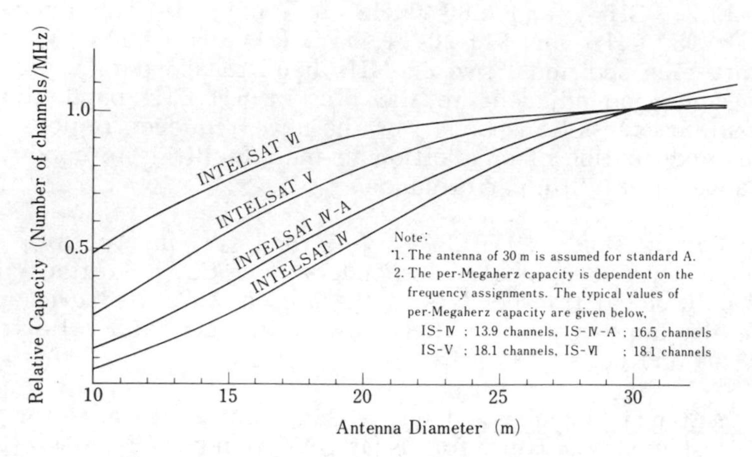

Fig. 8 Capacity decrease due to reduced antenna size.

station antenna gain resulted in an increase in the carrier-to-noise power ratio by the same amount. Later progress by INTELSAT in spacecraft technology has enabled the incorporation of multibeam antennas and higher-e.i.r.p. transponders. The introduction of frequency reuse techniques by means of orthogonal polarization or spatially separated beams permits multifold reuse of the communications bandwidth to enhance the traffic capacity of the satellites. One impact of frequency reuse is the increased cochannel interference noise due to imperfect isolation between multiple beams. In fact, the noise budget of the satellite links has become more and more interference limited, as shown in Table 15.

Considering that the use of large-size antennas can reduce only the thermal noise, the advantage of large dishes is becoming less significant. Figure 8 shows the per-Megahertz traffic capacity relative to that obtained with the Standard A Earth station, with a parameter of INTELSAT satellite series and the antenna size of Earth stations. The reduced incentive of using large-size antenna in the recent satellites may invite a discussion on the reconsideration of the G/T of the Standard A station in the future.

III. Present Technologies of Standard Earth Stations[2]

A. General Configuration

The facilities of INTELSAT Standard Earth stations are composed of 1) antenna subsystem, 2) high-power amplifier subsystem, 3) low-noise receiver subsystem, 4) communication

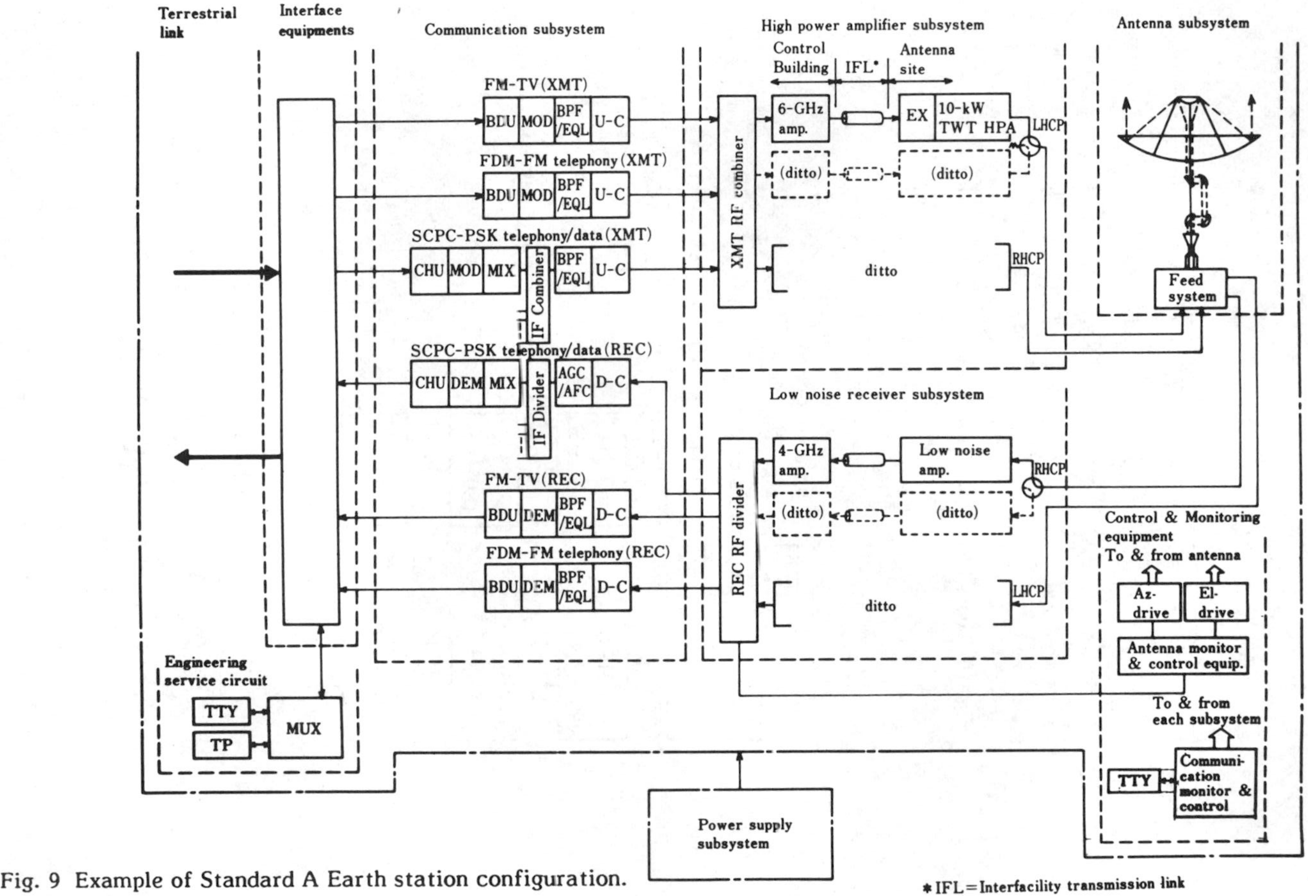

Fig. 9 Example of Standard A Earth station configuration.

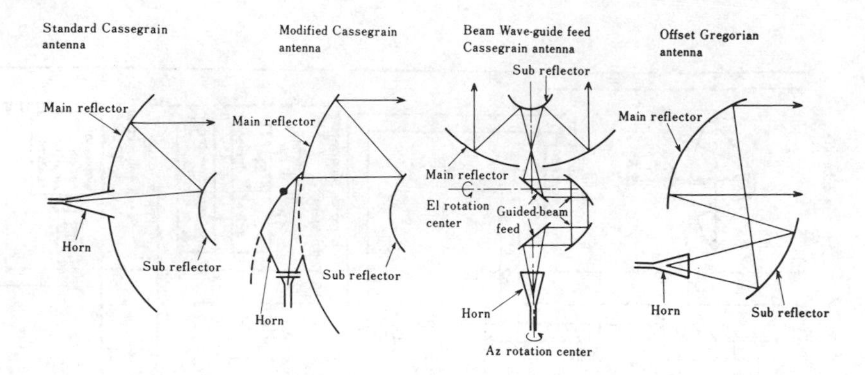

Fig. 10 Types of antennas for Standard Earth stations.

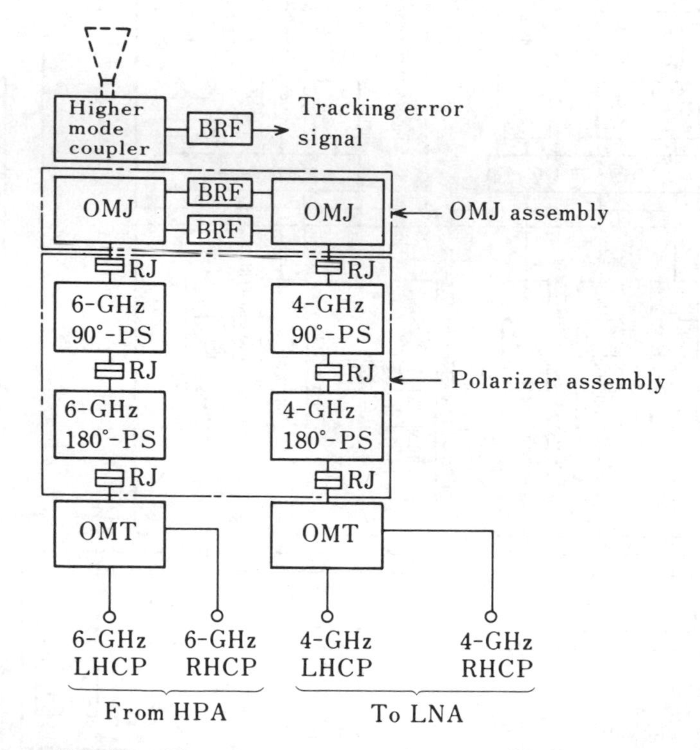

BRF : Transmit band rejection filter
RJ : Rotary joint
OMJ : Ortho-mode junction
OMT : Ortho-mode transducer
PS : Phase shifter

Fig. 11 Block diagram of the orthogonal polarization feed assembly.

subsystem, 5) control and monitoring equipment, 6) engineering service circuit, 7) interface equipment with terrestrial links, and 8) power supply subsystem. Figure 9 shows a typical configuration of the Standard A Earth station.

The following sections review the present technologies that are applied to typical Standard A, B, and C Earth stations.

B. Antenna Subsystem

Antennas, with their remarkable appearance, are fitting symbols for Earth stations and play an important role in determining the scale of a station.

1. Types of Antenna. All of the current Earth station antennas in the INTELSAT system are of the reflector type, which can achieve a high-gain performance at reasonable cost. Four types of reflector antennas have typically been used for Earth station antennas: namely, 1) the modified Cassegrain antenna, 2) the standard Cassegrain antenna, 3) the beam waveguide feed Cassegrain antenna, and 4) the offset antenna of the Cassegrain or Gregorian type (see Fig. 10).

The modified Cassegrain antennas have been used widely for Standard A Earth stations, particularly during the early period of the INTELSAT system. This antenna is relatively less expensive than the beam waveguide feed Cassegrain antenna, and a radio equipment room behind the main reflector rotates only in the direction of the azimuth angle. Most of the Standard B Earth stations use the standard Cassegrain antennas for economic reasons. The standard and modified Cassegrain antennas are usually mounted on an elevation-over-azimuth, York-and-Tower-style or King-post-style structure.

The beam waveguide feed Cassegrain antennas[6] were developed to offer convenience of maintenance and operation, as all of the high-power amplifiers and low-noise receivers can be installed in rooms on the ground. Excellent electrical performances are the result of higher aperture efficiency and better polarization purity due to its symmetrical feeding system over a wide bandwidth both in the 6- and 4-GHz bands or both in the 14- and 11-GHz bands. Thus most of the lately constructed Standard A and C Earth stations have adopted the beam waveguide feed Cassegrain antenna. This antenna is mounted on an elevation-over-azimuth, wheel-and-track-style structure.

The offset antenna of Cassegrain or Gregorian type[7] has been developed most recently. Because a subreflector and its supporting struts are not placed in front of a main reflector, the sidelobe's performance can be improved by 3 dB or more than those of the modified Cassegrain and the beam waveguide feed Cassegrain antennas. The disadvantage of the offset antenna is a requirement of the peculiar driving system and supporting structure for an asymmetrical reflector. Therefore its application would not be suitable to fully steerable antennas for economic reasons.

The present Earth station antennas have been constructed for operation in single transmitting and receiving frequency bands of 6/4 GHz or 14/11 GHz. Recently, INTELSAT has developed a feeder system that can work simultaneously in the 6/4 GHz and 14/11 GHz bands. This new technique will make it possible to use a single antenna for operation in both the 6/4 GHz and 14/11 GHz bands.

2. Feed Subsystem. The feed subsystem must provide the interconnection of low loss and high polarization purity between the antenna primary radiator (horn) and the high-power amplifier/low-noise receiver for simultaneous transmission and reception in the 6/4 GHz or 14/11 GHz band. Figure 11 is a typical example of the feed subsystem for dual-polarization use in the 6/4 GHz band. A polarizer that converts the linear polarization signal to circular polarization signal and vice versa is built by the 90-deg phase shifter or the hybrid-type phase shifter. For an Earth station that needs to restore the polarization degradation due to rain, a polarization compensator must be implemented in the feed subsystem.

For 14/11 GHz links in the INTELSAT V system, the linear polarization is adopted, and it is desirable to implement the rotatable 180-deg phase shifter to adjust the orientation of polarization planes.

3. Tracking System. It is desirable that the tracking accuracy of an Earth station antenna is maintained in less than one-tenth of the half-power beamwidth of the antenna radiation pattern. The

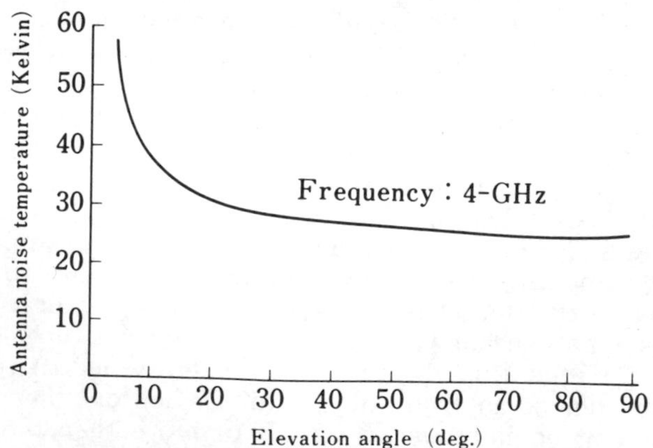

Fig. 12 Antenna noise temperature vs elevation angle measured at Yamaguchi Earth station.

required tracking accuracy of the Standard A antenna is less than about 0.01 deg rms. The following two autotracking methods are used today at most of the INTELSAT Earth stations: 1) the monopulse tracking system by using a higher mode and 2) the step-track system.

The monopulse tracking system has a better tracking accuracy performance and has been adopted for most Standard A and C Earth stations. The step-track system is simple and less expensive and is used typically by Standard B Earth stations.

4. Electrical Performances. Gain: The gain G of a reflector antenna is given by $G = (\pi D/\lambda)^2 \eta$, where D is the diameter of the main reflector; λ is the wavelength; and η is the aperture efficiency. To achieve a higher gain with a given aperture size, the aperture efficiency must be improved as much as possible. The main elements that contribute to the degraded performance are 1) nonuniformity of amplitude and phase distributions of the aperture plane, 2) phase error due to surface tolerance of the reflector, 3) spillover from the edges of the sub- and main reflectors, and 4) blocking of radiation power by the subreflector and its supporting struts. By careful design and construction of antennas, an aperture efficiency of 70%-80% is demonstrated today.

Noise temperature: The antenna noise temperature is attributed to the sky noise and ground noise picked up by the antenna sidelobes. Improvement of antenna sidelobes is essential to obtain a favorable figure of antenna noise temperature, particularly at Earth stations located at the low elevation angle. Figure 12 shows the antenna noise temperature vs the elevation angle measured at the Yamaguchi Earth station.

Sidelobes pattern[8,9]: To satisfy the requirement of the sidelobes pattern described in Sec. II, Earth station antennas must be carefully designed and constructed by 1) reducing the illumination level at the edge of reflectors, 2) improving the surface accuracy of reflectors, and 3) reducing the blocking area of the subreflector. A reflector shaping technique and the use of corrugated horn as a primary radiator are effective for the improvement of items 1 and 2. Figure 13 shows the sidelobes pattern measured at the Yamaguchi Standard A Earth station. An offset-type antenna also gives a good sidelobe performance by improving item 3.

Polarization[10]: All Earth station antennas that access the frequency reuse transponders of INTELSAT V must have a high polarization purity, such as described in Sec. II. Polarization purity is measured by the axial ratio or by cross-polarization discrimination (XPD). Reflectors and supporting struts do not introduce significant cross-polarization. To achieve high polarization purity, the techniques of a primary radiator with high polarization purity, such as a corrugated horn and a

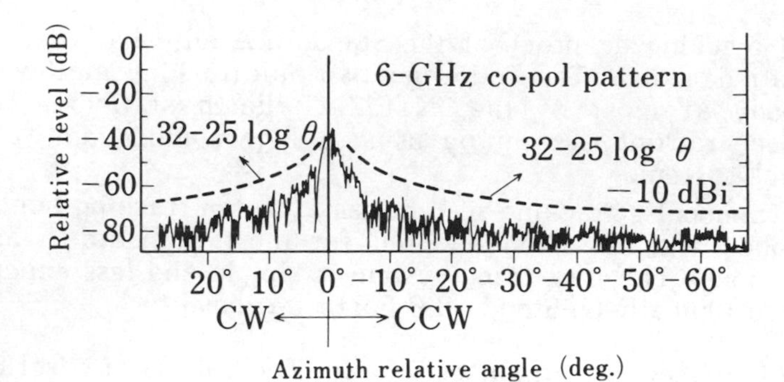

Fig. 13 Radiation pattern of Yamaguchi Earth station antenna (azimuth pattern).

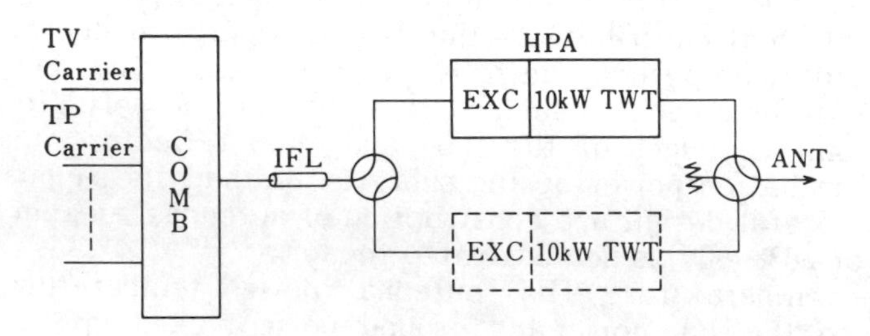

a) single amplifier configuration with redundancy

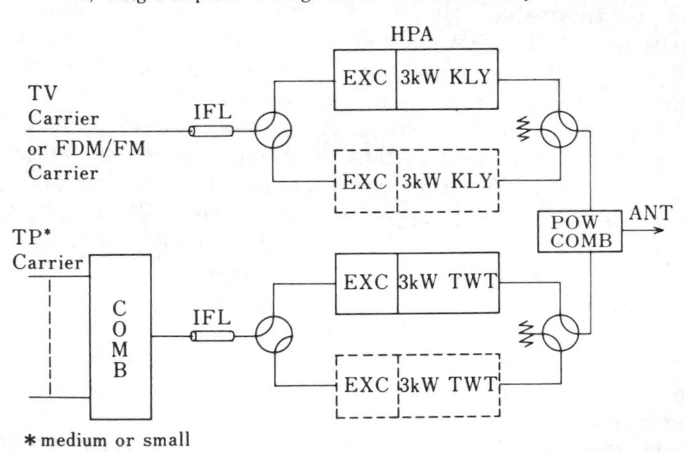

b) example of multiamplifier configuration with redundancy.

Note : [dashed box] redundancy.

Fig. 14 HPA configurations. a) Single amplifier configuration with redundancy; b) example of multiamplifier configuration with redundancy.

Table 16 Typical antenna performance of Standard A, B, and C Earth stations

Items	Standard A		Standard B		Standard C	
Type of antenna	Beam waveguide feed Cassegrain		Standard Cassegrain		Beam waveguide feed Cassegrain	
Type of mount	Elevation-over-azimuth, wheel-and-track		Elevation-over-azimuth, King post		Elevation-over-azimuth, wheel-and-track	
Autotracking method	Monopulse tracking using higher mode		Step-track		Monopulse tracking using higher mode	
Diameter, m	34		11		18	
Frequency, GHz	4	6	4	6	11.20	14.25
G/T (clear sky), dB	42.1 (at EL=5 deg)	...	32.7 (at EL=10 deg)	...	41.7 (at EL=20 deg)	...
Gain,[A] dBi[B]	61.9	65.6	52.0	54.8	64.3	66.2
Noise temperature,[A] K	48 (at EL=5 deg)	...	45 (at EL=10 deg)	...	64 (at EL=20 deg)	...
Axial ratio[A,C]	1.03	1.03	1.05	1.04	31.6	31.6

[A]Including feed subsystem.
[B]Gain relative to an isotropic antenna.
[C]Within tracking accuracy.

high-performance polarizer, are important design considerations. The lowest axial ratio practically achievable by a single 90-deg phase shifter is about 0.5 dB over the required bandwidth in both the 6- and 4-GHz bands. When the transmit and receive signals are separated prior to the input of polarizers, as shown in Fig. 11, the axial ratio of the individual polarizer that works over a 500-MHz bandwidth in the 6- or 4-GHz band can be reduced to as low as 0.07 dB. By carefully designing and constructing the total feed subsystem, polarization purity of 0.1-0.15 dB and insertion loss of 0.1-0.25 dB are achievable as the total feed subsystem at present.

INTELSAT has demonstrated that the existing antennas, which were designed for access to INTELSAT satellites prior to INTELSAT V, can be modified to improve their polarization purity to the specified levels for access to INTELSAT V by replacing the conventional primary radiator and existing feed with an improved primary radiator and a frequency reuse feed subsystem of high-performance purity.

5. Typical Performance of Standard Earth Stations. Table 16 shows the typical antenna performance of Standard A, B, and C Earth stations.

C. High-Power Amplifier Subsystem

The high-power amplifier (HPA) is the final stage of transmitting communication equipment prior to the input of the Earth station antenna.

1. Configuration of HPA. The configurations and the required output power levels of HPA depend on the number, type, and size of transmit carriers allocated to individual Earth stations. The two typical configurations of HPA in present Earth stations are 1) single amplifier configuration--a single HPA amplifies all of the multiple transmit carriers; 2) multiple amplifiers configuration--multiple amplifiers are installed and individual HPA amplifies a single transmit carrier or a part of the multiple transmit carriers.

In the former configuration, a single HPA must cover the total transmit bandwidth of 500 MHz in the 6- or 14-GHz band, and the output level of HPA should be large enough to maintain the levels of IM products below the specified limits. In the latter configuration, individual HPA can be of narrow bandwidth in the range of 40-80 MHz but the output of HPAs must be combined prior to the input of transmit antenna.

In the case of multiple amplifiers configuration, one of the following power combiners is required. A filter-type power

Table 17 Performance of typical TWT and klystron amplifiers

Type	10-kW TWT HPA	3-kW klystron HPA
Frequency range	5925 to 6425 MHz	5925 to 6425 MHz
Power output	10 kW, minimum	3 kW, minimum
Tube type	Coupled cavity	(with preset tuner)
Cooling system	Water cooling	Air cooling
Bandwidth	500 MHz (-6 dB points)	45 MHz, minimum (-1 dB points)
Gain,[A] dB	75 dB, minimum at 1-kW output	80 dB, minimum
Gain slope	0.07 dB/MHz maximum at 10 dB output backoff	0.07 dB/MHz maximum over the center 3/4 of bandwidth
Harmonic output	-60 dB at 1-kW output	-60 dB, maximum below fundamental
Spurious output	-60 dBW/4 kHz (5925 to 6425 MHz) -130 dBW/4 kHz (3700 to 4200 MHz)	-60 dBW/4 kHz (5925 to 6425 MHz) -130 dBW/4 kHz (3700 to 4200 MHz)
Intermodulation	-21 dB/2 kWx2	-22 dB/500 Wx2
Primary power	AC, 3-phase 200-240, 380-480 V, 50 or 60 Hz	
	95 kVA maximum	10 kVA (typical)
HPA configuration	FET + 6 GHz PWR amplifier (FET) + TWT	FET + FET + klystron

[A]Includes gain of intermediate power amplifier.

combiner has a low insertion loss of about 1 dB but has a slight dead band not available for the transmit frequency in the 500-MHz bandwidth. A hybrid power combiner suffers a 3-dB insertion loss, but full 500-MHz bandwidth is available for transmission. A ratio power combiner developed recently has a variable insertion loss in the range of 1-7 dB for one port and 7-1 dB for the other port without the dead band. These ratio power combiners are most commonly used at Earth stations of multiple HPA configuration.

Figure 14 shows typical examples of HPA configuration.

2. Type of HPA Tube. Two types of microwave tubes that have been used as HPA are a traveling wave tube and a klystron.

The traveling wave tube (TWT) is a wide-band amplifier that covers the entire bandwidth of 500 MHz in the 6- or 14-GHz band and is suitable for HPA, which simultaneously transmits multiple carriers. A TWT with coupled cavity can provide an output power level in the range 1-10 kW and a helix-type TWT can provide an output power level of 100 W-3 kW. In high-power TWT HPA, the reduction of power consumption is desirable and the depression method of collector voltage is adopted for TWT designing, resulting in an improvement of the overall power efficiency of TWT HPA. TWTs with an output power of 3 kW or less are now operated under air-cooling conditions, and TWTs with an output power of more than 3 kW are operated under water-cooling conditions.

In the 6-GHz band, TWT amplifiers with an output power level of 10 kW are now available and currently used at some Standard A Earth stations, with a single HPA configuration for the transmission of FDM/FM, TV, and other carriers simultaneously. TWT amplifiers with a 3-kW output power level are also used,

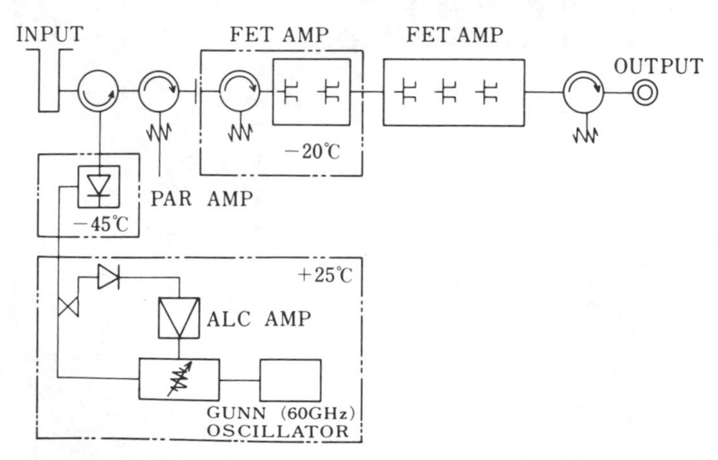

Fig. 15 Block diagram of 4-GHz thermoelectric cooled parametric amplifier.

coupled with the narrow-band HPA of a klystron for TV transmission at Standard A Earth stations of multiple HPA configuration. In the 14-GHz band, air-cooled TWT amplifiers with an output power level of 3 kW are now available for Standard C Earth stations.[11]

A linearizer is sometimes implemented to improve the nonlinearity of TWT amplifiers, thereby decreasing the level of intermodulation products produced in HPAs of multicarrier operation.[12]

The klystron is an amplifier with a narrow passband of about 40 MHz in the 6-GHz band or about 80 MHz in the 14-GHz band. These narrow passbands are tunable over a 500-MHz bandwidth, and a transmit passband can be automatically tuned by selecting one from the external five preset tuners. Most klystron HPAs used at INTELSAT standard Earth stations have output power levels of 100 W–3 kW with air-cooling and have the advantages of high efficiency, longer tube lifetime, simple maintenance/operation, and low cost resulting from relatively low supply voltage.

Because of the narrow bandwidth characteristic of the tube, the klystron HPAs can deal only with a single transmit carrier of TV, FDM/FM, or with SCPC carriers allocated within a 36-MHz transponder. For those Earth stations that transmit multiple FDM/FM and TV carriers, the multiple amplifier configuration becomes a necessary choice.

Further development is needed to expand the bandwidth of 6 GHz klystron to apply for the transmission of a TDMA carrier with an 80-MHz band unit.

3. Performance of Present HPA. Table 17 shows the performances of TWT and klystron amplifiers that are commercially available now as HPAs. HPAs should be selected taking into account not only current allocations of transmit carriers but also possible changes in Earth station requirements in the future.

D. Low-Noise Receiver Subsystem

To satisfy the G/T requirements of the standard Earth station, the total system noise temperature T at receiving frequency bands (4 and 11 GHz) must be sufficiently small. T is the sum of the antenna noise temperature under clear sky conditions (see Sec. IIIB) and the noise temperature of front-end low-noise receivers. Therefore it is desirable that the noise temperature of low-noise receivers at Earth stations be sufficiently small, on the order of magnitude almost equal to or less than the antenna noise temperature.

1. Types of Low-Noise Receiver (LNR).[13] LNRs that have been used at Earth stations for satellite communication are 1) the MASER, 2) parametric amplifier, and 3) GaAs FET.

The MASER is a quantum mechanical amplifier that operates by induced emission and is almost noiseless but can operate only at a liquid helium temperature of near 0 K. MASER was used during the early experimental period of satellite communications and has a performance of the lowest noise temperature, about 10 K, among the above three candidates. But due to a narrow bandwidth amplification in the range of 40-120 MHz, complicated maintenance, and the high cost, it was replaced by a parametric amplifier at INTELSAT Earth stations around the mid-1960's.

The parametric amplifier is a degenerative amplifier with a varactor excited by external pumping power (see Fig. 15) and has been widely adopted as a LNR at INTELSAT Earth stations due to excellent characteristics such as wide bandwidth amplification of about 500 MHz and relatively low-noise temperature in the 4- and 11-GHz bands. To achieve the low-noise temperature, 1) use of a varactor of high dynamic Q (capacitance variation/diode loss); 2) selection of reasonably high pumping frequency; 3) use of circulators with low insertion loss and high isolation; and 4) high gain enough to suppress noise contributions from following devices, must be carefully considered in the design. Even though this device can be operated at room temperature, cooling of the varactor and associated circuits of the parametric amplifier improves the noise temperature.

There are two types of cooling methods. One method is to use the cryogenic cooling system by gaseous helium, where the parametric amplifier in a vacuum vessel is cooled down to about 20 K. In the early days, parametric amplifiers could achieve a low-noise temperature of only about 30 K. Those early cryogenic systems were widely used as LNR's in the 4-GHz band, particularly at Earth stations constructed before the mid-1970's. The disadvantages of this cooling system are the complicated maintenance and operation and the high cost. The other method is to use the thermoelectric cooling system by Peltier effect, where the parametric amplifier is cooled down to the ambient temperature of -40 C. During the 1970's, there was remarkable progress in 1) high Q varactors (30 or more); 2) pumping Gunn oscillators of higher frequency (60 GHz) and power; and 3) circulators of high quality. Thus a parametric amplifier with a thermoelectric cooling system can now achieve a noise temperature of less than 40 K. This cooling system is more reliable and simpler to maintain, so most of the newly constructed Earth stations have implemented parametric amplifiers with thermoelectric cooling as the LNR in the 4- and 11-GHz bands. The above two cooling systems are also contributing to the good gain stability of parametric amplifiers by maintaining the devices in constant ambient temperature.

The gallium arsenide field effect transistor (GaAs FET) amplifier is the newest device developed recently as a more simple and less expensive LNR with good performances of wide bandwidth amplification and low-noise temperature.[14] Coupled with the thermoelectric cooling system, the noise temperature of the GaAs FET device is almost equal to that of a parametric amplifier operated at room temperature. Some Earth stations have implemented this device as a LNR in the 4- and 11-GHz bands.

Figure 16 shows trends of LNR noise temperature improvement in the last 20 years.

2. Performance of Present LNRs. Tables 18 and 19 show the performance of parametric amplifiers and GaAs FET amplifiers that are commercially available now as LNRs for Earth stations. Selection of these devices and cooling systems is determined by the required G/T value and antenna performances of gain and noise temperature. It should be noted that the antenna noise temperature at 11 GHz is relatively high, over 100 K, owing to the influence of atmosphere, and this fact reduces the relative effect of the noise temperature of the LNR to the total system noise temperature, compared to the situation in the 4-GHz band.

E. Communication Subsystem

The communication subsystem includes equipment that converts baseband signals into FM or PSK carriers of designated RF frequencies and vice versa. On the transmit side, output carriers of this subsystem are combined and sent to the HPA through interfacility links (IFL). On the receive side, output carriers from LNR through IFL are divided into individual received carriers and then fed to this subsystem (see Fig. 9).

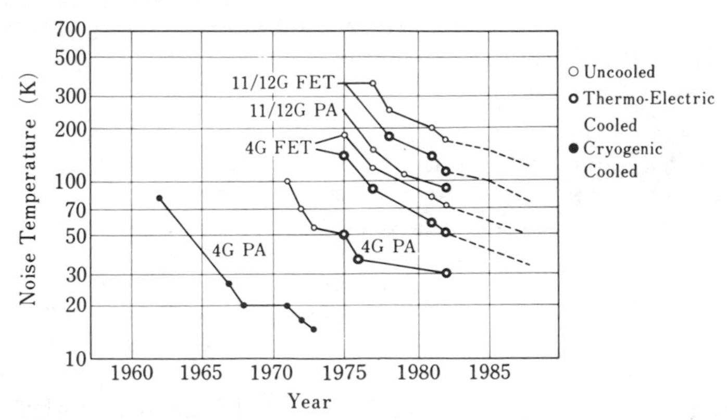

Fig. 16 Improvement of LNR noise temperature performance.

Table 18 Performance of typical parametric amplifier

Type	Paramp Cooled	Paramp Uncooled	Paramp Cooled	Paramp Uncooled
Frequency	3.7 - 4.2 GHz		11 GHz: 10.95-11.7 GHz or 12 GHz: 11.7 -12.2 GHz	
Noise temperature (typical)	30 K	40 K	90 K	100 K
Gain over 500 MHz	60 dB	60 dB	60 dB	60 dB
Configuration	Paramp (one-stage) with GaAs FET (five-stage)		Paramp (one-stage) with GaAs FET (six-stage)	
Cooling	Only Paramp is cooled by TE[A] devices	Temperature controlled by TE devices	Only Paramp is cooled by TE devices	Temperature controlled by TE devices

[A]Thermoelectric.

Table 19 Performance of typical GaAs FET

Type	FET LNR Cooled	FET LNR Uncooled	FET LNR Cooled	FET LNR Uncooled
Frequency	3.7 - 4.2 GHz		11 GHz: 10.95-11.7 GHz or 12 GHz: 11.7-12.2 GHz	
Noise temperature (typical)	50 K	75 K	125 K	170 K
Gain over 500 MHz	60 dB	60 dB	60 dB	60 dB
Configuration	GaAs FET (five-stage)		GaAs FET (eight-stage)	
Cooling	Only first GaAs FET is cooled by TE[A] devices	No cooling devices	Only first amp module (two-stage) is cooled by TE devices	No cooling devices

[A]Thermoelectric.

1. FM Equipment. In the transmit chain, baseband signals of FDM telephony or TV from the multiplexing equipments (MUX) pass through the FM modulator, the i.f. amplifier, the transmit filter and equalizer to the up-converter, where i.f. carriers are translated to RF carriers of designated frequencies and then fed to HPA. In the receive chain, RF carriers from LNR pass through the down-converter, the i.f. amplifier, the receive filter and equalizer to the FM demodulator, where FM carriers of i.f. are converted to baseband signals of FDM telephony or TV and then fed to the demultiplexing equipment (DEMUX). An individual transmit and

receive chain must be implemented for each FM transmit and receive carrier. Because of the multidestination transmit carriers adopted in the INTELSAT system, the number of transmit chains is not so large but the number of receive chains required at the Earth station is equal to or more than the number of communicating Earth stations in the network.

Modulators are designed to apply to all sizes of FDM/FM and TV carriers by replacing the baseband low-pass filters and pre-emphasis network.

Two types of demodulators are being used at Earth stations, namely, the conventional demodulator and the threshold extension demodulators of FM feedback type, phase locked type, and composite FMFB and phase detector type. It is recommended that the threshold extension demodulator be used for FDM/FM carriers with baseband configurations of up to and including 252 channels in order to overcome the marginal carrier-to-noise power ratio of satellite links. Demodulators can be made applicable to any channel size by replacing the baseband low-pass filters, de-emphasis networks and out-of-band noise detection filters. In the case of the threshold extension demodulator, filters in the feedback loop also have to be replaced to meet the change of channel size.

Two types of frequency converters are available, that is, a single conversion type and a double conversion type.

In the single conversion type of transmit chain, i.f. carriers of 70 MHz are directly translated to RF carriers of allocated radio frequencies of the 6-GHz band and vice versa for the receive chain. When the allocated RF frequency and size are changed, the RF bandpass filter must be replaced, and the local oscillator frequency must also be changed. In the double conversion type for transmit chain, i.f. carriers of 70 MHz are converted by the first converter to second i.f. carriers (for example, of 1.7 GHz) and then pass through the transmit filter and equalizer to the second converter, where the second i.f. carriers are translated to RF carriers of allocated radio frequencies. Carriers for the receive chain are processed in a similar manner. By this conversion type, reallocation of transmit and receive RF frequencies can be easily met by replacement or readjustment of only the second local oscillator, but replacement of the RF bandpass filter is not necessary. If a synthesizer is used as the local oscillator instead of a crystal oscillator, the frequency change can be made more easily without replacement of the oscillator.

Bandpass filters at the i.f. stage, which have been specified by INTELSAT, are required in transmit and receive paths to remove interference into and from adjacent carriers. Equalizers are required at the transmit i.f. stage to equalize the group delay in station transmit equipment and in satellite transponders and also required at the receive i.f. stage to equalize the group delay in the station receive equipment. Amplitude equalizers may also be required to compensate the amplitude nonlinearity of the transmit and receive equipment.

Communication equipment for the companded FDM/FM is identical to that for FDM/FM except for the additional requirement of a compressor and an expander prepared for each voice channel at the transmit and receive sides, respectively. The compandors (compressors and expandors) can be located either at the Earth station or international transit centers.

TV-associated audio programs are transmitted by means of a FM subcarrier in the video baseband. On the transmit side, an audio program channel from terrestrial links is processed through the 15-kHz LPF, the pre-emphasis circuit, the limiter, and the FM modulator with a subcarrier of 6.6 or 6.65 MHz, and then its output signal is combined with video signals. On the receive side, the FM subcarrier of audio programs is separated from composite TV and audio signals and then processed through the FM demodulator, the 15-kHz LPF, and the de-emphasis circuit to derive the audio program channel.

2. SCPC Equipment (BG-9-21 Rev. 3).[1] The SCPC equipment includes 1) channel units, 2) common equipment, and 3) frequency converters, as shown in Fig. 17.

For voice mode operation, a single voice channel is directly interfaced with the PCM Codec. The PCM Codec output signals are processed in the channel synchronizer of a transmit chain where timing buffering and framing functions are performed. The

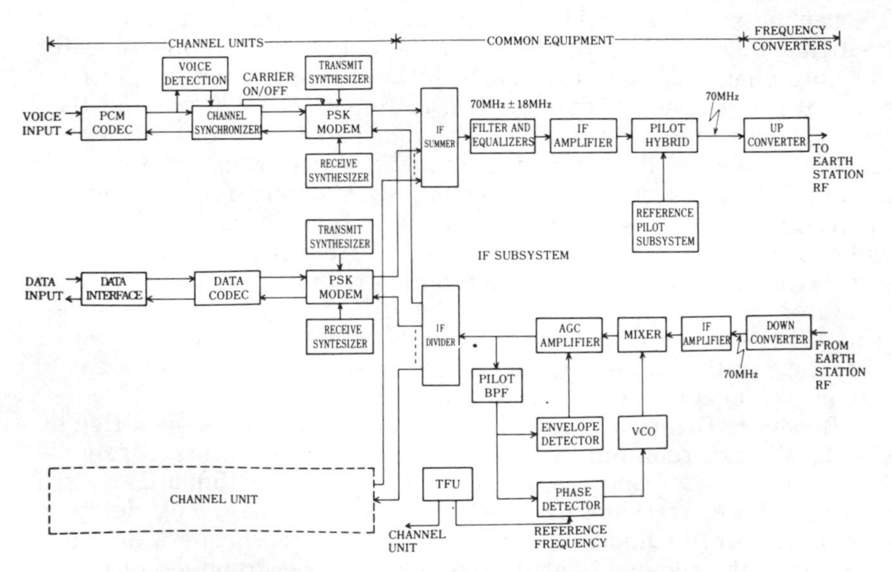

Fig. 17 Typical configuration of SCPC/PSK terminal.

channel synchronizer of the receive chain establishes frame synchronization and rebuffers PCM signals, and then they are sent to the PCM Codec. The four-phase PSK Modem modulates outgoing PCM signals and coherently demodulates the return PSK carrier. The transmit carrier is activated by the output control signal of the voice detector to conserve satellite power. Appropriate frequencies required for the transmit carrier and the receive carrier are supplied by a transmit and receive synthesizer, respectively. For the voiceband data channel unit for data above 4.8 kbit/s (120, 112), FEC is added to the channel synchronizer.

For digital data (48, 50, 56 kbit/s) operation, a data channel is directly interfaced with the data Codec where 3/4 or 7/8 FEC and synchronization are performed. The data Codec interfaces with the PSK Modem in a manner similar to that for the voice mode.

The common equipment is shared by all channel units and includes the i.f. subsystem, the timing and frequency unit, and, if required, the reference pilot subsystem.

The transmit i.f. subsystem accepts modulated carriers from the individual channel unit, combines them by the i.f. summer into a single spectrum that has a frequency of 70 MHz ± 18 MHz and passes them through the filter/equalizer and the i.f. amplifier to the frequency converter.

The receive i.f. subsystem accepts receive carriers of i.f. and passes them through the i.f. amplifier and the filter to the i.f. divider, where divided receive carriers are supplied to individual channel units for channel selection and demodulation. To perform an AFC function to center the received spectrum precisely at the center of the terminal i.f., an AFC loop using the reference pilot signal supplied by a reference Earth station is implemented.

The timing and frequency unit (TFU) provides the reference frequencies required by the channel units and AFC loop of the i.f. subsystem.

The reference pilot subsystem is installed at the SCPC terminal designated as a reference station and has the function of transmitting the reference pilot or the second pilot carrier.

In the up-converter, the spectrum of transmit i.f. carriers is converted to transmit RF carriers in the 6-GHz band. In the down-converter, received RF carriers of the 4-GHz band are converted to the i.f. spectrum.

3. SPADE (BG-14-30 Rev. 1).[1,15] SPADE is a SCPC system with demand-assignment capability. A typical configuration of communication equipment for SPADE is shown in Fig. 18. The main feature of the SPADE terminal is the requirement of the terrestrial interface unit (TIU), the demand-assigned signaling and switching unit (DASS), and the maintenance center (MC) in addition to the channel units, the common equipment, and the frequency converters that are included in the SCPC terminal.

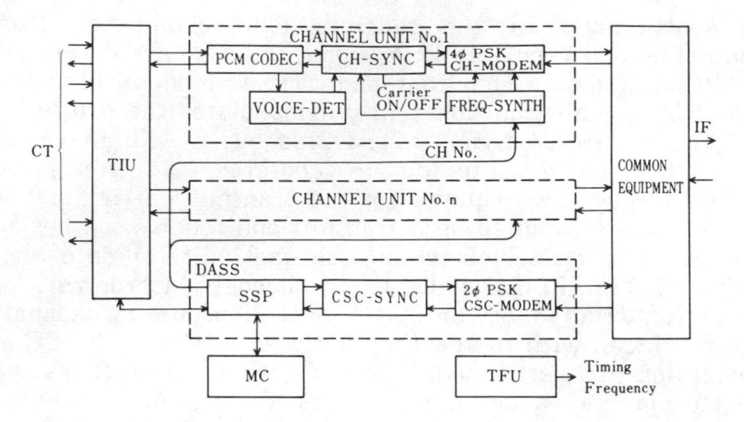

Fig. 18 Typical configuration of SPADE terminal.

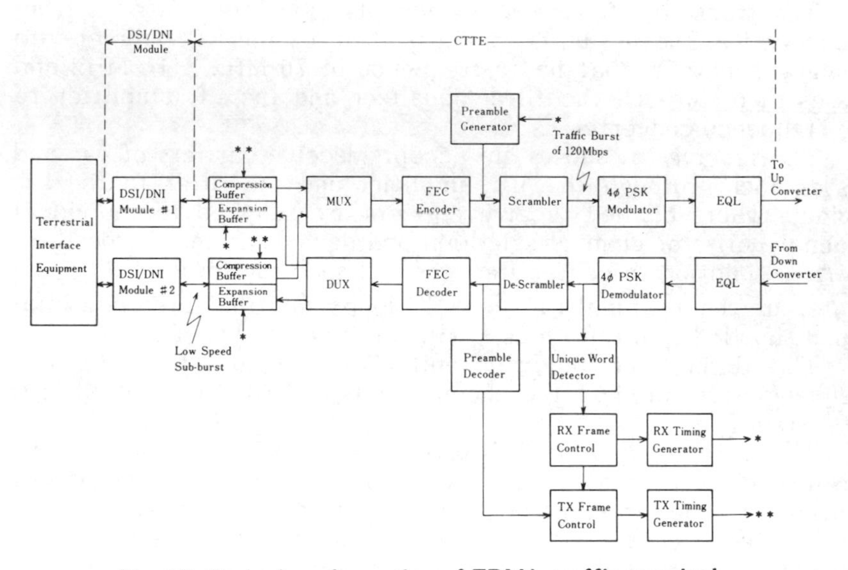

Fig. 19 Typical configuration of TDMA traffic terminal.

The functions of TIU are the connection of terrestrial voice channels with proper satellite channels by an access request and the conversion of the signaling between the terrestrial link and the SPADE network. DASS provides the capability to control the signaling and switching of each voice channel. All intersignalings between many SPADE terminals in the network are conducted by the common signaling channel (CSC) generated and received by DASS. MC provides a centralized arrangement of the terminal supervisory, maintenance, testing, and order wire capability.

4. TDMA/DSI Terminal (BG-42-65 Rev. 1).[1] The INTELSAT TDMA/DSI system is operated with two types of TDMA terminals. One is the reference terminal provided by INTELSAT to transmit a reference burst and the other is the local traffic terminal provided to transmit traffic bursts by accessing Earth stations.[16,17]

Figure 19 shows a typical configuration for the traffic TDMA/DSI terminal to be implemented at an Earth station.

DSI/DNI modules process incoming signals from the terrestrial facilities in the DSI mode or DNI mode and put them in a form of sub-bursts of DSI/DNI modes to interface with the common TDMA terminal equipment (CTTE), and also the inverse operation.

Voice signals of PCM/TDM forms from the terrestrial facilities pass through the DSI submodule shown in Fig. 20. On the transmit side, the presence of voice spurts is detected by the voice detector, and only active signals of voice channels are stored in the transmit speech momory and read out to CTTE in the sub-burst form. The transmit channel assignment processor controls the connection between the terrestrial channels and the satellite channels. On the receive side, the connection between the satellite channels and the terrestrial channels is processed through the receive speech memory controlled by the receive channel assignment processor.

Nonvoice signals of TDM form from the terrestrial facilities pass through the DNI module or the DSI module with a continuous channel assignment condition.

Common TDMA terminal equipment (CTTE) is a key equipment of the traffic terminal and performs the major functions of 1) traffic burst generation, 2) centralized timing control for the

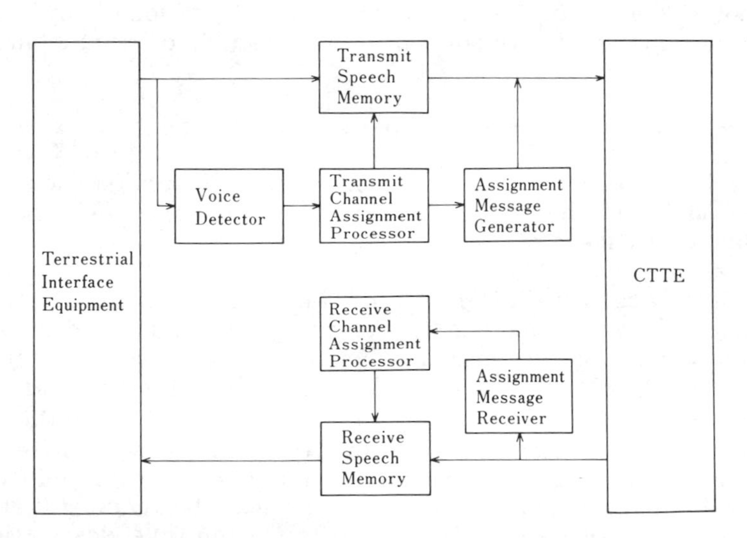

Fig. 20 Functional block diagram of DSI module.

initial acquisition and the steady-state synchronization of the bursts in the given time slots, and 3) four-phase PSK modulation/demodulation.

On the transmit side, the sub-bursts from the DSI/DNI modules are processed through the compression buffer to the multiplexer, which combines multiple sub-bursts to form the data portion of the traffic burst. The traffic data are passed to the FEC encoder, if necessary, and then combined with the preamble portion to generate the traffic burst of 120 Mbit/s. The four-phase PSK modulator accepts the scrambled traffic burst and feeds the PSK burst carriers of i.f. frequency (typically 140 MHz) to the i.f. amplifier and the equalizers. The timing of burst transmission is derived simply by applying a delay to the detection timing of the unique word of the reference burst. The delay time is periodically adjusted according to synchronization information supplied from the reference station.

On the receive side, the received PSK burst carriers of i.f. are coherently demodulated by the PSK demodulator and then pass through the descrambler, the FEC decoder if necessary, the demultiplexer, and the expansion buffers to the DSI/DNI modules. The preamble portion of demodulated signals are fed to the unique word detector, and the information of the detected signals is used to derive the timing of burst transmission and received clock signals.

Because transmitted bursts from other Earth stations are incoherent to each other, the four-phase PSK demodulator needs to be carefully designed to recover the reference carrier and clock timing from the short CBTR symbols in each preamble portion of received bursts.

Engineering service circuits are provided by the voice/telegraph order wire portion in the preamble of traffic bursts.

F. Terrestrial Interface Equipment

Satellite channels are extended from Earth stations to international switching centers through terrestrial networks in individual countries.

1. Interface with FDM Terrestrial Links. Until recently, terrestrial networks have been provided by microwave and coaxial cable links of FDM systems. The particular features of satellite links of FDM/FM mode are the use of different baseband allocations from those of terrestrial links and the adoption of multidestination transmit carriers (multiple channels of different destinations are combined into one transmit carrier, while multiple carriers from many other Earth stations must be received). This difference in baseband allocations and multiple-destinations

carrier system necessitates the implementation of the FDM multiplexer/demultiplexer (MUX/DEMUX) equipment at Earth stations for interface on the basis of voice grade channels and/or group signals and/or super-group signals.

In the event of TDMA introduction, TDMA terminals at Earth stations will have an interface with terrestrial networks of FDM systems and the conversion from FDM signals to TDM signals and vice versa at Earth stations is needed. One approach is to disassemble FDM signals from terrestrial networks into voice grade channels by FDM Demux and then pass through the channel distributor, the PCM encoder, and TDM multiplexer to supply TDM signals to the TDMA terminal on the transmit side. On the receive side, TDM signals of satellite channels are disassembled into voice grade channels and then passed through the channels distributor and the FDM multiplexer to derive FDM signals (part a of Fig. 21). The other approach, which is more economical, is to implement the digital transmultiplexer,[18] which converts directly from FDM to TDM and vice versa (part b of Fig. 21). A 120-channel transmultiplexer is now commercially available.

2. Interface with TDM Terrestrial Links. Recently, several countries have been introducing terrestrial networks of a digital type, which will enable them to have a direct connection between terrestrial networks and digital satellite links of a TDMA system by direct digital interface (DDI) equipment (part c of Fig. 21).

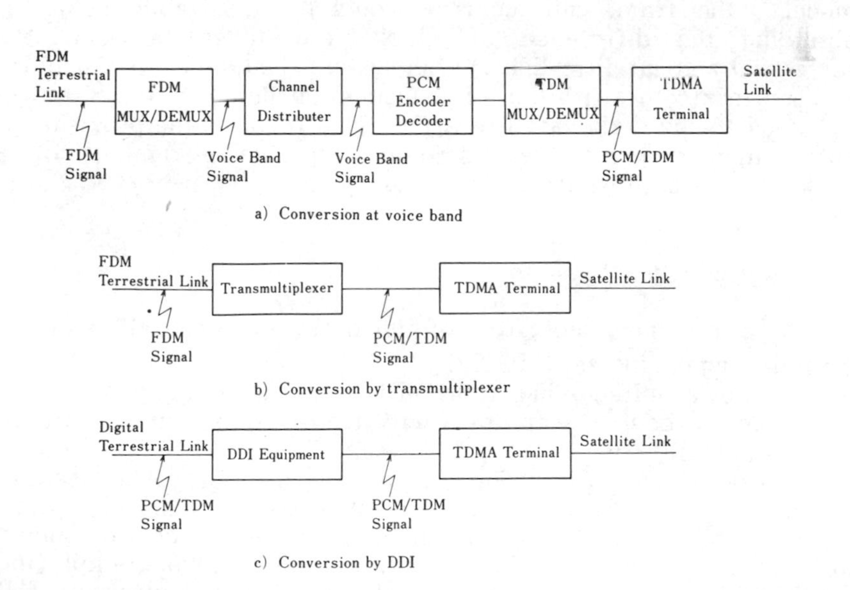

Fig. 21 Interface configurations for connection between the TDMA satellite link and terrestrial link: a) Conversion at voiceband; b) conversion by transmultiplexer; c) conversion by DDI.

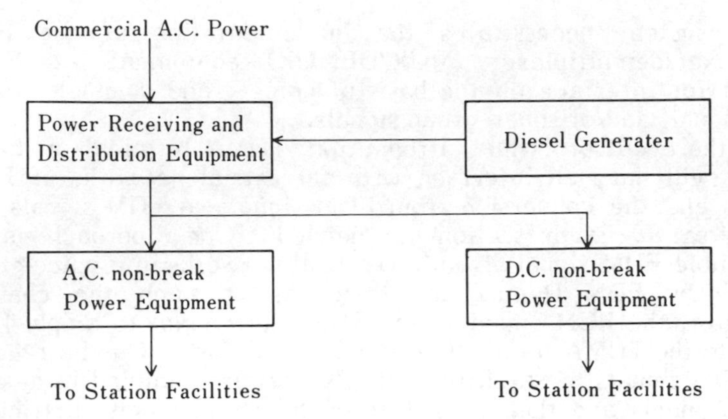

Fig. 22 Example of power supply subsystem.

Using DDI, Earth stations will no longer require the FDM Mux/Demux equipment, and the cost of interface becomes more economical. INTELSAT has specified the two DDI systems as follows:

1) Digital terrestrial network is synchronized to the TDMA system.

2) Digital terrestrial network is independently synchronized from the TDMA system, and the timing accuracies of both networks are of the order of 10^{-11}. With this plesiochronous mode,[19] the frame slip control would be a suitable means of absorbing the difference of clock frequencies between the terrestrial and satellite links. This system should have a capability to compensate the variation of transmission delay in the satellite links due to the daily satellite drift. This requires the provision of buffer memories that can accommodate the equivalent of a peak-to-peak path length variation of 2.2 ms.

G. Power Supply Subsystem

A power supply subsystem of Earth stations generally includes the following equipment (Fig. 22):

1) Power Receiving and Distribution Equipment; The equipment to receive, step down with transformers, and distribute the commercial ac power.

2) Diesel Generator; The ac power generator to be used in case of emergency when the commercial ac power fails.

3) ac Nonbreak Power Equipment; The equipment to supply nonbreak ac power of regulated voltage and frequency employing the thyristors, the dc batteries, and the inverter and to ensure full continuous operation irrespective of interruption of the commercial ac power.

4) dc Nonbreak Power Equipment; The equipment to supply nonbreak dc power with the rectifiers and the floating batteries to provide the stable voltages of output dc power and to ensure full continuous operation irrespective of interruption of the commercial ac power.

HPA and LNR are usually supplied with ac nonbreak power. The power consumption of a 10-kW TWT HPA, a 3-kW TWT HPA, and a 3-kW klystron HPA is about 100, 25, and 15 kVA, respectively. The power consumption of a LNR with a thermoelectric cooled parametric amplifier is about 500 VA.

The FDM/FM equipment in the communication subsystem is generally supplied with nonbreak dc power. The power consumption of a nonredundant FDM/FM equipment is estimated to be about 100 W for each transmit or receive carrier. SCPC/SPADE and TDMA/DSI equipment is generally supplied with the nonbreak ac power. The power consumption is estimated to be in the range of 2 kVA for the SCPC terminal (16 channels) and 5-10 kVA for the TDMA/DSI terminal.

The total power consumption of general Standard A, B, and C Earth stations is in the range of 150-500 kVA, including air-conditioning and lighting facilities for buildings.

H. Others

1. Control and Monitoring Equipment. For the satisfactory operation and maintenance of Earth station facilities, the station staff needs to monitor operational parameters of transmit and receive carriers, the quality of the circuits, and the operating condition of the subsystems and equipment installed at the Earth station. When failures or malfunctions are observed, adjustments of operation parameters, switchover to redundant subsystems, or the repairing of failed equipment must be carried out in a short time. Most Earth stations have centralized control and monitoring equipment so that the station staff can watch the operational parameters and the operating conditions of main subsystems and equipment to take prompt countermeasures when needed.

2. TV Standard Converter. There are two monochrome TV systems (625/50 and 525/60) and three color TV standard systems (NTSC:525/60, PAL:625/50, and SECAM:625/50) in the world. For INTELSAT TV transmission between countries employing different TV standard systems, the conversion of TV standards is carried out at receiving Earth stations.

The conversion of TV standards is processed through 1) the conversion of the line number (525 or 625 lines per frame) and the field frequency (60 or 50 Hz), and 2) the conversion of color systems. For the conversion of the line number and field frequency, the incoming TV signals are processed by deleting or repeating lines and fields according to the specific rule of signal conversion. For the conversion of color systems, incoming composite TV signals are decoded into a luminance component and two color difference components, and then encoded again into the outgoing composite TV signals of a different color standard system.

Acknowledgment

Because of the variety of standard Earth stations and their applications in the INTELSAT system, the author was able to provide just a general survey of the Earth segment but hopes that this material will help the reader to recognize the important role of the Earth segment in the international satellite telecommunications system.

The author wishes to express his gratitude to Takuro Muratani, Kinji Iwasaki, and many other colleagues at Kokusai Denshin Denwa Co. for their kind support in the preparation of this chapter.

References from IEEE publications have been excerpted with the kind permission of the Institute of Electrical and Electronics Engineers, Inc.

References

[1]INTELSAT documents for performance specifications: BG-28-72 Rev. 1 (1982), BG-28-74 Rev. 1 (1982), BG-28-73 Rev. 1 (1982), BG-52-75 (1982), BG-53-86 (1982), BG-35-78 (1978), BG-42-65 Rev. 1 (1981), BG-9-21 Rev. 3 (1982), BG-46-92 Rev. 1 (1983).

[2]Miya, K. "Satellite Communication Technology," KEC, Tokyo, 1981.

[3]Yamada, M., Ogawa, A., Furuta, O., and Yuki, H., "Rain Depolarization Measurement Using IS-IV Satellite in 4 GHz Band at Low Elevation Angle," Annales des Telecommunications, Vol. 32, pp. 524-529, Nov.-Dec. 1977.

[4]Yasukawa, K. and Yamada, M., "11 GHz Rain Attenuation and Site Diversity Effect at Low Elevation Angles," URSI 20th General Assembly, pp. 362, Washington D.C., Aug. 1981.

[5]Dorian, C., Potts, J. B., Rheinhart, E., and Weiss, H., "World Administrative Radio Conference and Satellite Communications," COMSAT Technical Review, pp. 1-26, Spring 1980.

[6]Mizusawa, M. and Kitsuregawa, T., "A Beam-Wave Guide Feed Having a Symmetric Beam for Cassegrain Antennae," IEEE Trans. Antennas and Propagation, Vol. AP-21, pp. 884-886, November 1973. © 1973 IEEE.

[7]Krichevsky, V. and Difonzo, D. F., "Beam Scanning in the Off-Set Gregorian Antenna," COMSAT Technical Review, pp. 251-269, Fall 1982.

[8]Kreutel, R. W., "Wide-Angle Sidelobe Envelopes of Cassegrain Antenna," COMSAT Technical Review, pp. 71-88, Spring 1976.

[9]Price, R., "RF Test in the Etam Standard C Antenna," COMSAT Technical Review, pp. 69-92, Spring 1982.

[10]Diffonzo, D. F. and Trachtman, W. S., "Antenna Depolarization Measurements Using Satellite Signals," COMSAT Technical Review, pp. 155-185, Spring 1978.

[11]Hamada, S., Tutaki, K., Horigome, T., and Fujii, G., "14/30 GHz High Power Coupled-Cavity Travelling Wave Tubes for Satellite Earth Stations," Record of IEEE International Electron Device Meeting, pp. 124-127, December 1977. © 1977 IEEE.

[12]Sato, G., Shibazaki, H., Asai, T., and Kurokawa, T., "A Linearizer for Satellite Communication," International Conference on Communication, pp. 33.3.1-33.3.5, Seattle, Washington, 1980.

[13]Kajikawa, M., Haga, H., Fukada, S., and Akinaga, W., "Development Status of Low Noise Amplifiers," Record of AIAA 8th Communication Satellite Systems Conference, pp. 309-316, Orlando, Florida, 1980.

[14]Akinaga, W., Fukada, S., Haga, I., Hayashi, H., and Kajikawa, M., "FET Low Noise Amplifier for Satellite Communications," Record of AIAA 7th Communication Satellite Systems Conference, pp. 544-551, San Diego, Calif., 1978.

[15]Edelson, B. I. and Werth, A. M., "Spade System Progress and Application," COMSAT Technical Review, pp. 221-242, Spring 1972.

[16]Pontano, B. A., Dicks, J. L., Colby, R., Forcina, G., and Phiel, J., "The INTELSAT TDMA/DSI System," IEEE Trans. Selected Areas in Communications, Vol. SAC-1, pp. 165-173, Jan. 1983. © 1983 IEEE.

[17]Muratani, T., Ohkawa, M., Koga, K., Mizuno, T., and Yamasaki, N., "Satellite Field Test of INTELSAT/DSI Terminal," 6th International Conference on Digital Satellite Communications, pp. II.1-II.8, Phoenix, Arizona, Sept. 1983.

[18]Takahata, F., Hirata, Y., Ogawa, A., and Inagaki, K., "Development of TDM/FDM Transmultiplexer," IEEE Trans. Communications, Vol. COM-28, pp. 726-733, May 1978. © 1978 IEEE.

[19]Inagaki, K., Hirata, Y., Takahata, F., Ogawa, A., and Niwa, F., "International Connection of Plesiochronous Network via TDMA Satellite Link," IEEE Trans. Selected Areas in Communications, Vol. SAC-1, pp. 188-198, Jan. 1983. © 1983 IEEE.

8. Transmission Techniques

Giuseppe Quaglione*
TELESPAZIO, Rome, Italy

I. Introduction

Transmission systems for satellite communications include baseband processing and multiplexing, modulation methods, and multiple-access methods. This chapter will discuss the major characteristics of the techniques most commonly used in operational communication satellite systems and the new techniques under study and investigation, with particular emphasis on the INTELSAT system.

II. Baseband Processing and Multiplexing

A. Analog Processing

The two major forms of voice signal analog processing are syllabic companding and voice activation.

1. Syllabic Companding. The principle of syllabic companding is to reduce at the transmission side the dynamic of the voice signal, according to Recommendation G.162 of the International Telephone and Telegraph Consultative Committee (CCITT), and to implement the opposite operation at the receiving side. The overall result of the two operations, compression and expansion, is an improvement in the quality of the voice signal due to an objective as well as a subjective reason. The objective reason is related to the operation of the companding device without a signal. This is due to the fact that the noise power in a telephone circuit is added only after the compression, at a level well below the unaffected reference level, and is expanded at the receiving side with a significant reduction of the noise power level relative to the average power of the voice signal. The subjective reason is

---

Invited paper received October 11, 1983.

*Director of Planning and International Relations.

related to the operation of the companding device with a signal, although, in this case, the overall gain of the expansion device will be driven by the signal level and will have the same value for the signal as well as for the noise, with no advantage on the signal-to-noise ratio. However, during pauses in conversation, the effect of the reduction in the noise level is a significant subjective improvement of the quality of the circuit. The overall improvement of the signal-to-noise ratio due to a compandor has to be tested experimentally and depends on several factors such as the performance parameters of the compandor, the noise level on the circuit, the speech activity factor, and even the language of the talkers. Improvement values in the signal-to-noise ratio on the order of 16 dB have been measured, resulting in a capacity improvement of about 2 for high-capacity carriers and about 2.5 for low-capacity carriers, assuming that the companding is added to an existing link so that the radio frequency bandwidth B_{RF} and the carrier-to-noise ratio C/N remain unchanged. INTELSAT decided in June 1981 to introduce companded frequency-division-multiplex/frequency-modulation (FDM/FM) carriers as a supplementary transmission technique between Standard A and Standard B Earth stations when the equivalent CFDM/FM carrier capacity approaches or exceeds 24 channels.

INTELSAT also decided on the use of CFDM/FM technique on links between Standard A Earth stations on selected carriers on a temporary basis for either expanding the capacity of an existing FDM/FM carrier or for reducing the bandwidth requirements of one or more existing FDM/FM carriers in order to release bandwidth for the introduction of new carriers.

The transmission parameters for companded FDM/FM carriers for the INTELSAT IV-A and V satellites have been derived assuming a 9-dB compandor gain.

Not currently included is partial companding of channels in a carrier (mixed mode operation) between Standard A Earth stations as well as between Standard A and Standard B Earth stations. Furthermore, there is a limitation on the introduction of more than 252 channels on any one carrier.

Recent studies have shown that CFDM/FM operation between Standard B Earth stations can also be approved when the antenna main beam gain and antenna sidelobe performance are somewhat better (up to 3 dB, depending on the carrier size involved) than the minimum mandatory requirements in order to meet the off-beam emission level recommended by the International Radio Consultative Committee (CCIR) Recommendation 524-1.

2. Voice Activation. Voice activation[1] is a method of achieving some bandwidth/power advantage through a device that withdraws the transmission channel from the satellite network during pauses in speech.

The fundamental parameters of the voice-operated switch are: 1) the threshold level above which speech is assumed to be present (typically -30 to -40 dBm0 for a fixed threshold level switch); 2) the time taken by the voice switch to operate on detection of speech (typically 6 to 10 ms); and 3) the time during which the voice switch remains activated after the cessation of a speech burst on a speech channel (typically 150 to 200 ms).

The threshold level needs to be set fairly low to avoid missing large portions of the start of a speech burst and low speech levels. This makes the voice switch susceptible to high noise levels, thus reducing the net system advantage. Voice activation has found practical application in the INTELSAT single channel per carrier (SCPC) systems, where its use yields a power saving of about 4 dB.

B. Digital Processing

Digital processing techniques include coding techniques for analog/digital conversion and error control, companding, voice activation, and speech interpolation.

1. Coding for Analog/Digital Conversion.

The first objective of coding is the digital representation of analog signals that are intended to be transmitted by digital techniques. The following coding systems are used for analog/digital (A/D) conversion[1] of analog messages.

(a) Pulse code modulation (PCM): In PCM, the analog signal is sampled periodically and quantized in an agreed upon manner. In order to decrease the inherent error occurring in the process of quantization (quantization noise), the signal may be previously subjected to companding (see Sec. IIB3).

CCITT Recommendation G.711 specifies the PCM format to be used in international connections of voice signals. The sampling rate is 8.000 samples per second, and each sample is coded into 8 bits, giving a total bit rate of 64 kbit/s per speech channel. However, a coding standard of 7 bits per sample is used in the INTELSAT SCPC system, which gives an information transmission rate of 56 kbit/s.

Some advantages of PCM are summarized in the following statements:

1) PCM presents a small but definite performance advantage over FM at the lower signal-to-noise ratios.

2) A PCM system designed for analog message transmission is readily adapted to other input signals.

3) PCM has the capability of regeneration and is very attractive for communication systems having many repeater stations.

PCM is generally used for the transmission of voice signals in connection with time-division multiplexing (TDM). However, another use of PCM is the direct encoding/decoding of a frequency-division-multiplexed (FDM) signal.

(b) Delta modulation (DM): In delta modulation, the analog signal is sampled at a rate greater than that for PCM (typically 24 to 40 kHz for good quality speech). Then, a 1-bit code is used to transmit the change in the input level between samples.

DM is very simple to implement. Subjective tests and noise measurements have shown that a DM system operating at 40 kbit/s is equivalent to a standard PCM system operating with a sampling rate of 8 kHz and 5 bits per sample. The effect of transmission errors in DM is not as serious as in PCM systems. This is because all of the transmitted bits are equally weighted and are equivalent to the least significant bit of a PCM system. The main drawback of DM systems is the difficulty of interconnecting them in a general way in an international network using analog and digital links.

(c) Differential pulse code modulation (DPCM): In DPCM, the difference between the actual sample and an estimate of it, based on past samples, is quantized and encoded as in ordinary PCM and then transmitted. To reconstruct the original signal, the receiver must make the same prediction made by the transmitter and then add the same correction. Thus DPCM uses the correlation between samples to improve performance, as compared to PCM. In the case of voice signals it is found that the use of DPCM may provide a saving of 8-16 kbit/s (1 to 2 bits per sample) over standard PCM. For monochrome television, a saving of about 18 Mbit/s (2 bits per sample) can be obtained.

(d) Adaptive PCM, DPCM, and DM : In PCM, DPCM, and DM, the quantizer has been assumed to have a fixed step size. It is possible to improve the dynamic range of these encoders at the expense of increased circuit complexity by using an adaptive quantizer, wherein the step size is varied automatically in accordance with the time-varying characteristics of the input signal. In ADM, for example, the variable step size increases during a step segment and decreases during a slowly varying segment of the input signal. Experimental results have indicated that, with these techniques, a telephone signal can be transmitted at a rate of 32 kbit/s with the same subjective quality achieved by a standard PCM signal at the rate of 64 kbit/s.

2. Coding for Error Control. There are two main classes of error control[1] that may be used to improve the performance of digital systems: automatic repeat request (ARQ) and forward error correction (FEC). ARQ uses an error-detecting code together with an auxiliary feedback channel to initiate a retransmission of any block of bits received in error. It is not suited for telephony

purposes but can be used in other applications, such as data transmission. FEC codes have the capability of correcting a certain amount of transmission errors without the need for retransmission. Among the FEC codes considered for application in noisy channels with power limitation and available frequency bandwidth are BCH codes and convolutional codes.

3. Special Signal Processing Techniques. Special processing techniques,[1] involving baseband signals, may also be used in satellite communications to provide additional gain in performance. Some of these are discussed below.

(a) Instantaneous companding: Practically, whereas syllabic companding is often implemented for analog transmission (refer to Sec. IIA1), instantaneous companding is the conventional technique implemented for digital transmission. Instantaneous compressors and expanders are zero-memory nonlinear devices. The so-called μ-law and A-law are functions defining the instantaneous companding in PCM and other coding schemes for A/D conversion. CCITT Recommendation G.711 prescribes that the conversion of the companding law should be performed by the countries using the μ-law.

Nearly instantaneous companding (NIC) is a source coding technique for reducing the bit rate in PCM. This is achieved by rescaling the quantizing range of a block of PCM samples according to the magnitude of the largest sample in the block. A control word is transmitted and used by the receiver to convert the received bit stream to the normal PCM format. It has been reported that NIC makes possible the transmission of good quality speech at bit rates of 35 kbit/s.

(b) Voice activation: Voice activation is used for achieving some power advantage by using the transmission channel only during voice activity periods. Further information on this subject is given in Sec. IIA2.

(c) Speech interpolation: Speech interpolation techniques are based on the fact that, in a normal two-way conversation, each speaker actually engages the telephone circuit for only about half the total time taken up by conversation. Furthermore, gaps between syllables, words, and phrases add to the idle time, so that a normal one-way telephone channel is active, on the average, only for 35 to 40% of the time during which the circuit is connected.

Accordingly, if the same transmission channel can be assigned to different speakers, on a voice-activated basis, the transmission channel is better utilized. This brings a significant gain in the number of conversations that can be routed over the same channel in a given time period, thus leading to a gain in traffic capacity. The theoretical speech interpolation gain is defined by the ratio between the number of speakers (input trunks) and the number of transmission channels required to service them.

The digital speech interpolation (DSI) gain is discussed in Sec. IVC1, and experimental values achieved also during the INTELSAT TDMA field trials are around 2.5. DSI techniques can also be used in conjunction with adaptive PCM or similar techniques, giving an overall gain of about 5.

C. Multiplexing Techniques

Multiplexing is the reversible operation of combining two or more information-bearing signals to form a single, more complex signal. The signals that are combined in a multiplexer usually come from independent sources, such as subscribers in a telephone network. Prior to multiplexing, each signal travels over a separate electrical path, such as a pair of wires or a cable, whereas the multiplexed signal can be transmitted over a single communication medium of sufficient capacity. The reversibility of the multiplexing operation permits recovery of the original signals, which often have different final destinations, at the receiving end of the transmission. This inverse operation whereby the original signals are recovered is called demultiplexing.

The most commonly used mutiplexing techniques are frequency-division multiplex (FDM) and time-division multiplex (TDM).

In a frequency-division-multiplex (FDM) system, a number of channels are arranged side by side in the baseband in the form of single sideband (SSB) signals with suppressed carriers at 4-kHz spacing, and thus many narrow bandwidth channels can be accommodated by a single wide bandwidth transmission system. An important technical requirement of FDM concerns the accuracy and coherence of SSB carrier frequencies, which are usually derived from stable master oscillators.

As the name implies, time-division multiplexing (TDM) is simply the sharing of a common facility in time. Some of the digital bit streams arriving in parallel at a satellite communications terminal or intermediate switching center originated from different sources but may have the same destination. It is usually convenient to time-division-multiplex (time interleave) bit streams that have the same ultimate destination into one single serial bit stream for transmission on a single RF carrier. However, the bit streams to be multiplexed may have slightly different bit rates. Even if these bit streams are nominally at the same bit rate, they have imperfect clock stabilities and hence are not exactly synchronous with one another or with the Earth terminal clock. Furthermore, some of these bit streams may have arrived via a satellite link in a multiple transmission where the satellite-Earth station propagation time is changing because of satellite motion. Many of these bit streams may have the same nominal frequency but are not necessarily synchronous. In order to time-multiplex these bit streams together, they must be converted to the same rates or multiples of the same rates. If the bit streams are nonsynchronous, one or all of the bit streams should be buffered and have "stuff" bits added to synchronize them (justification bits).

A simpler type of multiplexing problem occurs when the two bit streams are either frequency-locked or plesiochronous (the clocks may not be locked but should be close, with the frequency accuracies better than or equal to 10^{-10} or 10^{-11}. In these cases, multiplexing can be performed without using "stuff" bits. For example, a frame slip will be used at the plesiochronous interface (see CCITT Rec. G.811).

As with the frequency-division multiplex hierarchy of group, supergroup, etc., two kinds of TDM hierarchy are recommended by CCITT (see Fig. 1).

The problem of the conversion of the TDM standard is still under discussion. A possible solution is that of a mixed hierarchy based on 2, 6, 45, 140 Mbit/s.

III. Modulation Methods

The most common methods of modulation used in fixed satellite service are frequency modulation (FM) for analog signals and phase shift keying (PSK) for digital signals. In the following paragraphs we will briefly describe the main characteristics of these modulation methods, as applied to satellite communications, together with some information on other modulation methods that appear to be promising for the future.

A. Frequency Modulation

Frequency modulation[1,2,3] has been largely used for terrestrial radio relay links as well as for satellite communications. The possibility of compromising between power and bandwidth utilization has suggested the use of large values of the frequency deviation for satellite communications, normally more power-limited. Frequency modulation is generally used in the two well-known systems, FDM/FM and SCPC/FM. In a single carrier per transponder mode of operation, the constant envelope of the FM signal allows the power amplifiers to operate at saturation, thus making maximum use of the available power.

On the other hand, in a multicarrier per transponder mode of operation, a set of intermodulation products formed by combination of these carriers, which arise due to the amplitude and phase nonlinearities of the high-power amplifier, becomes a source of noise in the transmission frequency band and reduces the saturation output power of the transponder.

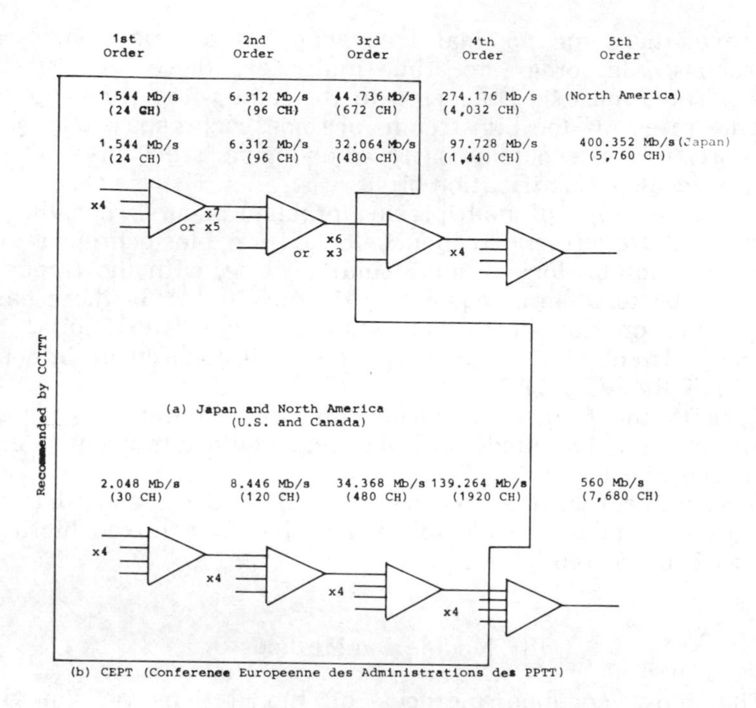

Fig. 1 Digital hierarchy.

Frequency-modulation systems have a characteristic threshold of operation, above which the carrier-to-noise density ratio C/N_o at the input of the demodulator is proportional to the signal-to-noise ratio S/N at the output. However, if the C/N_o decreases below the threshold value, the S/N does not keep the 1:1 relationship and deteriorates rapidly. It is assumed that, for conventional demodulators, the FM threshold occurs at a carrier-to-noise ratio of about 10 dB. However, the threshold level corresponding to the radio frequency bandwidth B_{RF} can be reduced by about 4 dB by such techniques as frequency feedback demodulators. It is interesting to analyze the distribution of noise in the INTELSAT FDM/FM carriers.

The INTELSAT space segment allocates 8000 pW0p to up- and down-path thermal, transponder intermodulation, Earth station RF out-of-band emission, cochannel interference within the operating satellite, and interference from adjacent satellite networks. Within this 8000-pW0p allocation, an allowance of 500 pW0p is reserved for Earth station RF out-of-band emission caused by multicarrier intermodulation from other Earth stations in the system.

The total INTELSAT system noise budget is therefore divided into four major portions:

Space segment	7,500 pW0p
Earth stations (excluding thermal noise at the receiver)	1,000 pW0p
Earth stations out-of-band emission	500 pW0p
Terrestrial interference	1,000 pW0p
Total (CCIR Rec. 353-4)	10,000 pW0p

Uplink thermal, intermodulation, cochannel interference, and downlink thermal noise levels are determined through the use of a computer program using many of the proven subroutines that have been carried over from one satellite series to the next. For example, uplink and downlink thermal noise levels are calculated for the top channel in each FM carrier using the familiar equations shown below. In all cases, the best available measured values for fixed parameters (e.g., ψ_s, G/T, P_s) are employed with appropriate corrections for geographic location.

$$\frac{S}{N}\text{ weighted, with pre-emphasis} = \frac{C}{N} \cdot \frac{B_{if}}{b} \cdot \left(\frac{f_{tt}}{f_{max}}\right)^2 \cdot P \cdot W$$

where B_{if} is the carrier IF (or Carson's rule) bandwidth; b the channel noise bandwidth (= 3100 Hz); f_{tt} the rms 0-dBm0 test tone deviation; f_{max} the top baseband frequency; P the pre-emphasis improvement; W the psophometric weighting; and C/N the total link carrier-to-noise ratio.

In the uplink, C/N is calculated from

$$\frac{C}{N} = \psi_s - G - BO_I + \frac{G}{T_s} + 228.6 - 10 \log B_{if}$$

where ψ_s is the single carrier saturation flux density; G the effective gain above isotropic of a 1-m^2 antenna; BO_I the transponder input backoff due to the carrier in question (see Sec. IVA); and G/T_s the satellite gain-to-noise temperature ratio.

In the downlink, C/N is calculated from

$$\frac{C}{N} = P_s - BO_o - PL + \frac{G}{T_E} + 228.6 - 10 \log B_{if}$$

where P_s is the e.i.r.p. of the saturated satellite TWT; BO_o the transponder output backoff due to carrier in question (see Sec.

IVA); PL the downlink path loss; and G/T_E the gain-to-noise temperature ratio of the receiving Earth station.

The distribution of the 7500-pWp (picowatts psophometric) space segment noise among the various contributions differs, depending on the technical characteristics of the specific satellite generation. Table 1 shows distribution variations for typical telephone carriers operating similarly on the three satellite generations INTELSAT IV, IV-A, and V.

Therefore, consideration has been given to carriers operating on transponders connected to the spot beams of INTELSAT IV vs carriers operating on the interzone transponders of INTELSAT IV-A and on interhemispheric transponders of INTELSAT V. The values shown in Table 1 have been derived according to the following considerations. The specific carrier-to-noise ratios for the uplink, downlink, and intermodulation have been determined by computer, using appropriate values for the fixed satellite and Earth station characteristics [e.g., single carrier saturation flux density, gain-to-noise temperature ratio of the satellite and Earth station, and equivalent isotropically radiated power (e.i.r.p.) of the saturated satellite TWT].

The C/N values, representing the interference from transponders reusing the same frequency bands by spatial discrimination (east-west) or polarization discrimination, are based on the isolation performance of the satellite antennas. The 24 dB of C/N for the east-west cochannel interference of INTELSAT IV-A are derived from the power combination of the two C/N values of 27 dB, referred to the isolation specifications, which are separately applicable to the up- and downlink between one main beam (e.g., east) and the interfering sidelobe of the other beam (e.g., west).

For INTELSAT V, the calculation is complicated by an additional interference source attributable to frequency reuse by

Table 1 Space segment typical noise budget for INTELSAT IV, IV-A, and V-A

Noise source	INTELSAT IV (spot) C/N, dB-N	pWp	INTELSAT IV-A (zone) C/N, dB-N	pWp	INTELSAT V (interhemispheric) C/N, dB-N	pWp
Uplink thermal	23.7	3270	25.4	1030	27.4	370
Downlink thermal	27.3	1430	20.1	3460	18.3	3050
Intermodulation	24.3	2800	23.5	1590	20.6	1800
Cochannel east-west interference			24.0	1420	22.2	1240
Cochannel cross-polarization interference					22.9	1040
Total	20.1	7500	16.8	7500	14.4	7500

[A] Valid only in nominal clear sky conditions.

polarization discrimination; hence it is also necessary to consider the polarization isolation performance of the Earth stations. In addition, INTELSAT V uses the frequency interleaving carriers planning technique extensively to limit cochannel interference. The advantage of this technique must be properly assessed and quantified. Therefore, for example, the C/N value of 22.2 dB for the east-west cochannel interference has been calculated with the following procedure: The uplink and downlink contributions have been separately calculated; each has been derived from the combination of the two contributions, due to interference from copolarization and cross-polarization channels (but always irradiated from a spatially discriminated beam).

These last contributions have been derived based on the fundamental hypothesis relative to the isolation performance of the satellite antennas and the evaluation of the advantage of the frequency interleaving technique.

When the partial C/N values given in Table 1 for the uplink and downlink thermal noises have been determined, assuming the noise bandwidth as the total useful bandwidth of a transponder, the total C/N has been calculated. Then, it is always possible to establish a correspondence between the total C/N and the total noise budget for the space segment (7500 pWp) by a proper choice of the noise bandwidth and, therefore, of the modulation index of the various carriers. Finally, the noise in pWp corresponding to each partial C/N can be calculated from the difference between the total C/N and the partial C/N. For example, the comparison of the 18.3-dB value of C/N for the cochannel interference due to rain depolarization for 0.01% of the year with the 22.9-dB value of C/N in clear weather corresponding to 1040-pWp noise makes it easy to calculate a noise figure of about 3000 pWp in degraded weather conditions.

Table 1 shows that, during the transition from INTELSAT IV to INTELSAT IV-A and to INTELSAT V, it has been necessary to reduce the sum of the noise contributions due to the uplink, downlink, thermal noise, and intermodulation noise to allow the allocation of noise contributions for interference due to the frequency reuse techniques. Therefore the trend, which is expected to accelerate, is from satellites limited by thermal noise to interference-limited satellites.

The uplink thermal noise reduction from INTELSAT IV to IV-A is justified by the improved satellite gain-to-noise temperature ratio (G/T); the additional reduction of this contribution on INTELSAT V is due to the higher saturation flux density and to the decreased incidence of the uplink C/N relative to the total C/N. The downlink thermal noise is lower for INTELSAT IV, due to the higher e.i.r.p. of its spot beams, while the minor differences between INTELSAT IV-A and V are mainly due to the fact that the output backoff of the satellite TWTA is slightly higher for INTELSAT V than for INTELSAT IV-A; for INTELSAT IV, this contribution is higher even with a higher backoff because of the higher incidence relative to the total C/N.

B. Phase Shift Keying

Phase shift keying (PSK)[1] modulation is an efficient method of modulating a carrier with digital information. In a two-phase system, one phase of the carrier is used to represent one binary state, and phase 180 deg apart from the first one, is used for the other binary state. Higher-order PSK schemes exist. Theoretically four-phase PSK requires the same power, but half the bandwidth, as two-phase PSK for a given link performance. Higher-order (greater than four-phase) PSK systems are more susceptible to noise and therefore need more power than either two- or four-phase systems to achieve the same standard of performance. For most purposes four-phase PSK gives the best power-bandwidth compromise.

In general, the bandwidth efficiency of PSK modulation, expressed in bits/Hz, varies according to $\log_2 \phi$, where ϕ is the number of states, while the power required, expressed by the ratio E/N_o (bit energy-to-noise density ratio), increases with the number of states for the same bit error probability P_{be}.

Two basic types of PSK modulation exist: the coherent phase shift keying (CPSK) and the differential phase shift keying (DPSK).

The DPSK system is based on a differential demodulation realized by comparing the signal received in a given signaling interval with the signal received in the previous signaling interval. Therefore, DPSK, contrary to CPSK, does not require any carrier recovery circuit but only requires a device that can introduce a delay equal to the signaling interval on the received signal.

The bit energy-to-noise density ratio E/N_o corresponding to a bit error probability of 10^{-4} is given in Table 2 for coherent and differential detection of PSK signals.

E is the energy per bit in W-s, and N_0 is the noise power in a bandwidth of 1 Hz. Table 2 accounts for a Gaussian noise signal disturbing the information signal.

In CPSK, however, there is the problem of resolving at the receiving end which of the n possible states of the recovered

Table 2 E/N_0 for CPSK and DPSK signals

No. of states	E/N_0 ($P_{BE} = 1\times10^{-4}$)	
	CPSK	DPSK
2	8.4	9.3
4	8.4	10.7
8	11.8	14.8
16	16.1	19.1

reference carrier is the reference state. The most suitable way to resolve this problem is to differentially encode the data bit stream before modulation. This implies an additional power requirement that, for example, in the case of four-phase differentially encoded CPSK, amounts to 0.4 dB.

The performance of PSK systems is therefore characterized by the parameters E/N_0 of Table 2, which, however, in the practical cases have to be increased by an amount M_t, defined as the total margin needed to compensate for various sources of degradation. These sources can be grouped in two categories: the modem implementation margin and the system margin. The modem implementation margin is essentially due to the intersymbol interference performance tied to the type of filtering used. The system margin is essentially due to the impairments introduced by the onboard filtering, by the AM-AM and AM-PM nonlinearities of the transponder, by the adjacent channel and cochannel interferences, and by multipath effects. The order of magnitude of the total margin for a two-phase CPSK system is about 2 dB, while it is 3 or 4 dB for a four-phase CPSK system and even larger for a higher number of states.

INTELSAT has adopted four-phase CPSK modulation with absolute encoding (not differential encoding) for its SCPC, SPADE, and TDMA systems (see Secs. IVA2 and IVC1).

C. Single Sideband Amplitude Modulation

Single sideband (SSB) transmission[1] is simple in concept and has the merit of greatest economy in bandwidth. Unlike frequency modulation, it is not subject to the threshold effect. On the other hand, single sideband modulation has certain disadvantages relative to FM and PSK; e.g., it requires higher transmitter power than wide deviation frequency modulation; it is more sensitive to nonlinearity; and it requires highly accurate frequency synchronization, which is particularly difficult to achieve with nongeostationary satellites. It is also both more susceptible to interference than wide deviation frequency modulation and will produce more interference when its signals fall within one baseband-bandwidth of the carrier of other systems.

Similar noise allocation can be considered for single sideband as for frequency modulation, although for single sideband, it may be more difficult to provide the necessary linearity.

In a single sideband transmitter, it is necessary to consider both mean power and peak power. Peak-to-rms voltage ratios for SSB transmitters can be on the order of 10 to 13 dB.

Recent investigations have been performed, also under an INTELSAT R&D contract, on amplitude companded single sideband (ACSSB) modulation, which uses methods of voiceband processing such as 12 dB per octave emphasis, ratio-2 companding, pilot signal insertion, and a second stage of ratio-2 companding. Preliminary results show, theoretically, that there is a possibility of transmitting 9000 channels per 36-MHz transponder. However, further investigation will be needed on subjective evaluation of speech quality.

D. Hybrid Modulations

Hybrid modulations[3-5] are combined amplitude and phase modulations (apm) that can be applied to digital signals and to analog signals as well. In a graphical form, these modulation schemes can be represented by a constellation of points or segments in a plane, respectively, for digital and analog signals, which indicate the states that the carrier can assume during the modulation process.

The Nyquist bandwidth of the modulated carrier is given by the following formula:

$$B_N = \frac{f_c}{\log_2 S}$$

where B_N is the Nyquist bandwidth, in Hz; f_c the transmission bit rate, in bit/s; and S the number of states of the constellation.

As shown by the formula, the bandwidth is reduced as the number of states increases. On the other hand, a higher number of states means a higher value of the signal-to-noise ratio needed to achieve a given error probability, more complex modulation and demodulation circuits, and increased sensitivity of the system to the linear and nonlinear distortions and to the implementation imperfections of the circuits.

Studies and experimental investigations performed by INTELSAT concentrated originally on one-pulse amplitude-and-phase modulation (1P-apm) and two-pulse apm (2P-apm). Both were designed specifically for efficient FDM/FDMA telephone transmission and, later on, for four-dimensionally coded quadrature amplitude modulation (4D-QAM) for the transmission of digital signals. Another approach has also been suggested for digital TDM/PCM/FDMA telephony that combines 2P-apm and four-phase PSK in one modem.

The modem would employ 2P-apm for PCM digits representing voice channel samples (channel digits) and four-phase PSK for service digits.

Figure 2 compares the ideal performance of various modulation systems, including one- and two-pulse apm. Bandwidth and power requirements are shown relative to the requirement for single sideband modulation. Also shown are contours of constant RF C/N ratio and Shannon's rate-distortion bound for a Gaussian baseband noise power ratio (NPR) of 32 dB, which is appropriate for large FDM telephone multiplex assemblies. All realizable systems lie above this bound.

The following assumptions are used in Fig. 2. For FM, apm, and AM, a 10-dB peak factor is assumed, and NPR = 32 dB. The FM curve includes a 4-dB pre-emphasis gain and shows the Carson's rule bandwidth. The apm and PSK bandwidths are the noise bandwidths obtained by sampling at the Nyquist rate. The 1P-apm performance is based upon the signal mapping described in Ref. 4, while the PSK performances are based on 8-bit coding and a bit-error rate of 10^{-4}. The peak power values for PSK give the approximate envelope power at the output of a Nyquist filter, which can be exceeded with a probability of 10^{-5}.

It should be noted that 2P-apm uses less power and less bandwidth than any of the angle modulation systems. Figure 2 does not include practical system degradations, which vary among modulation systems. As a matter of fact, detailed link analysis together with simulations performed to take into account the interference environment and the linearity performance of the INTELSAT V satellite indicated that hybrid modulation suffers from severe impulse noise during degraded conditions.

IV. Multiple-Access Methods

Multiple access[6] is the ability of a large number of Earth stations to interconnect their respective transmission links

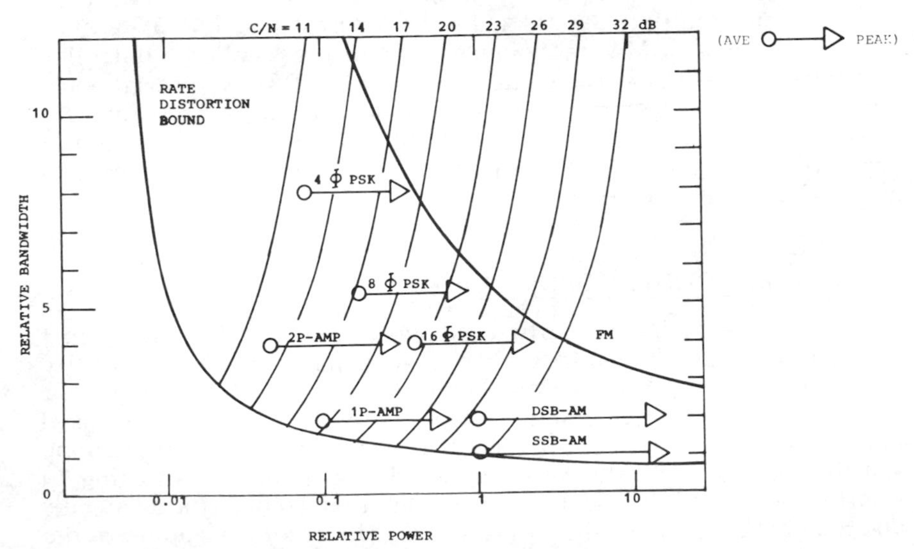

Fig. 2 Modulation systems comparison.

simultaneously through the same satellite resource. This capability is very important in a satellite system because of the large coverage area achievable by the satellite antennas. Transmissions from one or more Earth stations can be received at many different locations, resulting in an interconnectivity that is a peculiar characteristic of satellite communications systems.

Generally, the information transmission capability available per transponder is greater than that needed by most Earth stations for their individual use. Therefore, in order to fully utilize the available repeater capability, more than one Earth station is allowed to access the same transponder with either single destination or multidestination transmissions. This use of one or more transponders by more than one transmitting Earth station is called multiple access.

Two different categories of multiple-access techniques can be identified according to speech channel allocation or the type of transponder sharing. The first way is broadly divided as follows:

1) Preassigned multiple access, in which the channels required between two Earth stations are assigned permanently for their exclusive use.

2) Demand-assigned multiple access, in which the channel allocation is changed in accordance with the originated call. The channel is automatically selected from a common pool and is connected only while the call is continued. This system remarkably increases the efficiency of the satellite transponder in comparison with preassigned multiple access.

Secondly, the multiple-access system can be classified according to the type of transponder sharing, as follows: 1) constant envelope carriers (frequency-division multiple access and spread-spectrum multiple access) and 2) pulsed carriers (time-division multiple access and pulse-address multiple access).

FDMA and TDMA techniques can also be called controlled multiple-access systems, while SSMA and PAMA are random multiple-access systems. This categorization is useful for discussion because it groups together systems with similar properties. Figure 3 illustrates the manner by which this categorization is reached.

A. Frequency-Division Multiple Access

Frequency-division multiple access (FDMA)[1] is the most commonly used multiple-access technique in the satellite communication system. FDMA is such that, by giving a different frequency to respective Earth stations, satellite resources are used in common. One detrimental effect of such a system is that many signals pass through the satellite at the same time, resulting in interference noise caused by intermodulation among these signals due to nonlinearity of the transponder. In order to suppress the

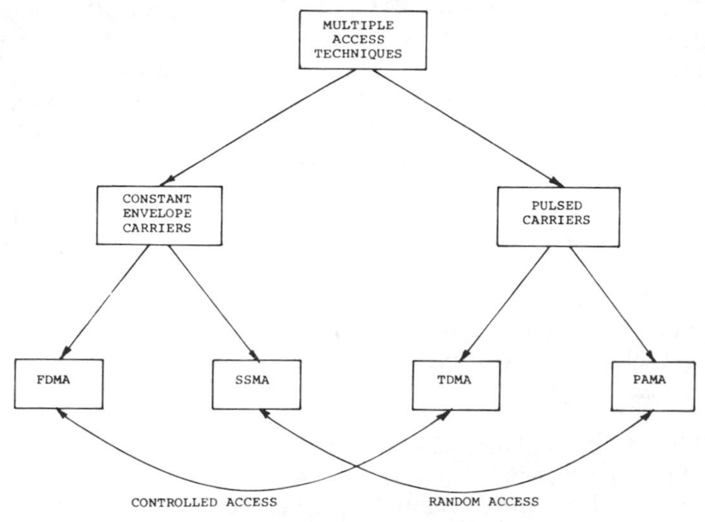

Fig. 3 Categorization of multiple-access techniques.

influence of intermodulation, it is necessary to keep the transmitting power output level considerably lower than the saturation point. This is called "backoff." Furthermore, the transmitting power output from each Earth station has to be precisely controlled.

FDMA employs many baseband multiplexing and modulation combinations. Typical systems used by INTELSAT are FDM-FM and single channel per carrier (SCPC). The former is the most common, using carriers that are frequency modulated by a frequency-division multiplexed baseband signal. The latter uses an RF carrier in each telephone channel, with PSK modulation, and it is particularly suited for Earth stations with relatively few channels.

1. FDM-FM-FDMA. In FDMA, each signal is assigned a separate nonoverlapping frequency band, and power amplifier intermodulation products are controlled to acceptable levels by appropriate frequency selection and/or reduction of input power levels to permit quasilinear operation. Attention is focused on the satellite transponder effects, since this power is more critical and costly than Earth terminal power. Typically, one might reduce the satellite average output power by 50% or more to reduce intermodulation products to an acceptable level.

The guard band between adjacent frequencies must be decided also, taking into account the frequency drifts of the oscillators controlling the signal center frequencies at the satellite and Earth station frequency translators. Doppler shifts of satellite can also be significant for very low data rate transmission.

Computation of the effects of intermodulation products created by satellite and Earth station transmitters' nonlinearities must account for changes in the relative signal strength received at the satellite. These signal levels can change because of Earth station transmit e.i.r.p. fluctuation, localized rain losses that can significantly attenuate the signal received from one Earth station, and antenna pointing losses at both the Earth station and the satellite.

FDM/FM is the most widely used multiple-access technique in the INTELSAT system. Table 3 shows some examples of carriers used in the INTELSAT system.

2. Single Channel per Carrier (SCPC). In a single channel per carrier (SCPC) system,[7-9] each carrier is modulated by only one voice channel. The voice channel can be processed in a number of ways. Typical SCPC systems use companded frequency modulation (CFM), delta modulation with PSK modulation of the RF carrier, or PCM with PSK modulation of the RF carrier.

In this type of system, the carrier is usually voice activated, and this fact causes 60% of power savings in the satellite transponder (the carriers are active only 40% of the time on average).

SCPC/CFM systems may have RF bandwidths of 22.5, 30, or 45 kHz per carrier; SCPC/DM/PSK systems transmit 32 kbit/s in a 22.5-kHz bandwidth; and SCPC/PCM/PSK systems use 64 kbit/s in a 45-kHz RF bandwidth.

The assignment of transponder channels to Earth stations may be fixed or variable. In the first case, we have a preassigned system, in which each channel slot of the transponder is dedicated to the use of a particular station. In the second case, we have a demand-assigned system, in which the channel slots of the transponder are assigned to different Earth stations according to their instantaneous needs.

Table 3 Examples of INTELSAT FDM/FM carriers

No. of channels	Maximum baseband, kHz	rms test tone deviation, kHz	Occupied bandwidth, MHz	Allocated bandwidth, MHz	C/T at operating point, dBW/K
24	108	164	2.00	2.5	-153.0
60	252	270	4.00	5.0	-149.9
132	552	376	6.75	7.5	-145.9
252	1052	358	8.50	10.0	-139.9
432	1796	616	18.00	20.0	-139.9
972	4028	802	36.00	36.0	-135.2

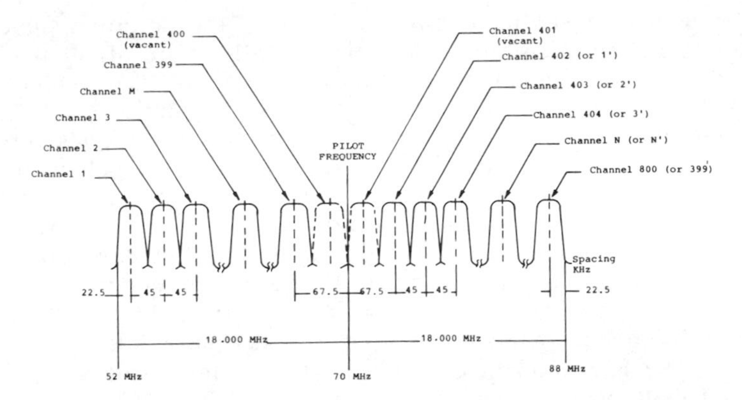

Fig. 4 SCPC frequency plan for full transponder operation.

Both systems have been implemented in the INTELSAT network, using 64-kbit/s PCM/PSK modulation in an RF bandwidth of 45 kHz. A 36-MHz transponder can therefore accommodate 800 simultaneous SCPC channels.

Figure 4 shows the frequency plan for full transponder operation of the INTELSAT SCPC system. In case of the demand-assignment operation (SPADE), channels 1, 2 and 1', 2' are not usable because of the common signaling channel (CSC) carrier. Channels 400 and 401 are not usable in all frequency plan arrangements, to guard against interference with the pilot. Considering the 40% activity factor, the transponder would only be loaded with 320 carriers at any instant.

The demand-assignment feature of SPADE may be thought of as a channel concentrator. It has been calculated that, for traffic intensity of 0.5 erlangs (30 channel-min/h) to each destination, 40 preassigned or 11 demand-assigned channels would give the same grade of service. Experience has shown 900 min per day on some SPADE channels, while 150 min is normally considered good usage on an international trunk.

The SPADE terminal consists of four basic hardware subsystems. These are 1) a terrestrial interface unit (TIU); 2) a demand-assignment signaling and switching unit (DASS), 3) individual channel units up to a total of 60 per terminal; and 4) an i.f. system.

The TIU is connected to the transit center (CT) for international calls. It is designed to accommodate whichever CCITT signaling and switching system is in local use, whether automatic or manual. It orders the DASS to set up a circuit and tells the DASS when the call is completed. The TIU also connects to the channel units to pass along the call.

The DASS maintains a continuously updated table of active frequencies and randomly selects an unused pair of frequencies when it wants to set up a circuit. It uses the common signaling channel to inform all other terminals that the pair of channels is engaged. The i.f. system takes the output of the PSK modem and the DASS and sends it to the Earth station's i.f. panel.

When the DASS sets up a call, the conversation goes through the channel units that are selected for the call. The channel unit has a PCM coder-decoder (codec). Its digital output goes to a voice detector and to a channel synchronizer. These subsystems, along with a frequency synthesizer, feed into the PSK modem of the channel unit. The frequency synthesizer can generate all 800 frequencies of the frequency plan and is instructed by the DASS as to which frequency pair will be used for the particular call. The output of the PSK modem goes to the i.f. system, where all of the outputs of all of the channel units are combined with the DASS burst of the common signaling channel (CSC). On the receive side, these systems work in the reverse direction in a similar manner.

The system sets up calls through a TDMA CSC, which all stations access and monitor and which is used to update a table of channel assignments that are engaged. The CSC's bit rate is 128 kbit/s and uses two-phase PSK. The bit error rate at threshold is 10^{-7}. Since any terminal can reserve channel assignments and all terminals monitor the channel assignments, SPADE has a decentralized control. A recording terminal that records all system events for analysis and billing purposes can be installed at any station of the network.

In those cases where dedicated point-to-point service is required, SCPC may be used to provide one or more channels. The modulation method and the transmission parameters are the same as for SPADE, but the common control facility (DASS) is not needed, so the terminal is simpler and less expensive.

For light traffic requirements, it is certainly more efficient to use SCPC rather than a 24-channel FDM/FM carrier occupying 2.5 MHz of RF bandwidth. Current frequency planning now reserves blocks of bandwidth for SCPC assignments on all satellites.

Table 4 Typical SCPC/companded FM characteristics

Characteristic	Value	Unit
Channel spacing	45.0	kHz
Allocated channel bandwidth	45.0	kHz
i.f. noise bandwidth	25.7	kHz
C/N_0 per channel at nominal operating point	53.9	dB-Hz
C/N in i.f. bandwidth at nominal operating point	10.0	dB
Pre-emphasis improvement	6	dB
Total weighted, uncompanded test tone/noise ratio	34.3	dB
Compander advantage	17.0	dB
Companded equivalent test tone/noise ratio for average talker	51.3	dB

A number of countries now lease space segment capacity on INTELSAT satellites in the form of a full transponder or fractional part of a transponder. The type of modulation utilized depends on the range of services required, and INTELSAT's only restriction is in the form of specific technical performance criteria that must be adhered to in order to minimize mutual interference between the various services provided by INTELSAT satellites, be they leased or within the regular global network.

However, SCPC/CFM has emerged as the most widely used transmission system on leased transponders, and typical parameters are shown in Table 4.

B. Spread-Spectrum Multiple Access

Although this multiple-access system is not currently being used in the INTELSAT network, it might be usefully employed in the future to permit communication of message information under difficult conditions of very low signal-to-noise ratio, such as may be encountered due to high interference from radio-relay links.

The spread-spectrum multiple access (SSMA),[10,11] sometimes also called code-division multiple access (CDMA), is a random access mode with constant envelope carriers that utilizes modulated signal spectra that deliberately employ large emission bandwidths to send information requiring a bandwidth much less than is transmitted.

Since wide-band FM produces a spectrum much wider than that required for the transmitted information, it might be considered a spread-spectrum technique. However, it differs from the spread-spectrum techniques subsequently to be described in that these latter techniques use some signal or operation other than the information being sent to broadband (spread) the transmitted signal, whereas in the FM case the information signal itself is used to broadband the signal transmitted. However, like wide-band FM, spread-spectrum techniques have a processing gain, permitting reduced carrier-to-noise, or low transmitted power levels, to produce satisfactory system performance.

Three general types of spread-spectrum techniques are possible: 1) pseudorandom sequences, in which a carrier is modulated by a digital code sequence having a bit rate much higher than the information signal bandwidth; 2) frequency-hopping, in which the carrier is frequency shifted in discrete increments in a pattern determined by a digital code sequence; and 3) frequency-modulation pulse compression, or "chirp," in which a carrier is swept linearly over a wide-band of frequencies during a given pulse.

The easy generation of discrete pseudorandom codes allows a large number of accesses. Link synchronization is required, but not central timing. Control of uplink power is required to avoid the "capture effect" which is a characteristic of constant envelope carriers and which occurs when one signal at the satellite repeater input dominates the sum of all the other repeater input signals, "capturing" the limiter or bringing to saturation the output traveling wave tube amplifier (TWTA).

SSMA signals occupy the full channel bandwidth, unwanted signals appearing as Gaussian noise. The mutual interference limitation occurs when the number of multiple accesses generate sufficient noise to reach the power-to-noise density ratio required for the type of modulation. The system is essentially bandwidth-limited by the expression

$$\frac{P_r}{N_o} = \frac{\frac{\pi}{4} W}{M}$$

where P_r is the satellite power received at the terminal; N_o the receiver noise density; $\pi/4$ the hard-limiter suppression factor; W the system bandwidth; and M the number of accesses.

For example, using a P_r/N_o ratio of 42 dB-Hz required for a 2400-bit/s signal with a 10^{-3} error rate and a 10-MHz satellite bandwidth gives M equal to 490. This number of accesses can be supported before mutual interference becomes the limiting factor.

C. Time-Division Multiple Access

Time-division multiple access (TDMA)[1,12] is a digital multiple-access technique that permits individual Earth terminal transmissions to be received by the satellite in separate nonoverlapping time slots, thereby avoiding the generation of intermodulation products in a nonlinear transponder. Each Earth station must determine satellite system time and range so that the transmitted signals are timed to arrive at the satellite in the proper time slots.

There is no intermodulation caused by instantaneous nonlinearity, because only one signal enters the satellite transponder at a time. The bit rates of the transmitted bursts are generally many times higher than that of the continuous input bit streams to the Earth terminal.

Compared with the FDMA system, the TDMA technique has the following advantages:

1) Because of the absence of intermodulation effect, the satellite transponder can be driven nearly to saturation to yield efficient use of satellite power.

2) With a digital speech interpolation (DSI) technique, more than three times the channel capacity can be obtained compared with FDMA, and the capacity does not decrease steeply with an increase in the number of accessing stations.

3) Establishment and change of traffic is easily accommodated by changing burst length and burst position.

Each TDMA Earth station has parallel input digital bit streams, or analog streams that are digitized at the Earth station, which are addressed to separate receiving Earth stations. The signals addressed to separate receive terminals are allocated separate portions of the transmit TDMA burst following the TDMA burst preamble. The TDMA receiver demodulates each of the TDMA bursts from separate transmit terminals and multiplexers, then demultiplexes the appropriate portions of them into separate serial bit streams.

The input signal to a transponder carrying TDMA consists of a set of bursts originating from a number of Earth stations. This set of bursts is referred to as the TDMA frame. A small time gap, called guard time, is left between bursts to make sure that bursts do not overlap with one another. The first burst in the frame contains no traffic, and it is used for synchronization and network control purposes. This burst is called the reference burst. Each burst is periodically transmitted with a period identical to the TDMA frame period, which is a multiple of 125 μs (corresponding to a 8-kHz frequency).

The information bits are divided into groups called subbursts. Each subburst contains information originating from a terrestrial interface module located at the Earth station. The information bits are preceded by a group of bits referred to as preamble, which is used to synchronize the burst and assist in controlling the network. The reference burst contains only the preamble. In general, the burst preamble contains the following parts:

1) a carrier and bit timing recovery sequence that has the function of providing the carrier reference and the bit timing clock required for demodulation in the receive terminals;

2) a bit pattern called the unique word that is used to identify the starting position of the burst in the frame and the position of the bits in the burst;

3) order wire bits, carrying telephone and teletype information for interstation communications; and

4) control bits, carrying network management information (this section of the preamble is referred to as the control channel).

Several different synchronization problems are involved in operating the TDMA system.

To demodulate the input burst mode PSK carriers, it is necessary to recover the carrier and bit timing within the recovery sequence at the top of each burst. Thus the TDMA demodulator usually has very high-speed recovery circuitries made of filters and carrier and clock extraction circuits.

Another crucial synchronization problem arises in the burst transmission timing at each accessing station in order to prevent an overlap of bursts at the satellite transponder. This control is called burst synchronization. The burst synchronization is performed so that each burst maintains the predetermined timing difference referenced to the position of the reference burst at the satellite transponder. To achieve the burst synchronization, several methods--some of which follow--have been studied.

(a) Global beam synchronization: The transmit timing error is detected at each transmitting station just by examining the received signal sequence, which includes its own transmission as well as that from the reference station.

(b) Feedback synchronization: The detection of the timing error is performed at the destination station or the reference station, and the information on the burst position error is sent back to the transmitting station via the control channel.

(c) Open loop synchronization: Transmit timing is determined by knowing the range data from each Earth station to the satellite.

A particularly attractive extension of the TDMA technique in connection with onboard multiple beam antennas is that referred to as satellite-switched time-division multiple access (SS-TDMA). This technique is based on the concept of implementing a time-division switching device onboard the satellite that, operating in synchronism with the ground TDMA terminals, permits splitting the incoming TDMA bursts into multiple subbursts that are routed toward the appropriate downlink beam.

1. System Parameters in TDMA Systems. In TDMA systems, preamble requirements cause a decrease in the available telephone capacity so that the frame efficiency (ξ), defined as the usable fraction of the frame capacity, decreases as the number of stations in the system increases, as shown in Fig. 5. A suitable method of limiting this loss of efficiency is to lengthen the time frame by using buffer store circuits. This method of operation, called the "multiframe" mode, offers increasing advantages as the number of stations per frame and preamble length per station increase. Alternatively, a system employing clock coherent operation among the time slots allocated to a given Earth station allows shorter preambles and thus increases the frame efficiency. Generally, the frame efficiency increases with increasing TDMA frame length, as shown in Figs. 5 and 6. It is seen that the point at which this increase becomes insignificant is a function of the preamble length and the number of stations in the system. In practice, the optimum TDMA frame length for a particular system will be determined by the operational and Earth segment equipment requirements as much as by the need to have a high frame efficiency.

Calculations summarized in Fig. 6 have been performed using the following formula:

$$N_C = \frac{RL_f - n_p N_{st}}{L_f \cdot R_C}$$

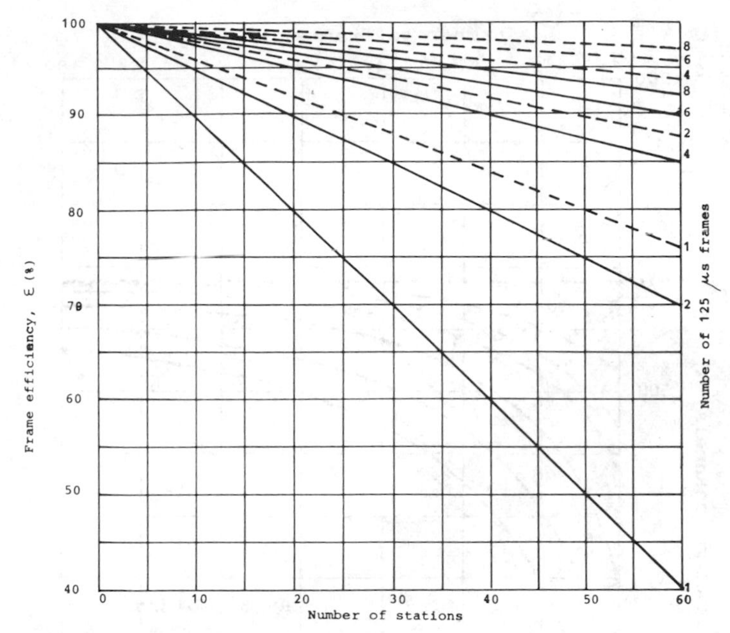

Fig. 5 Frame efficiency in TDMA systems. (——) Preamble length: 40 symbols (80 bits) (clock asynchronous operation). (-----) Preamble length: 16 symbols (32 bits) (clock coherent operation).

where N_c is the number of useful channels; R the link bit rate; R_c the channel bit rate; L_f the frame length; n_p the preamble bit number; and N_{st} the number of stations.

The capacity of a TDMA system can be significantly improved, incorporating a digital speech interpolation (DSI) facility either in a "point-to-point" or "multidestination" configuration.

In the DSI system, the theoretical capacity gain has its limit set by the inverse of the mean activity factor of the input trunks. DSI, however, may introduce some systematic or sporadic degradation of the voice signal. The system degradations are due to the voice detection process and to the internal signaling message and can be reduced or even partially eliminated by employing suitable hardware solutions. The sporadic degradations are due to the statistical behavior of traffic, which causes system overload to occur when the number of active input trunks is higher than the number of available transmission channels. The degradations introduced in this case are directly related to the DSI gain and therefore to the system capacity.

When the system goes into overload, clipping is introduced--the average value of which is referred to as freeze-out (Φ). Practical tests on users' reactions to the presence of speech degradation introduced by DSI show that the speech degradation is unnoticeable by an average listener as long as the

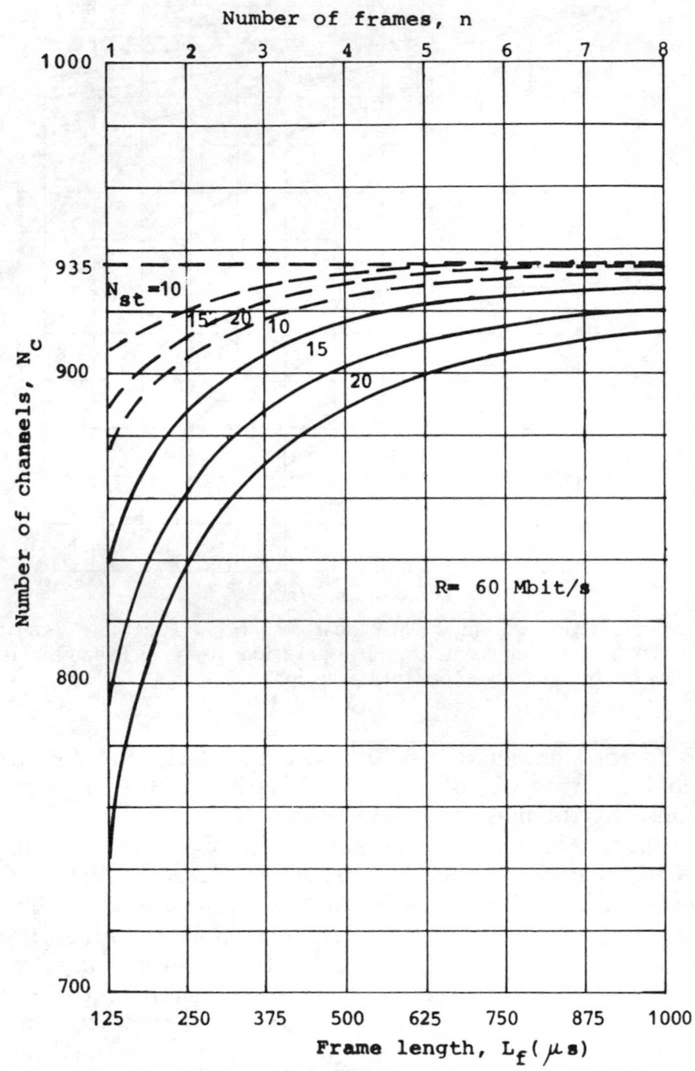

Fig. 6 Net channel capacity of a TDMA system on a 36-MHz satellite transponder vs frame length (L_f). (——) Preamble length: 40 symbols (80 bits) (clock asynchronous operation). (-----) Preamble length: 16 symbols (32 bits) (clock coherent operation).

freeze-out fraction is less than 0.5%. Theoretically, the DSI gain G and the freeze-out fraction Φ are related to the mean voice activity A, the number of input trunks N_t, and the number of transmission channels N_c.

Figure 7 shows G as a function of N_c when A = 35% and $\Phi \leq$ 0.5% and 1.0%. From this curve, it appears that the DSI advantage increases as the number of transmission channels increases.

A bit reduction strategy can be employed to reduce the freeze-out fraction further. With this strategy, the eighth bit of the various normal channels is used to yield additional transmission channels when all of the normal channels of 8 bits are overloaded. In this case, the degradation introduced consists of a decrease of the S/N ratio of the channels involved, which is limited to the overload time, and a possible residual freeze-out, which is qualitatively similar to that which occurs without the bit reduction strategy.

Subjective tests performed in Italy on equipment with and without the bit reduction strategy under the same operational conditions, i.e., with N_t, A, and G, revealed that the voice quality obtained by means of the bit reduction strategy is better than that achieved without the bit reduction strategy. For example, for the system with bit reduction, a freeze-out fraction of less than 0.05% can be expected, compared with 0.5% for the system with bit reduction but of the same DSI gain. Moreover, a higher DSI gain and therefore a greater total system capacity is achieved through the bit reduction strategy as compared with the normal

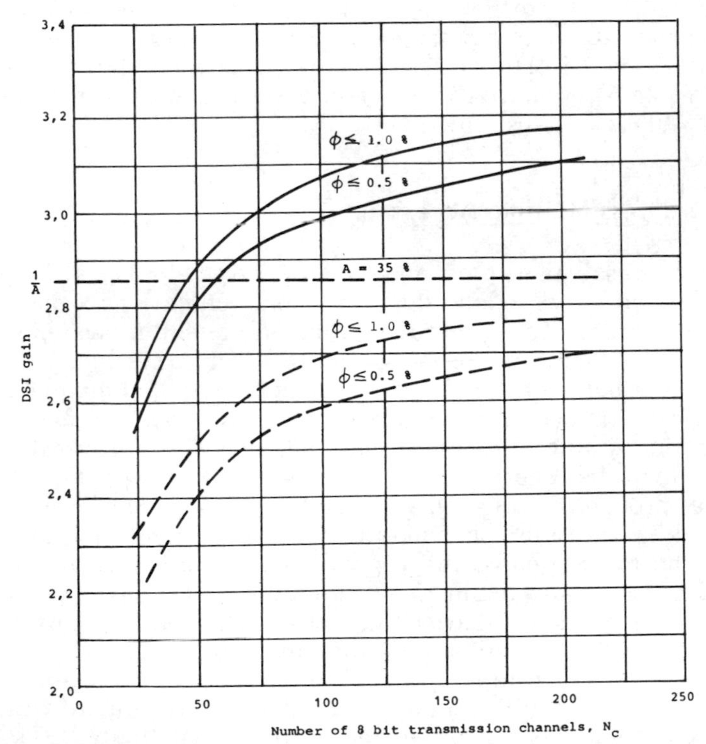

Fig. 7 DSI gain for different numbers of transmission channels (N_c). (—) Bit reduction strategy. (-----) Normal DSI.

DSI for the same voice quality, other conditions being equal. Figure 8 compares, as an example, the channel capacities using TDMA multiframe and DSI for the two strategies, assuming equivalent subjective quality of service.

A TDMA/DSI system will be used with INTELSAT V, EUTELSAT I, and later satellites. The INTELSAT TDMA/DSI system is designed to operate with 80-MHz hemi and zone beam transponders at 6/4 GHz and with spot beam transponders at 14/11 GHz. It may also operate with cross-strapped transponders. The system is also designed to be compatible with the satellite-switched transponders that will be present on INTELSAT VI satellites (SS-TDMA). Each DSI interface module can accommodate up to 240 terrestrial channels. The digital noninterpolated (DNI) modules can accommodate up to 128 terrestrial channels.

The TDMA system employs the use of separate reference stations for the acquisition and synchronization of the traffic bursts. Each reference station is equipped with sufficient redundancy to provide a high degree of reliability. In order to satisfy INTELSAT's continuity of service requirement, each reference station operating in a given coverage area will have one secondary reference station. Since many functions necessary for monitoring are inherently present in the reference stations, the TDMA system monitor function is carried out by the reference stations. Table 5 summarizes the main transmission parameters of the INTELSAT TDMA system.

D. Pulse-Address Multiple Access

Pulse-address multiple access (PAMA)[1,13] is a digital technique used for uncoordinated communications that is particularly suitable for interconnection of computer systems. It can also be thought of as un uncoordinated TDMA system. A user will transmit a burst of data (one or many packets), complete with address information at a time determined only by the need for communication at the originating end and regardless of whether or not there may be another user accessing the repeater. The acceptable probability of collision of bursts will determine the maximum data transmission capacity of the system. As can be expected, the utilization of the repeater capacity is low for such a system. This is weighed against the simplicity of the system.

An increase in utilization can be achieved by initiating transmission of a burst only at periodic time intervals, i.e., using a frame structure but retaining the unordered access to a frame. This will improve the utilization. A further improvement will be achieved by going to an organized (but uncomplicated) TDMA system.

Table 5 Main transmission parameters of the INTELSAT TDMA system

Transmission bit rate	120.832 Mbit/s
TDMA frame length	2 ms
Frequency	Transmit: 6 GHz or 14 GHz Receive: 4 GHz or 11 GHz
Acquisition method	Open loop
Synchronization method	Feedback closed loop
Modulation	Four-phase PSK (QPSK)
Demodulation	Coherent
Encoding	Absolute (not differential encoding)
Error correction coding	Rate 7/8 FEC BCH (128,112); code used on selected links
Nominal signal bandwidth	80 MHz
Nominal Earth station e.i.r.p.	89.0 dBW at 6 GHz 85.0 dBW at 14 GHz (west spot) 88.0 dBW at 14 GHz (east spot)
Bit error rate requirements	As per CCIR (10^{-6} long term, 10^{-3} short term)
Terminal transponder hopping capability	Up to four transponders
Terminal traffic capability	Transmit or receive up to 32 DSI and/or DNI subbursts per frame

The University of Hawaii's ALOHA system has pioneered in the use of random access techniques to connect by radio a large number of geographically distributed users to a central computer.

One of the basic aspects of a random-access packet transmission system (ALOHA) is the "collision" phenomenon between packets transmitted by different users sharing the same transmission channel. After transmitting a packet, the transmitting station waits a given amount of time for an acknowledgment; if none is received, the packet is automatically retransmitted. The price paid, however, is that the maximum allowable throughput drops considerably below the channel capacity. As a consequence, the total number of packets in the transmission channels will be made by type A and type B packets, where type A are newly generated packets and type B are retransmitted packets.

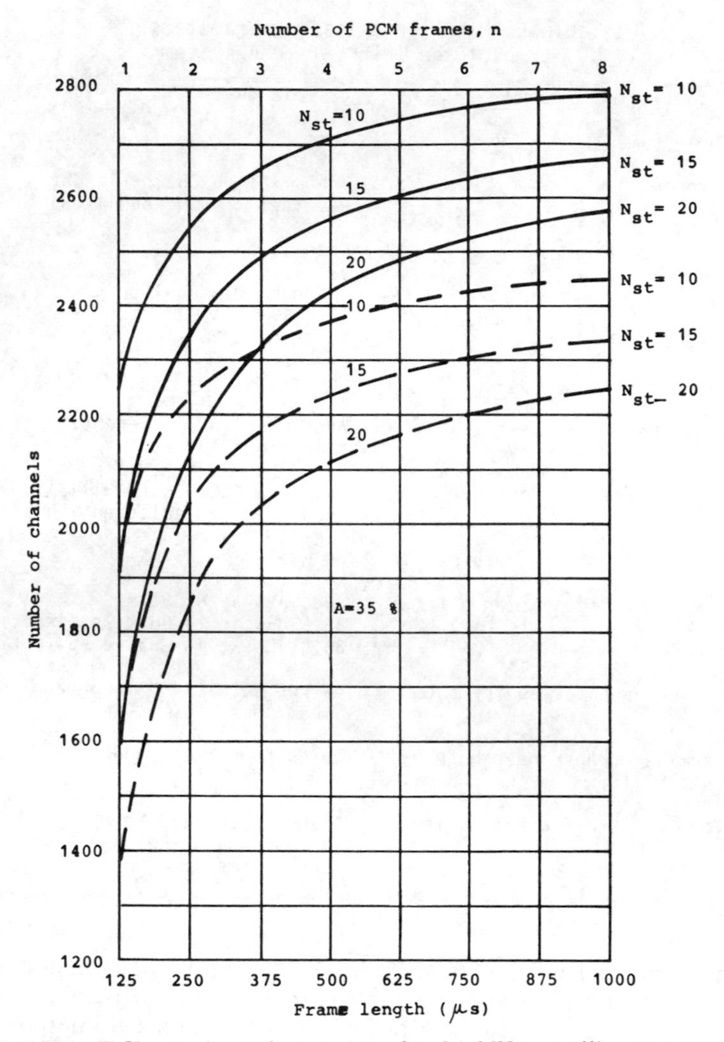

Fig. 8 TDMA/DSI net channel capacity of a 36-MHz satellite transponder. (-----) Concentrated freeze-out strategy; $\Phi \leqq 1\%$. (——) Bit reduction strategy; $\Phi \leqq 1\%$. N_{st} is the number of stations in the system.

If we call G the total number of packets in the transmission channel and S the number of useful packets (i.e., the real traffic), it will be

$$G = A + B$$

$$S = A$$

It is therefore clear that the ratio between G and S represents the number of times that a single packet must be transmitted in order

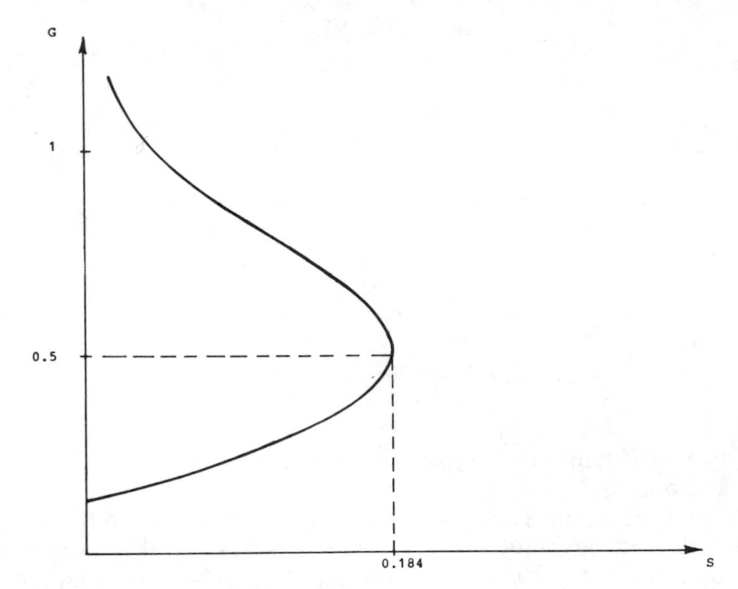

Fig. 9 Average number of useful packets S vs the average total number of packets G.

to be successfully transferred from one station to another. Assuming then that the total channel traffic obeys Poisson statistics with parameter G, it has been demonstrated that

$$S = G \bullet e^{-2G}$$

Such an equation is sketched in Fig. 9 for a pure ALOHA system. As has been mentioned, the ratio G/S represents the average value of retransmission of a packet, which is equal to 2.7 and is also a measure of the time delay in the system (exclusive of propagation delay). S/G represents the probability that no collision will take place and is equal to about 37%.

It should be noted that the maximum value of S that can be achieved with a pure ALOHA system cannot exceed 18.4% of the available channel capacity. The value of G corresponding to this value of S is 0.5; i.e., the total gross volume of data transmitted in the channel is 50% of the available channel capacity.

It is essential to note that the mentioned values are average values, due to the Poisson statistical distribution assumed both for the traffic of the various stations and for the total traffic of the network. Therefore the maximum value of retransmissions of a packet, which has an impact on the buffer size and on the time delay, is not a definite number and should be evaluated as a function of the particular signal to be transferred. Here, a basic distinction should be made between the signals of the

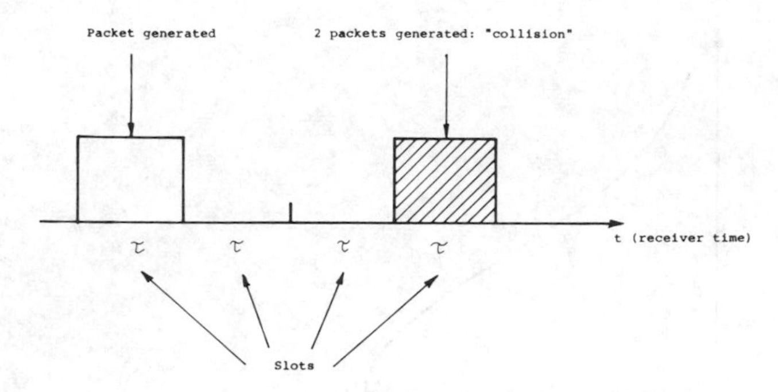

Fig. 10 Time slots in slotted ALOHA.

"computer-to-computer" type and the "real time" signals (such as video and audio).

As a matter of fact, for the computer-to-computer signals, the possible large number of retransmissions of the same packet does not have a significant impact on the buffer size because the number of bits to be transferred in a certain time interval is not very large (a few thousand bits). Also, the time delay is not very important, within certain limits, because there is no rigid "frame" requirement such as in the real time signals like audio or video.

The pure ALOHA scheme provides a maximum utilization of $1/2e = 0.18$. This is due to collisions with other packets possibly transmitted within a $\pm\tau$ or 2τ second (packet) interval of the one under consideration.

Consider now a system in which all users are synchronized to one common clock (presumably located at the common receiver). Let the time scale now be divided into specified packet intervals or slots τ seconds long. An example is shown in Fig. 10. Users are constrained to transmit in these time slots only (as referenced to the common receiver time). It is apparent that "collisions" will now occur only if two or more packets arrive at the receiver during the same time interval. Since there can be no overlap of packets from adjacent intervals, the capacity of this scheme is exactly twice that of the pure ALOHA system. In particular, it is apparent that the system throughput S (in packet/slots) obeys the equation

$$S = Ge^{-G}$$

and

$$S_m = \frac{1}{e} = 0.368$$

Here G is again the average channel traffic (new packets plus retransmissions) in packets/slots. Poisson statistics have again been assumed for both the newly arriving and retransmitted packets.

Slotted ALOHA thus provides substantial improvement in allowable throughput over the pure ALOHA technique but at a cost in complexity: All transmitting terminals must be synchronized to the receiver.

Considerable thought has been given to the possibility of improving the throughput characteristic of ALOHA systems, and a variety of schemes have been proposed. These include several reservation ALOHA schemes in which users request slots in advance of transmission (note that these schemes are then similar to contention-type accessing schemes) and carrier-sense ALOHA in which users monitor other users' transmissions and transmit only when they sense no other transmissions are taking place. The former schemes, of course, trade off increased throughput for increased time delay. (The reservation request requires an additional two-way propagation delay.) The latter schemes have, in theory, been shown capable of exceeding a throughput figure of 0.8. The requirement here, however, is that user stations be close enough together so that the delay involved in listening to all other transmissions is a small fraction of a slot interval. As the propagation delay increases, the throughput improvement deteriorates.

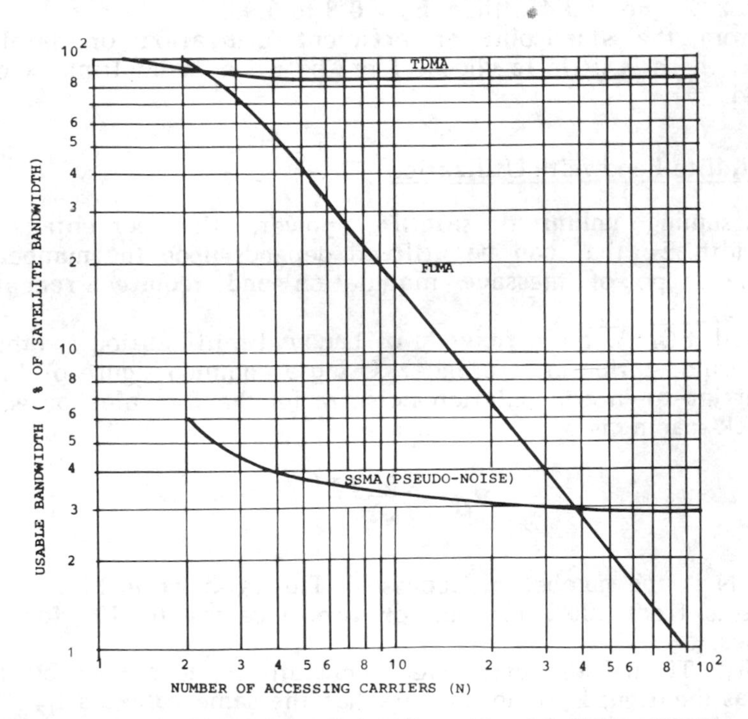

Fig. 11 Percentage of satellite bandwidth usable with FDMA, TDMA, and SSMA.

V. Comparison of Various Transmission Systems

The comparison among various transmission systems[11] must consider, in particular, the satellite power utilization, satellite bandwidth utilization, and vulnerability to interference.

A. Satellite Power Utilization

Assuming unlimited satellite bandwidth, the percentage of satellite output power E_P that can be used for information transmission is mainly dependent on the suppression effects in the hard limiter for FDMA and SSMA. On the assumption that the input level of any one carrier is not greater than the sum of all other carrier levels, about 20% of the satellite e.i.r.p. would be used for transmitting intermodulation products. If one carrier dominates, the wasted power is lower. Hence, for FDMA or SSMA, $E_P = 0.8$.

For TDMA, the problem of intermodulation does not arise. However, guard time between transmissions and link synchronization time is required. This is equivalent to a lost power of 5 to 20%. For TDMA, then, $E_P = 0.8$ to 0.95.

From the standpoint of efficient utilization of satellite e.i.r.p., there is little to choose from between the multiple-access methods.

B. Satellite Bandwidth Utilization

Assuming unlimited satellite power, the percentage of bandwidth E_B that can be utilized depends upon the number of carriers, type of message modulation and required reception quality.

With FDMA, the efficiency of bandwidth utilization is limited by intermodulation interference. Using a common figure of 14 dB for carrier-to-intermodulation-noise ratio, the E_B value based on Babcock spacing is

$$E_B = \frac{3.33}{N^{1.3}}$$

where N is the number of accesses. This is shown in Fig. 11. E_B decreases from 100% for one or two accesses to 1% for 100 accesses.

With TDMA, the effective bandwidth is decreased by lost time, as mentioned previously. E_B has the same value as E_P. For SSMA, the value of E_B is equal to the ratio of the hypothetical bandwidth occupied by accessing carriers with message modulation only to the satellite channel bandwidth. This is given by

$$E_B = \frac{3\pi}{16\ R} \bullet \frac{1}{1-1/N}$$

where R is the minimum acceptable C/I ratio, and N is the number of accesses. Assuming a 14-dB C/I ratio, E_B approaches 3% as shown in Fig. 11.

Figure 11 shows that TDMA gives the most efficient use of bandwidth and is power-limited. FDMA is power-limited if the number of accesses is small. It provides greater efficiency than SSMA if the number of accesses is less than about 40. SSMA is bandwidth-limited and its capacity tends to be independent of the number of accesses.

C. Resistance to Interference

From the point of view of mutual interference between accessing users, TDMA is superior to FDMA or SSMA, because TDMA, when properly synchronized, is free from mutual interference if each user is assigned his own time slot. However, from the point of view of intentional interference (jamming), SSMA provides the best protection.

FDMA is very vulnerable to selective jamming of a particular carrier if the jammer power P is equal to or greater than the terminal power P_g. The power required to jam all terminals is $P_{jt} = NP_g$, where N is the number of simultaneous accesses.

With TDMA, the jamming power for complete jamming is $P_{jt} = P_{gp}$, where P_{gp} is the peak power of each terminal in the case of peak-power-limited terminals. In the case of mean-power-limited terminals, $P_{jt} = NP_g$, where P_g is the mean power per terminal. Thus TDMA is about as vulnerable to jamming as FDMA.

For SSMA, the effective jamming power is reduced by the processing gain. If the information rate is reduced, the processing gain is increased. To jam SSMA completely would therefore require radiated powers much greater than NP_g. Selective jamming would be very difficult and would require knowledge of link synchronization timing and the pseudorandom code sequency.

D. Transponder Capacity Evaluation

Table 6 summarizes the evaluation of traffic capacity in terms of the number of telephone channels achievable in an INTELSAT V 80-MHz transponder with various combinations of multiplexing, baseband processing, modulation, and multiple-access (FDMA or TDMA) systems already in use or for future introduction in the INTELSAT system.

Table 6 INTELSAT V transponder capacity (number of channels in 80 MHz)

	Coding and modulation \ MUX	FDM	TDM	TDM/DSI
FDMA	FM	1000[A]	...	...
	CFM	2000[A]	...	...
	Hybrid modulation (2P-apm)	1700	...	...
	PCM/PSK	1500	1700	3300[C]
	ADPCM (32 kbits)/PSK	...	3600	6600[C]
TDMA	PCM/PSK	1500	1700	3300[B]
	ADPCM (32 kbits)/PSK	...	3800	6600[C]

[A]Techniques already in use.
[B]Techniques being introduced.
[C]Techniques under study for possible future introduction.

The main assumptions made for the above calculations are:

Transponder usable bandwidth	72 MHz
Total noise bandwidth	60 MHz
C/N available (clear sky)	15 dB
C/N_0 required for FDM/FM	63 dB-Hz
C/N_0 required for FDM/CFM	60 dB-Hz
C/N required for 2P-apm	17 dB
DSI gain	2:1
C/N required for TDM/PCM/QPSK (BER = 10^{-4})	15 dB
FEC rate	7/8
TDMA frame efficiency	99.5%
Actual bit rate	120 Mbit/s

It should be noted, however, that in selecting the appropriate combination of baseband processing/modulation/multiple-access techniques, other factors in addition to capacity per unit bandwidth must be taken into account, such as:

1) Transparency to various types of signals (voice, data, television, etc.).

2) Interface with terrestrial links.

3) Operational flexibility.

4) Overall system cost, including Earth station equipment cost.

5) Type of traffic (preassigned or on demand, occasional service, trunk to a single destination, or small streams to many destinations).

6) Satellite beam coverage (global or limited coverage).

Taking into account all previous factors, the INTELSAT voice traffic is expected to gradually migrate in the next few years from an analog system (FDM/FM/FDMA) to a digital system (TDM/DSI/TDMA, with SS/TDMA starting with the INTELSAT VI generation). It is interesting to note that most of the capacity increase from the INTELSAT V-A to the INTELSAT VI generation is due to the higher penetration expected for TDMA.

This will bring the nominal capacity from about 15,000 telephone circuits with approximately 20% TDMA penetration for INTELSAT V-A to about 35,000 telephone circuits with almost 70% TDMA penetration for INTELSAT VI. This capacity could be further increased with the introduction of 32 kbit/s ADPCM on the TDMA system, which might take place starting from 1986 or 1988. Small traffic streams will mostly be using SCPC/PSK carriers or companded FM carriers. Trunk telephone traffic could use in a very efficient way a TDM/DSI/ADPCM/PSK/FDMA transmission system.

The transmission of television signals through the INTELSAT satellites has been performed up to now using FM. However, the recent consideration of the introduction of video-teleconferencing as part of INTELSAT services will make attractive the use of digital coded video/PSK transmission systems at a transmission rate of 2 Mbit/s, while several techniques are under study to transmit high-quality television at data rate of 30 Mbit/s or lower.

Acknowledgments

The author wishes to acknowledge that the reproduction in this chapter of material taken from the publications of the International Telecommunication Union (ITU), Place des Nations, CH-1211 Geneva 20 Switzerland, has been authorized by the ITU.

Substanial use has been made of material from: Mischa Schwartz, COMPUTER-COMMUNICATIONS NETWORK DESIGN AND ANALYSIS, © 1977. Reprinted by permission of Prentice-Hall, Inc., Englewood Cliffs, N.J.

References

[1]CCIR, Vol. IV-1, Rep. 708-1, "Multiple Access and Modulation Techniques in the Fixed-Satellite Service", 1982.

[2]Quaglione, G., "Evolution of the INTELSAT System from INTELSAT IV to INTELSAT V," Journal of Spacecraft and Rockets, Vol. 17, March-April 1980, pp. 67-74.

[3]Welti, G.R., "Application of Hybrid Modulation to FDMA Telephony via Satellite," COMSAT Technical Review, Vol. 3, Fall 1973, pp. 419-429.

[4]Welti, G.R., "Pulse Amplitude and Phase Modulation," Second International Conference on Digital Satellite Communications, Paris, Nov. 1972, pp. 208-217.

[5]Welti, G.R., "TDM/PCM/FDMA Telephony Using 2P-APM," Fourth International Conference on Digital Satellite Communication, Montreal, Oct. 23-25, 1978, pp. 51-57.

[6]Schwartz, J.W., Aein, J.M., and Kaiser, J. , "Modulation Techniques for Multiple Access to a Hard-Limiting Satellite Repeater," Proceedings of the IEEE, May 1966, Vol. 54 no. 5, pp. 763-777. © 1966 IEEE.

[7]Edelson, B.I., and Werth, A.M., "SPADE System Progress and Application," COMSAT Technical Review, Vol. 2, Spring 1972, pp. 221-242.

[8]Dicks, J.L., and Brown, M.P., "Frequency Division Multiple Access for Satellite Communication Systems," Electronics and Aerospace Systems Conference, Oct. 1974, pp. 167-178.

[9]Dicks J.L., and Schachne, S.H., "The Use of SCPC Within the INTELSAT System for International and Domestic Telecommunications Services," Electronics and Aerospace Systems Conference, Sept. 1977, pp. 13-3A-13-3G.

[10]Utlaut, W.F., "Spread-Spectrum Principles and Possible Application to Spectrum Utilization and Allocation," Telecommunication Journal, Vol. 45, Jan. 1978, pp. 20-32.

[11]Webb, P.R.W., "Military Satellite Communications Using Small Earth Terminals," IEEE Transactions on Aerospace and Electronic Systems, Vol. 10, May 1974, pp. 306-318. © 1974 IEEE.

[12]Pontano, B.A., Forcina, G., Dicks, J.L., and Phiel, J., "A Description of the INTELSAT TDMA/DSI System," Fifth International Conference on Digital Satellite Communications, Genoa, Italy, March 23-26, 1981.

[13]Schwartz, M., Computer Communication Network Design and Analysis, Prentice-Hall, Englewood Cliffs, N.J., 1977.

9. INTELSAT's Operations

H. W. Wood*
INTELSAT, Washington, D.C.

Introduction

From the earliest days, satellite and submarine telephone cable systems were recognized as part competitive and part complementary. A practical demonstration of this came at the very outset of INTELSAT's service. On 18 June 1965, before regular commercial services started, the Early Bird satellite restored services interrupted by the failure of the transatlantic cable system. Such restorations are now just part of a day's work for INTELSAT operations.

Santiago Astrain
Director General (1976–1983)
Secretary General (1973–1976)
INTELSAT

The first of the INTELSAT II satellites was nominally a launch failure (because it was launched into a highly elliptical orbit rather than a circular geostationary orbit). On 24 and 25 November 1966, it became co-visible to the Australian station at Carnarvon (Western Australia) and the U.K. station in Goonhilly (Cornwall, England). Carnarvon was at that time a truly remote area without television interconnections with the rest of Australia, let alone the rest of the world. Three British families, who had settled in this remote area, were hurriedly rounded up to "speak to and see" relatives gathered in the BBC studios in London. More than a decade of isolation and silence was dramatically broken by families reunited by what seemed to them an electronic miracle.

Graham Gosewinckel
President of AUSSAT
Formerly of Overseas
Telecommunications
Corporation
(Australia)

Invited paper received October 11, 1983.

*Deputy Director General, Operations and Development. (Deceased: March 18, 1984.)

Early Bird was the first satellite to provide "live" transatlantic communications by establishing 66 telephone circuits working twelve hours a day starting in July 1965. As summed up in an understated press account at the time, "In the United States and abroad, public and press interest in the Early Bird was intense."

Joseph V. Charyk
Chairman and Chief
Executive Officer
COMSAT

Television via satellite has always been extra special to us. The first live commercial TV broadcast was a black-and-white transmission of a Le Mans race in France on 20 July 1965. After that, no one ever looked back. Perhaps the proudest moment has to be the worldwide distribution of the moon walk, telecast on 20 July 1969, when we established a new world record of a viewing audience of 500-million. Today, we have more than a billion people watching the Olympic Games and the World Cup Matches, but no one can ever duplicate the global excitement and thrill of that first moon walk "Live Via INTELSAT."

George Lawler
Vice President, Marketing
COMSAT

As the preceding observations of some of the pioneers in satellite communications make clear, the growth and development of INTELSAT's "conventional" services have been exciting, challenging and fast-moving. It is only a slight exaggeration to say that the growth of these services and their global impact are unique in the history of international telecommunications. Seldom has any innovation anywhere affected so many people in so little time.

INTELSAT's basic responsibility, simply stated, is to make the communications capability of its satellites available to users when and where they need it, and to do so in the most readily available and convenient form. The definition of Earth station performance characteristics, the establishment of associated transmission parameters, the setting of contractual terms and conditions, and the development of tariffs are actually ancillary, albeit essential, activities.

Now, to operate a successful bus company, one needs not only buses, but drivers, route planners, maintenance personnel, accountants, and administrators. All of these, however, are ultimately of little importance if there are no passengers on the buses. Telecommunications services are the passengers on INTELSAT's bus line. The history of INTELSAT's "conventional" services is, in many ways, akin to the building of bigger and better buses to accommodate a growing number of passengers more efficiently.

The operational techniques and practices that have contributed to the commercial exploitation of communications satellites on a global scale are critical to understanding how INTELSAT services developed. In the following section every key aspect of INTELSAT's operations is presented, while in the following chapter the rapid evolution of INTELSAT's exciting new services of the 1980's and 1990's will be explained.

INTELSAT Charges

INTELSAT's basic service unit is defined in the INTELSAT Tariff Manual as follows:

> The unit of satellite utilization is the measure of entitlement, secured through the allotment by INTELSAT of space segment capacity, to the use of such capacity for the establishment of one end of a two-way 4-kHz telephone circuit providing, as an objective, quality of service in accordance with appropriate CCITT/CCIR recommendations, by means of:
>
> (i) access to a satellite in the multi-channel FDM/FM mode, and
>
> (ii) an approved Standard A or C Earth station.
>
> Such a circuit may also be used for telephony, voice frequency telegraphy, facsimile telegraphy and data signals. Further, a combination of such circuit capacity can be used to derive one-way, two-way or multi-destination audio, stereo audio, or television services.

These, then, are the "conventional" space telecommunications services provided by INTELSAT, starting in 1965.

With the passing of time, and as new service requirements arose, new modulation techniques were introduced. The unit of utilization was used as the basis for developing differential charges for Earth stations of different sizes. Different types of Earth stations required different signal strengths to achieve comparable quality of service and, inevitably, Earth stations of different sizes became necessary. Those new developments of satellite technology and diverse service requirements led to the concept of "equivalence of satellite resource utilization" and the need to apply "rate adjustment factors" that were multiples of the basic unit.

Thus, the smaller the Earth station, with lower gain or effective performance, the larger the rate adjustment factor applied to establish equitable and "equivalent" charges based upon how much satellite power and bandwidth were required to provide service.

Figure 1 shows the mainstream services now carried in the INTELSAT system. (A mainstream service is one that contributes 1% or more to INTELSAT's total operating revenues.) The figure is of interest because it highlights the years when significant revenue shifts occurred, due either to the emergence of new types of service or to a significant increase in an existing service. Revenue from full-time telephony and data service constituted almost 100% of the revenue until 1968, when it dropped to 95% as a result of temporary services and television becoming "big business." Today however, a broad range of diverse new services is claiming an even greater proportion of total revenues.

Implementation of Services

Other chapters have described INTELSAT activities that led to the establishment of services in accordance with generally recognized international standards of quality. Those activities centered around satellite design and maintenance of satellites in orbit and standardization of Earth station interface with the space segment. It now remains to describe how satellite capacity is shared and the processes by which quality of service is maintained, and to demonstrate the practical achievements of the INTELSAT system.

Satellites are designed to provide certain permutations of interconnectivity broadly congruent with the traffic projections received from users. Switching elements in a specific satellite are set to connect the receive and transmit antennas on the satellite to provide appropriate proportions of capacity in the east/west, west/east, west/west, east/east, and global modes of access. Frequency-division-multiplexed/frequency-modulated (FDM/FM) carriers (which carry a significant but decreasing proportion of all international traffic in the system) are sized and geographically configured to meet the traffic requirements projected for the following two or three years. The link parameters for each carrier are iteratively optimized to ensure that the worst-located Earth station within the relevant area of beam coverage receives adequate power from the satellite. In practical terms, the noise level in any FDM/FM carrier is set to 7500 pW (out of the end-to-end budget of 10,000 pW), and the process of iteration achieves this objective in all carriers in the same transponder after making appropriate allowances for internal degradation [intermodulation, amplitude-modulation/phase-modulation (AM/PM) transfer], thermal noise in the uplink and downlink,

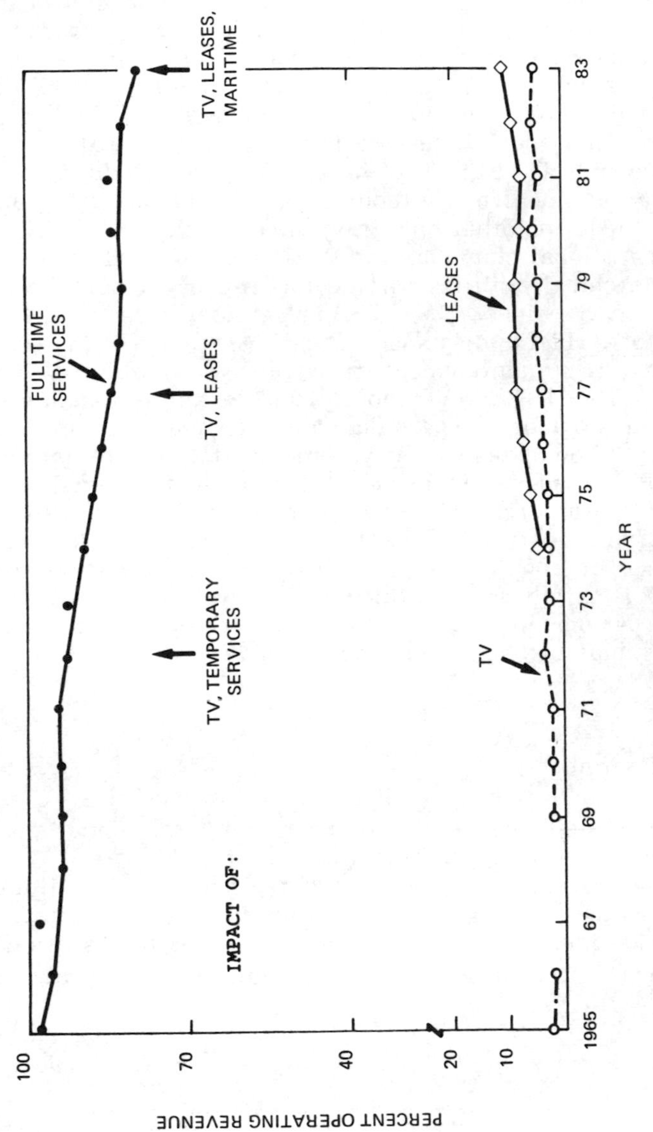

Fig. 1 INTELSAT mainstream services.

cross-polarization effects in feed systems at the Earth station, frequency reuse interference, and, to some extent, even interference from satellites of other satellite systems.

For purposes of this preoperational optimization, an extensive data base of satellite characteristics (prelaunch and postlaunch) and a corresponding data base of Earth station characteristics using nominal characteristics are used. During the course of practical operations, the actual measured characteristics of Earth stations, which are, in general, better than the nominal values, are invoked to exact the full service potential from a satellite. These processes enable INTELSAT to handle growth of individual services from different countries without frequent readjustment and with minimum impact on other operating Earth stations.

Preoperational planning of this kind is by itself not adequate without matching facilities for the monitoring of performance and midcourse corrections. INTELSAT services are utilized by a network of widely dispersed Earth stations with significant differences in maintenance facilities and on-site technical expertise. The INTELSAT monitoring system seeks information from Earth stations on periodic routine measurements, while a sophisticated communications system monitor (CSM) maintained by INTELSAT provides extensive data on measurements of all operational parameters. These data from the Earth stations enable INTELSAT to identify incipient faults and to track sources of interference or abnormal degradation. The information also helps in studying satellite performance. Preventive maneuvers are executed periodically to change satellite operating parameters (gain steps, bias of beams) in order to maintain the desired quality of service.

Quality of Services

Table 1 summarizes the measurements of signal quality in FDM/FM services during a typical month of 1983.

An indication of limiting performance is obtained by determining the number of carriers in the system operating with out-of-band noise (OBN) at the worst-located station. This performance criterion is very conservative, since each carrier

Table 1 Signal quality of FDM/FM services during typical month of 1983.

Number of operating satellites	6	
Number of operating FDM/FM carriers	489	
Number of receiving Earth stations	1944	
Number of Earth stations receiving signal quality better than the normal standard (-50 dBm0p)	1380	(71%)

serves a number of destinations, many of which, because of their geographical location within the satellite beam area, receive signal quality better than the required standard.

Table 1, a subset of monitoring data on the performance of the Atlantic Ocean region Major Path I satellite, shows that the percentage of Earth stations reporting out-of-band noise levels equal to, or better than, the specification level of -50 dBm0p is above 85%. Particularly noteworthy is the fact that, in transponders, the range of OBN even at the worst-located station drops significantly below the specified level of -50 dBm0p.

Reliability of Service

INTELSAT operational practices are structured to maximize service reliability. The concept of an in-orbit spare, for example, was adopted very early in INTELSAT's history, and, in the course of time, satellites of different generations served in regular operating positions--thus leading to the practice of maintaining a spare of each type in orbit.

The operating lifetime of a satellite is carefully controlled at each stage, and, in deployment plans, orbital shifts of satellites are minimized. During regular operation, maneuvers correcting orbital parameters are minimized by tailoring the corrections to service requirements (e.g., coverage area, stationkeeping). In some cases, satellite transponders not required for active service are switched off. Such measures have paid dividends in the form of useful service life extending beyond design life, as Table 2 illustrates.

Continuity of Service

In cooperation with Earth station operators, INTELSAT maintains records of service continuity in three areas: space segment, Earth segment subsystem outages, and path continuity. Path continuity represents the weighted impact of an outage in either the space or Earth segment on the circuits handled by the affected part of a satellite path. Figure 2 shows the dramatic decrease in satellite-related outages since 1971.

Figure 3 is a graphic representation of system outages in Earth station equipment in the INTELSAT system. The plots confirm the improvements in reliability (well known to industry participants) of Earth station equipment, and spotlight the significant reduction in the number of outages per Earth station with regard to every subsystem.

Figure 4 is a plot of the percentage path continuity of service. The path continuity of service steadily improved to 99.8% by 1971 and has remained between 99.8% and 99.98% ever since. This record is a combination of improved reliability of satellites,

Table 2 Satellite service life extending beyond design life.

Type	Number placed in orbit	Number in service	Achieved average life to end-of-life or to date (6/83)
INTELSAT IV	7	2 still in service	116 months
INTELSAT IV-A	5	all in service	78 months
INTELSAT V	6	all in service	16 months

[A]Nominal design life is seven years in all cases.

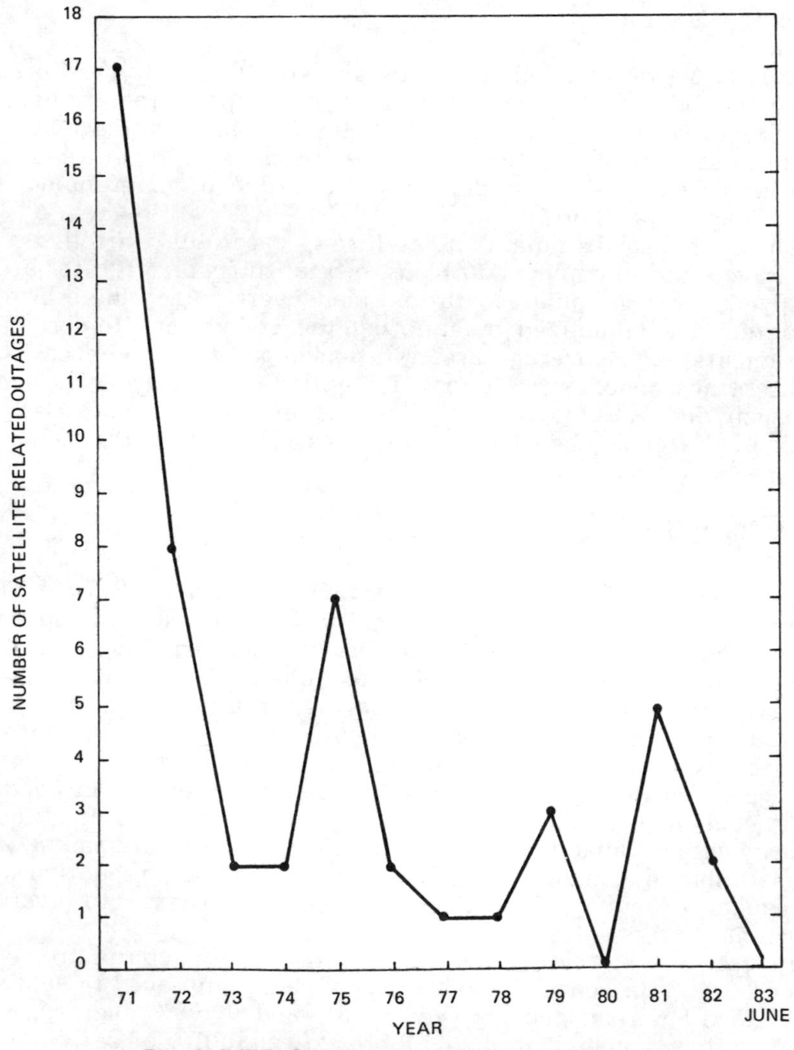

Fig. 2 INTELSAT system satellite-related outages.

as well as specific operational procedures to reduce, and even eliminate, the need to interrupt service. The most recent factor in reducing interruptions is the collocation of satellites with phased telemetry control of the two satellites involved in a planned change of satellite. This procedure significantly reduces interruptions due to pointovers of large numbers of Earth stations.

System Management

The objective of system management is to make maximum practical use of satellites so that services are established in accordance with approved policies and tariffs. System management in this context covers four areas: satellite operations, operational plans, operational coordination and implementation, and utilization and billing.

Satellite Operations

When postlaunch in-orbit tests on a satellite have demonstrated that the satellite is functioning properly, the satellite is maneuvered to its assigned operating location, which is part of a predetermined deployment plan that is based on the utilization of all satellites in orbit and on the proposed utilization of satellites due to be launched in a given time frame.

Conceptually, a deployment plan reflects the likely pattern of use of satellites currently under procurement. Conceptual plans are based on a set of assumptions, typical of which are 1) seven-to-ten-year design life in orbit of all satellites; 2) projected availability of launch dates; 3) rate of launch successes; 4) need to use multiple-purpose satellites (e.g., the Maritime Communications Subsystem) in some locations; and 5) projected dates by which existing satellites will be unable to accommodate further significant growth in user requirements (briefly referred to as "saturation dates").

Each of these assumptions could change with time for entirely independent reasons. The projected saturation dates could, in some cases, be extended by means of operational solutions (increased use of TDMA/DSI, for example), whereas unanticipated growth in services could bring the date forward. The launch success rate is an uncertainty, and the failure of a particular launch obviously could lead to the delay of the succeeding launches. The current conceptual deployment plan consists of plans for location of over a dozen new satellites (excluding launch failures) and an even larger number of relocations of satellites of three generations over the next ten-year period.

These plans are reviewed and updated every time a new satellite is launched or whenever a satellite must be relocated in order to fulfill a different role in the network. This practice

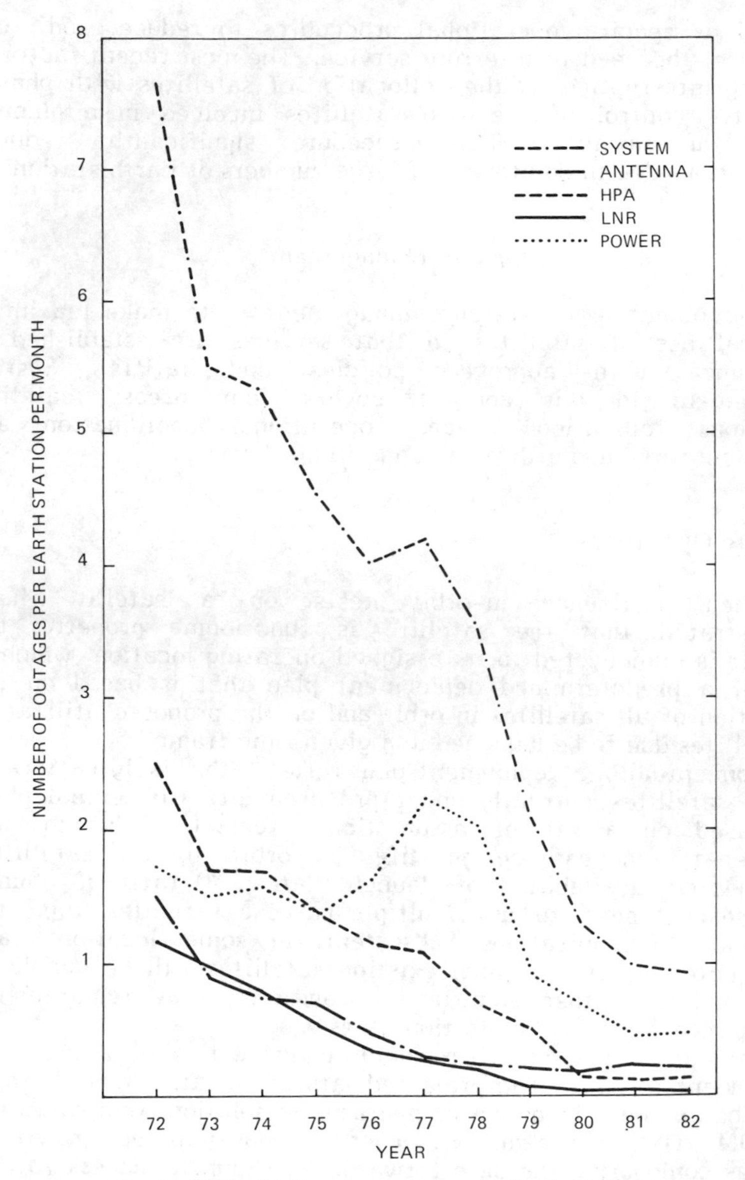

Fig. 3 INTELSAT system outages in Earth stations.

ensures that the use of satellites is optimized to meet global requirements and that users receive adequate advance information that may impact on their plans to construct matching Earth segment facilities (e.g., Earth stations, TDMA terminals).

Satellites are maintained in orbit within close tolerances of stationkeeping accuracy, and the telemetry systems are monitored

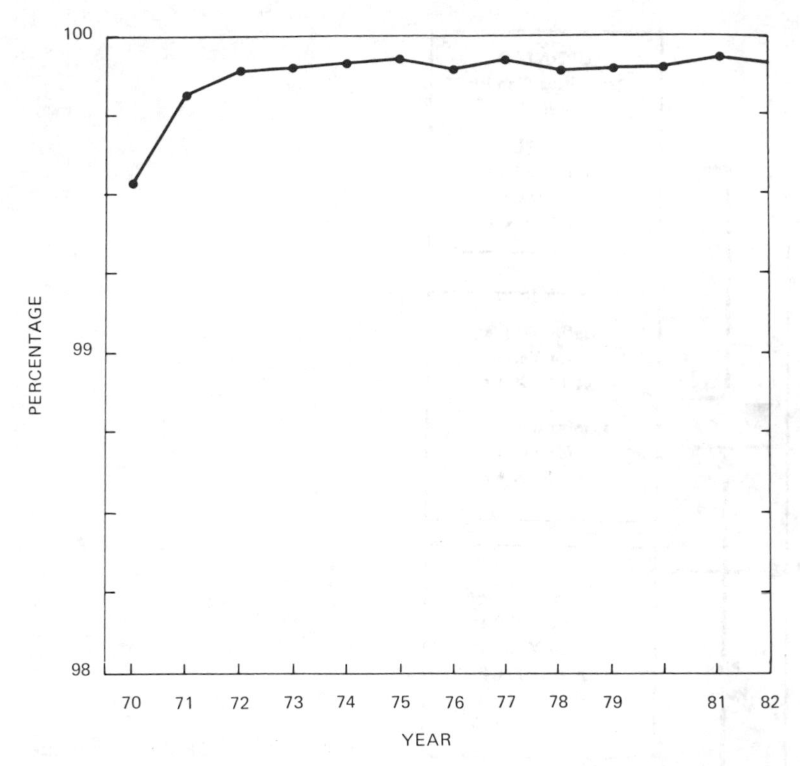

Fig. 4 INTELSAT services path continuity of service.

continuously to guard against unexpected anomalies. These operations are supported by the network of tracking, telemetry, command and monitoring (TTC&M) stations described in Chap. 6, and they are provided with sophisticated computer software for a variety of purposes, such as the generation of commands for routine maneuvers, real time monitoring of key parameters of satellite performance, and retrieval of historical performance and test data for the evaluation of anomalies.

During a recent month, the following types of spacecraft-related activities were performed as part of service maintenance: 1) one satellite was drifting at 1.5 deg/day to a new location; 2) 42 reorientation and velocity correction maneuvers were performed; 3) interconnectivity patterns in two satellites were completely altered and one less extensive change was made in another satellite; and 4) three changes were made in operating point or amplifier switching.

Operational Plans

Operational plans for the system deal with all aspects of satellite utilization. They summarize assignments of capacity in

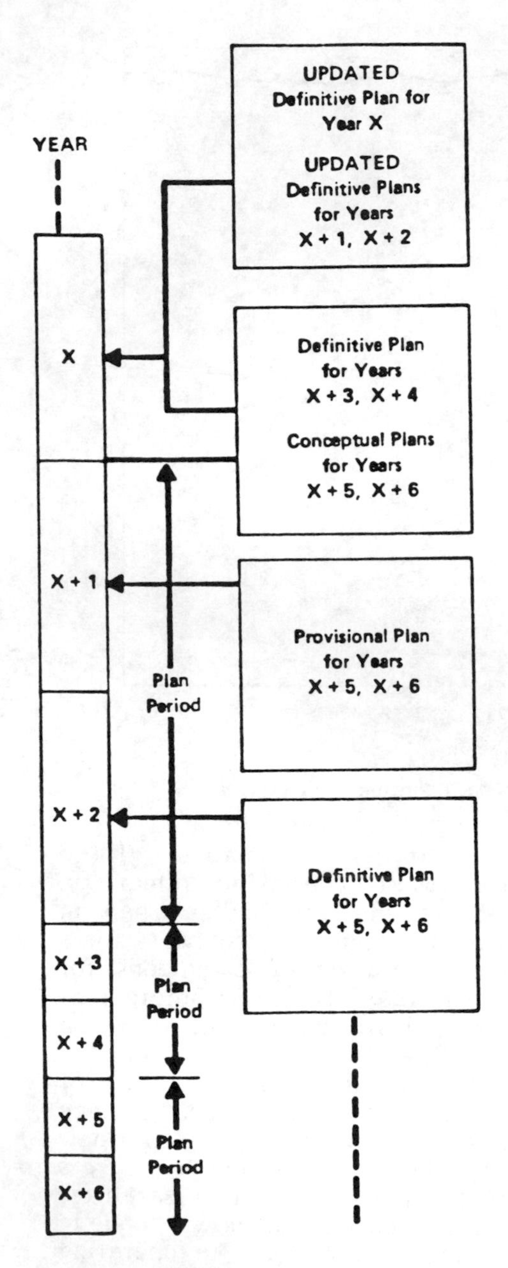

Fig. 5 Nominal planning cycle.

existing satellites in accordance with user forecasts of full-time services, approved allotments for leases and allocations for occasional use services, as well as other services that may arise from time to time. Estimates of "saturation" dates are an important byproduct of this activity. The operational plans have, as a subset, the conceptual satellite deployment plans to meet the requirements of all users, including INTELSAT's contractual commitments to INMARSAT for maritime services. Finally, the operational plans also consider the impact of service development to the extent that it requires the incorporation of new technical features or a choice of new orbital locations.

The operational plans are synthesized from component plans for each satellite and each region that determine the practical interconnectivity within each satellite, and are used to assign segments of transponder capacity to each accessing Earth station in accordance with the users' forecasts.

In practice, capacity assignments cannot be matched precisely to forecasts, and an accepted guideline is to size assignments to within 10% of requirements. It is equally unlikely that the forecast rate of overall growth is followed by all Earth stations. The operational plan for each satellite is normally designed to serve the system for at least two years, and operational entities agree on the implementation schedule about a year prior to implementation. Since each plan may require augmentation of equipment (additional or expanded carrier) or installation of new systems (e.g., TDMA), the planning process has to provide adequate lead time. For this reason, the planning cycle covers up to six years: the conceptual phase followed by the implementation phase. Conceptual plans are

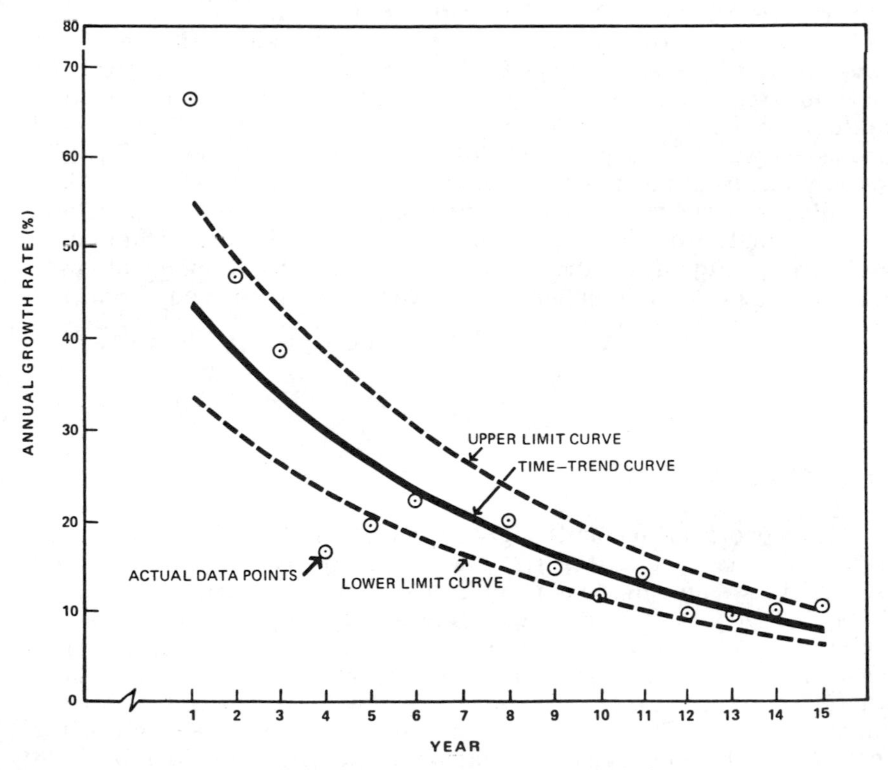

Fig. 6 Time–trend analysis of annual growth rates.

intended to help Earth station owners make arrangements for procurement of Earth station equipment and to complete other interface arrangements with the distribution network. Figure 5 is a schematic of this planning cycle.

Forecasts

Every year, INTELSAT invites the community of telecommunications user entities to agree on forecasts of service requirements for the ensuing five years.

Long-term requirements for the subsequent ten years are also solicited, either as agreed estimates or, if this is not achievable, on the basis of the demand preceived by each user entity. A 15-year forecast aggregate is thus built up and used for planning purposes.

This pragmatic approach to forecasting results in a planning tool that directly relates to the intended use of the system. The forecasts for each link are custom-tailored to the circumstances at each end and lend themselves to bilateral adjustments in timing. INTELSAT acts as a clearinghouse for this information and coordinates matching changes in capacity assignments.

Long-term estimates are less precise, but since they are also developed by link and category of service, a statistical analysis of built-in trends can yield indicators that do not rely on external factors, which themselves are hard to forecast. These trend analyses establish upper and lower bounds for probable traffic growth and are used to test satellite design against traffic growth.

Figure 6 shows the trend lines and the associated upper and lower bounds for long-term growth rates derived from user estimates. Figure 7 shows a comparison, by region, of user forecasts against the upper and lower bounds derived from the trend lines of Fig. 6.

System Operations

The operational plans agreed with Earth station operators indicate, for each satellite, the frequencies, capacities, and baseband assignments. Prior to implementation, these plans have to be individually analyzed for deriving the optimized operating parameters--Earth station e.i.r.p.'s transponder input and output backoff--using measured values of satellite performance and, wherever necessary, measured values of Earth station characteristics. The plans are then converted into schedules for test and implementation with supporting facilities provided by CSM stations wherever necessary.

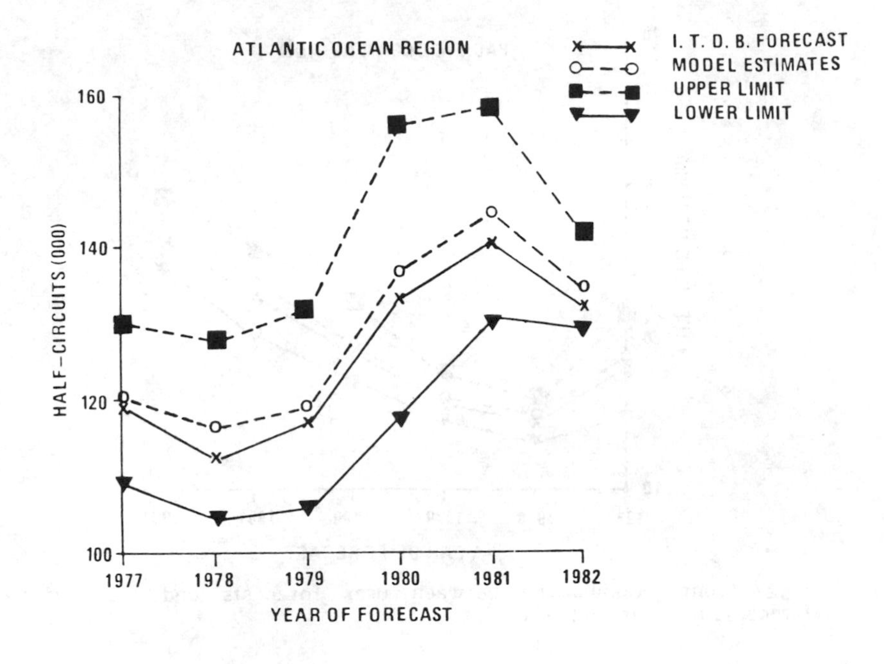

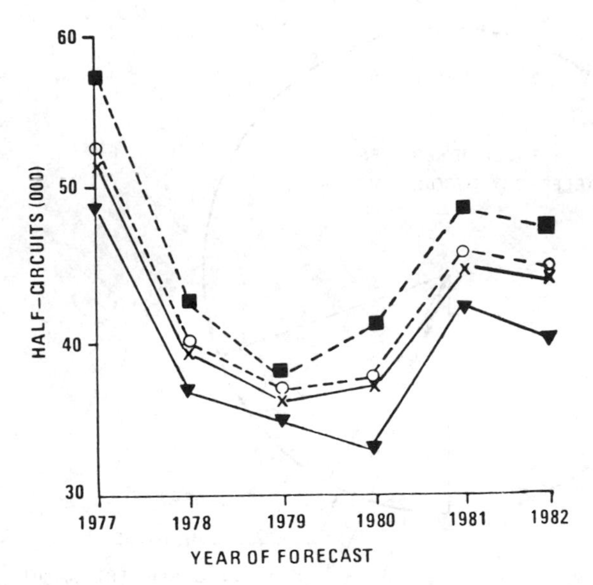

Fig. 7 Comparison between user forecasts and trend model forecasts for year-end 1990.

(Figure 7 continued on next page.)

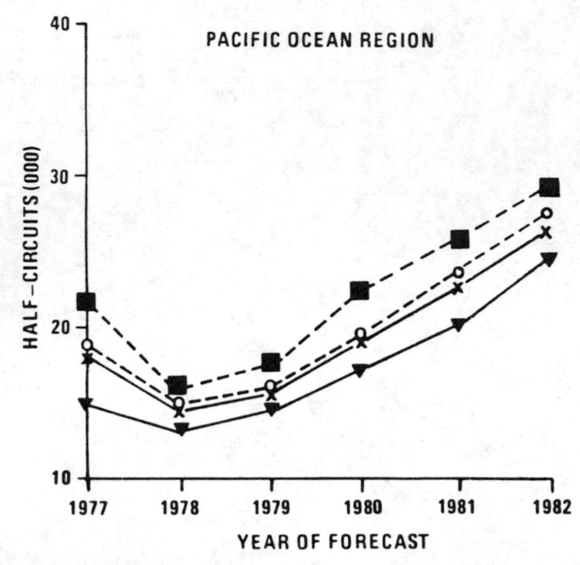

Fig. 7 (cont.) Comparison between user forecasts and trend model forecasts for year-end 1990.

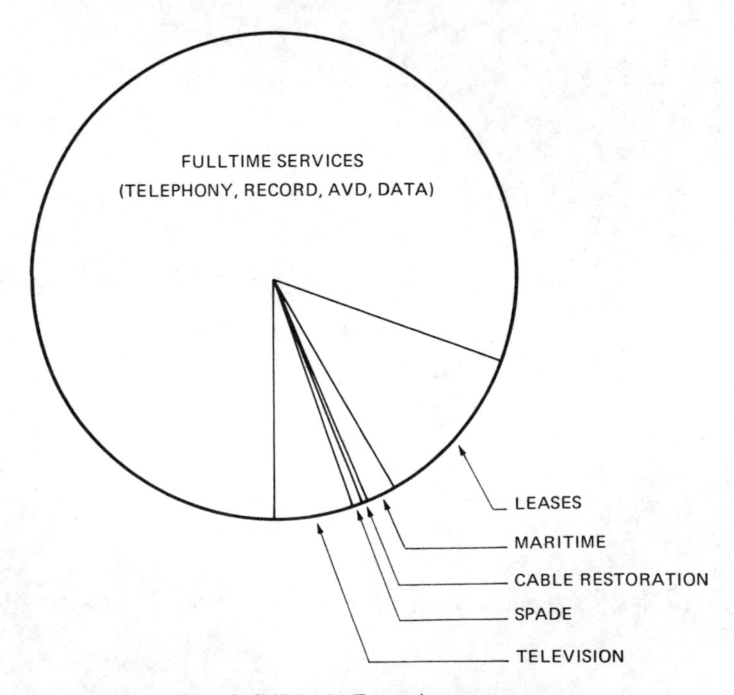

Fig. 8 INTELSAT services revenue.

Transition Plans

The implementation of an operational plan proceeds in a series of steps involving pairs of Earth stations. When a satellite is replaced by a satellite of a later generation, the changeover usually requires several intermediate steps, because the expansion of capacity splits the destinations served by a carrier into two or more and also because the frequency reuse capabilities in the new satellite provide beam interconnectivities different from those in the satellite that is being replaced. Since the planned replacement of a satellite generally takes place when its capacity is almost fully utilized, there is little spare capacity in it for intermediate steps. Two techniques are used to overcome this problem: the collocation of the relieving satellite and temporary transfer of part of the traffic to another operating satellite that may have some spare capacity. A typical transition plan could consist of more than 100 steps. A prime objective throughout is to eliminate, or at least minimize, interruptions to service.

Technical Operational Control and Coordination

The implementation of an operational plan requires the coordinated sequencing of carrier activations and testing prior to commencement of traffic. FDM/FM carriers, in general, are multi-destinational, and in order to minimize testing time several abridgement techniques are adopted, such as the conducting of a full range of tests for new carriers only and the testing with all destinations other than the prime one. The activities in each ocean region are coordinated by a technical and operational control center (TOCC), which schedules the test sequences and carrier

Table 3 Number of Earth stations tested between Jan. 1 and June 30, 1983.

New Standard A,B,C stations	15
Domsat system Standard Z	44
Nonstandard and coast stations	5
Retrofitted stations	9

Table 4 Full-time INTELSAT service paths as of Dec. 31, 1982.

	FDM/FM	SCPC
Atlantic Ocean region	479	95
Indian Ocean region	300	69
Pacific Ocean region	101	44
Total	880	208

activations well in advance and maintains close liaison with all Earth stations throughout the sequence of steps. The TOCC also reviews the test results in significant cases of departure from normal operations.

The INTELSAT Operations Center (IOC) is the focus of coordination during each stage of activation and testing. For this purpose, the IOC can access the spectrum display for each transponder on each satellite through a suitable gateway Earth station and receive measurements from a network of CSM stations on all parameters.

Earth Segment Facilities

INTELSAT's plans for meeting users' forecast requirements necessitate the use of multiple satellite configurations, frequency reuse techniques, and different kinds of modulation systems. The Earth segment facilities have to be correspondingly augmented, or retrofitted, from time to time. INTELSAT provides a continuous feedback to the user community regarding its implementation schedule and solicits information from users regarding the proposed plans for provision of corresponding Earth segment facilities.

Major augmentation of facilities includes activation of additional Earth stations or antennas, retrofitting of antennas for dual-polarization, and activation of TDMA terminals. In such cases, INTELSAT conducts a series of tests to verify that the Earth station equipment complies with the mandatory performance specifications. This precaution is essential to ensure that system integrity is not impaired due to improper operation of the new unit and that the interference caused to other satellite networks (including INTELSAT's own) is kept within prescribed limits. These verification tests do not relate to acceptance testing of the facility, for which the responsibility lies solely with the Earth station owner. INTELSAT, however, offers support facilities for testing through the satellites whenever requested by the Earth station owner. This support consists of 1) provision of carriers from a CSM station for test purposes; 2) measurement of antenna sidelobe levels; 3) reference station support for TDMA tests; and 4) evaluation of test results on a real time basis.

The level of this support activity to system management is demonstrated in Table 3.

INTELSAT provides a satellite system operations guide (SSOG) describing test procedures for common adherence by participating Earth stations. The test covers the verification of mandatory performance characteristics as well as select sets of system lineup tests. The SSOG generally contains test procedures in all cases that require INTELSAT support in minimizing operational delays in commencement of service.

Utilization of Full-Time Services

Figure 8 shows the major service categories contributing to INTELSAT revenues in 1983. As of Dec. 31, 1982, the INTELSAT system provided full-time service on 1088 paths (see Table 4).

The total global network of 493 Earth stations consisted of 178 6/4 GHz Standard A stations; 57 Standard B stations; three 14/11 GHz Standard C stations; and 255 Standard Z (DOMSAT) and nonstandard Earth stations.

The correlation between Fig. 9, showing the growth of Earth stations in the system, and Fig. 10, showing the growth of full-time half-circuits is striking.

Figure 11 shows the traffic interconnectivity in the Atlantic Ocean region and the number of Earth stations operating in each of the major subregions in this region.

Occasional Use Services

Among the occasional use services available to INTELSAT users, three are regularly and extensively used: SPADE, television, and cable restoration services.

SPADE

Demand for SPADE exists only in the Atlantic Ocean region. Thirty-four terminals are in operation, providing a demand for telephone service on some 450 paths. About 120,000 minutes of service is provided every day--representing an estimated 10,000 telephone calls.

Television

From Early Bird to INTELSAT V, television has played a conspicuous role in INTELSAT's public image. This service now provides about 6% of revenue, and in recent years has grown annually at a rate of 35-50%. Regular telecasts from the United Kingdom, the United States, Italy, Spain, and France to a large number of destinations are a daily feature. Approximately 2500 bookings are received each week, with the service being characterized by frequent and pronounced peaks of demand. The global interest in international events is directly evidenced in television activities: It is commonplace for INTELSAT to be involved in arrangements for major events up to five years in advance. It has also established special arrangements to ensure a response to any booking within two days and to assist in the advance planning of telecast requirements for events such as the Olympics, Wimbledon Tennis, and World Cup Soccer Matches.

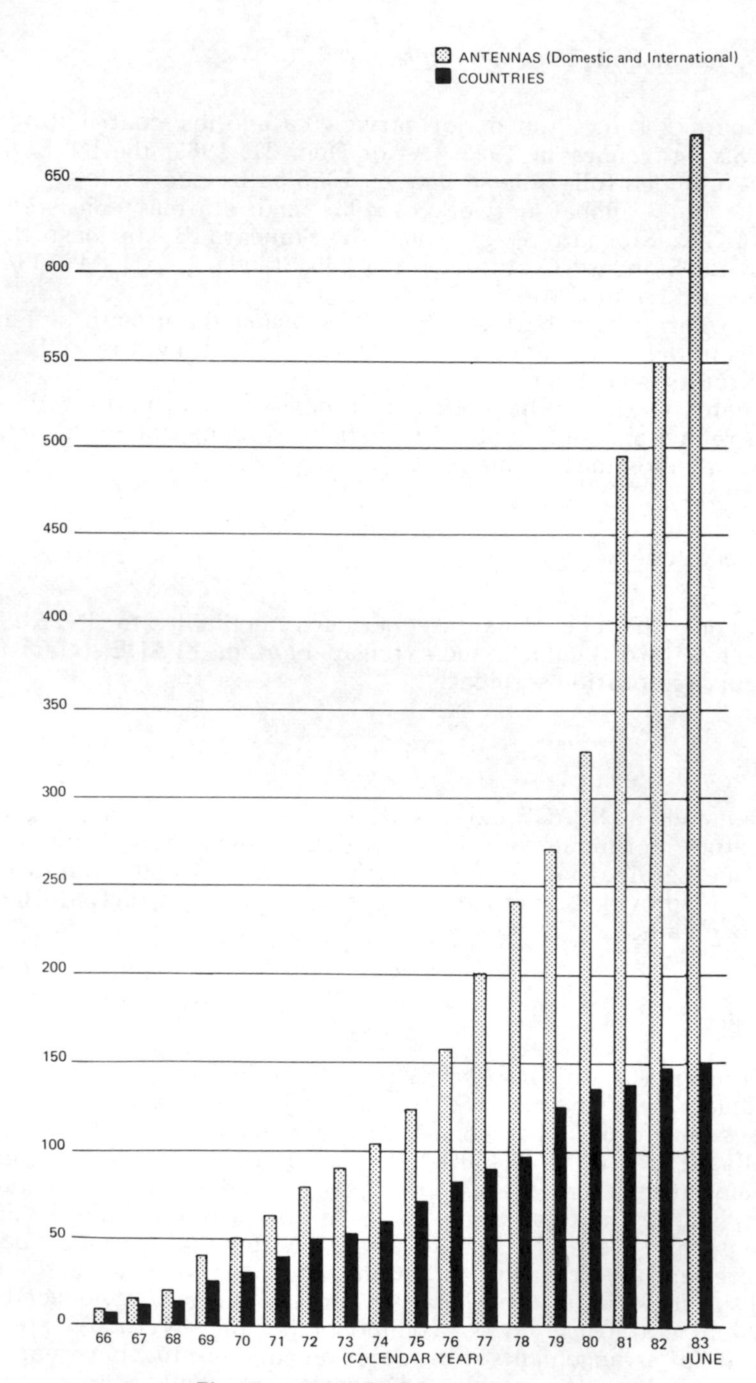

Fig. 9 INTELSAT system.

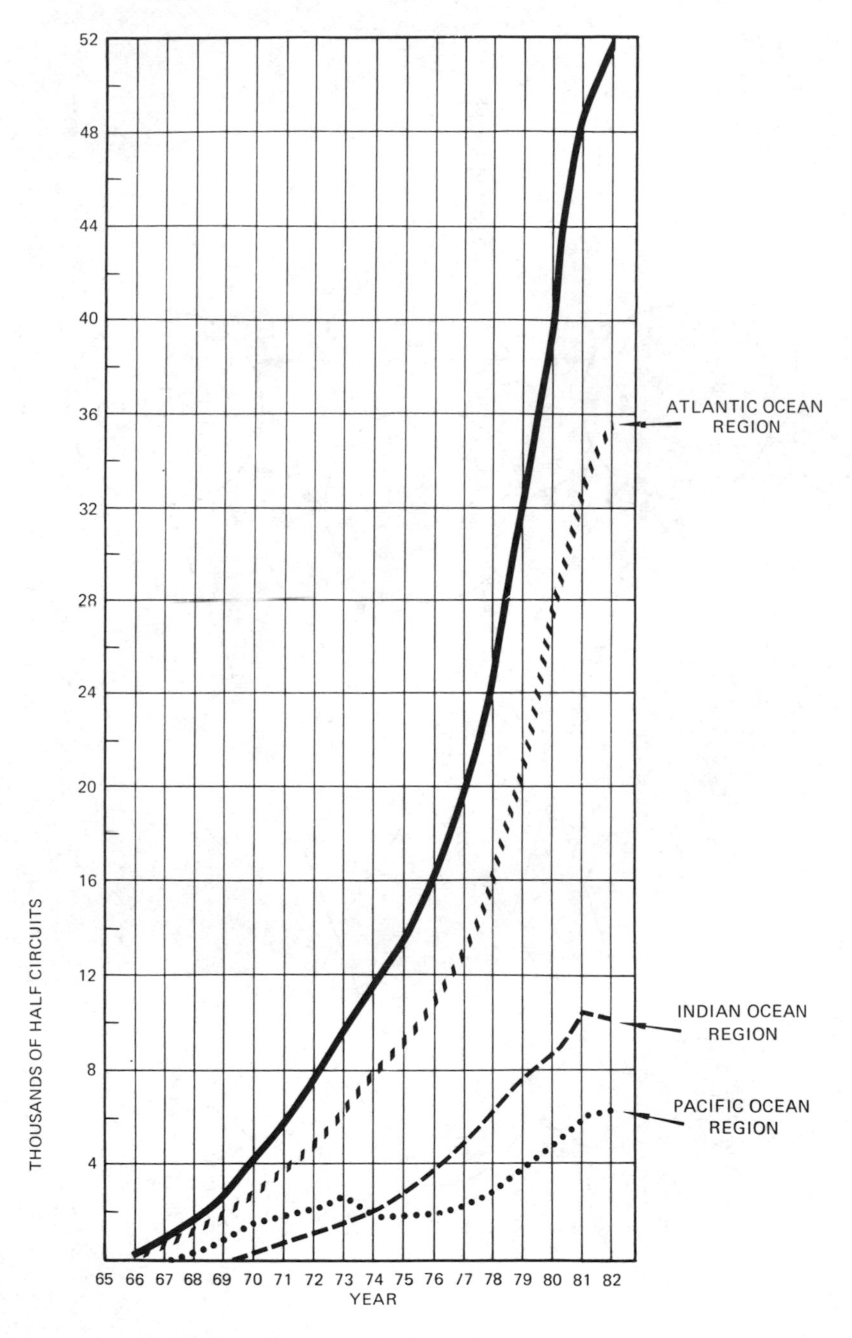

Fig. 10 Growth of full-time traffic in half-circuits.

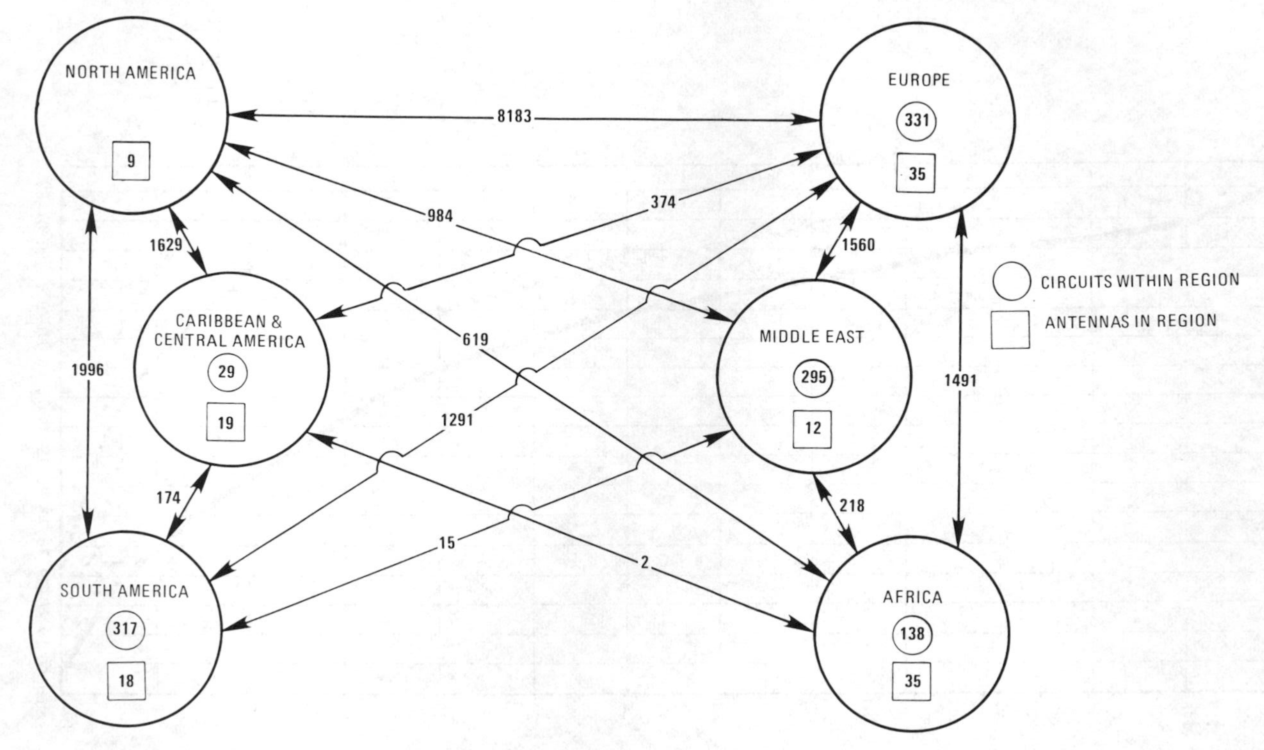

Fig. 11 Distribution of full-time circuits in Atlantic Ocean region (year-end 1982).

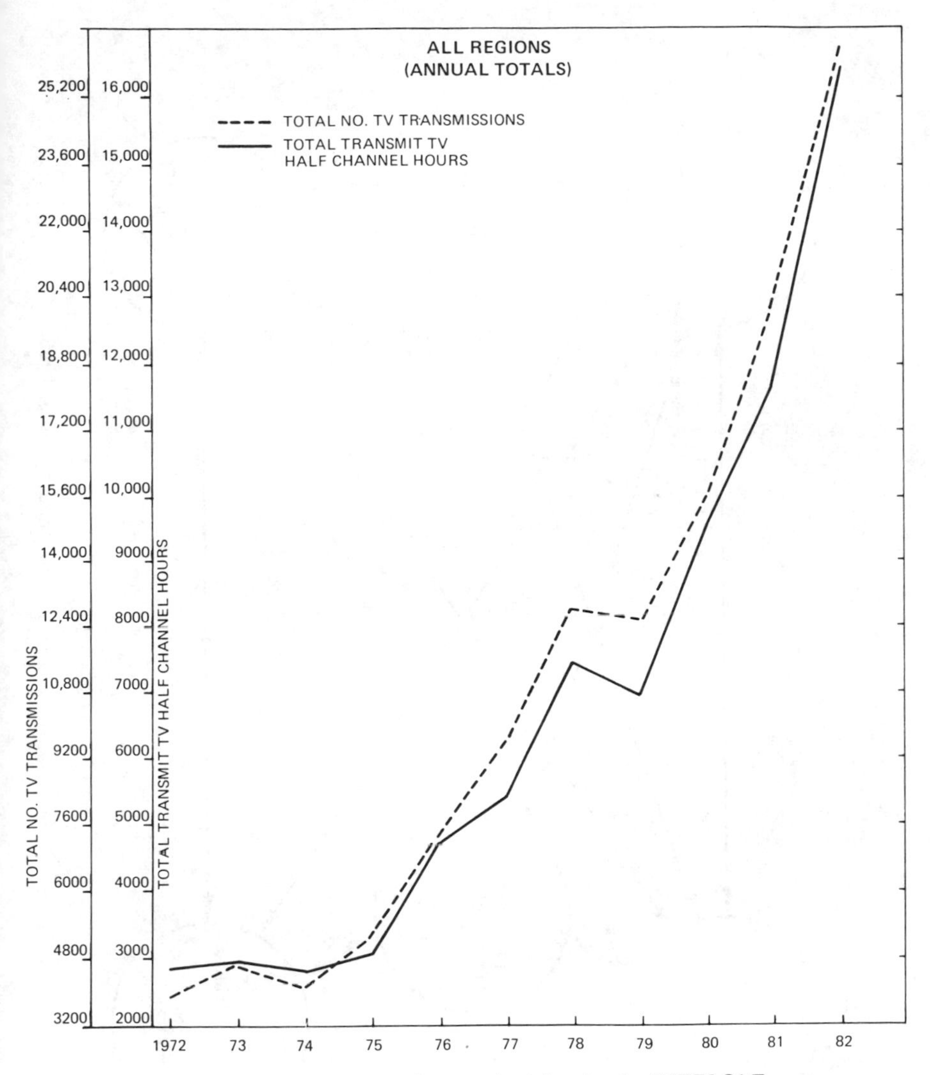

Fig. 12 Growth of occasional use television in the INTELSAT system.

Figure 12 shows the growth of television in recent years, and Figure 13 is an indication of global interest in television service.

Cable Restoration Services

INTELSAT maintains a close liaison with mutual aid groups established by owners of submarine cable systems. Plans exist for satellite restoration of major cable systems if any of them should

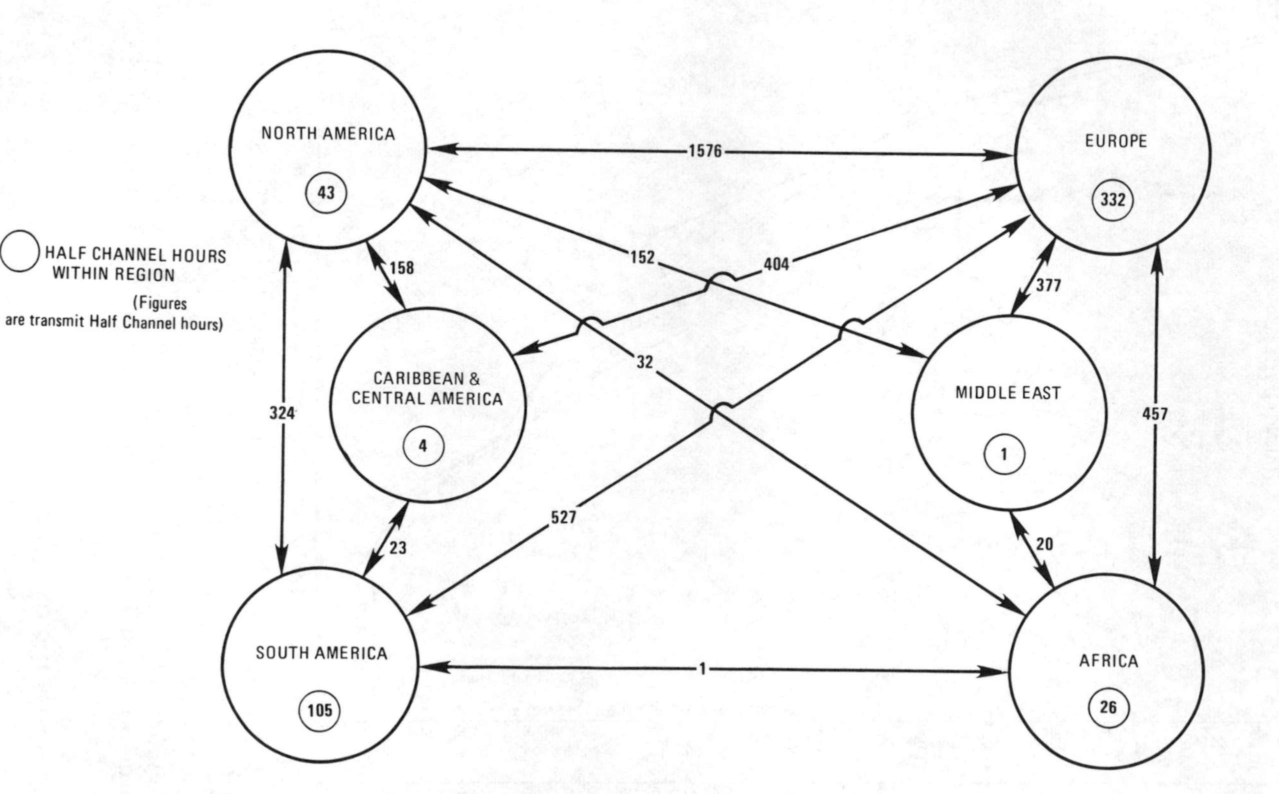

Fig. 13 Television service in the Atlantic Ocean region (January–June 1983).

suffer an unexpected interruption, with restoration usually achieved within an hour of receipt of the request by the INTELSAT Operations Center. There are also instances of other terrestrial systems or cable systems being restored on an ad hoc basis. INTELSAT's response to such contingencies is equally rapid, with the restoration arranged through spare capacity in existing links, or even by making temporary use of additional space segment capacity.

Conclusion

At the very first meeting of Operations Coordinators, it was emphasized that the Earth and space segments of the INTELSAT system are inseparable elements of an integrated system and that the utmost cooperation and coordination between the operators of the two segments would be essential if the full attainment of the benefits of satellite technology were to be realized. INTELSAT's outstanding record of service and reliability over the years is testimony that the two segments have indeed worked together in exceptional harmony.

10. Development of New INTELSAT Services

Marcel Perras*
INTELSAT, Washington, D.C.

I. Introduction

The emerging need for a global wideband telecommunications system was the prime motive for the establishment of INTELSAT when satellite communications technology became available in the early 1960's. Meeting this requirement still remains INTELSAT's prime objective, but increasingly INTELSAT has also successfully responded to other types of service requirements for which the satellite is well suited. Its service philosophy is to be fully responsive to user requirements, as voiced to it by its Signatories, and to anticipate new market trends. Thus, INTELSAT has been innovative in adapting satellite transmission techniques to the types of service requirements and to the demands for an expanding range of international services. In 1965, it began with just two services--black-and-white television and associated audio and telephony circuits that could be accessed by only two Earth stations at a time. Today, the INTELSAT Tariff Manual lists scores of services, and the number is still growing.

Full-time services such as telephony, telex, telegraph, and facsimile were prime revenue producers right from the start, but occasional services (particularly television) also were important. It actually was through the "live via satellite" captions on their television screens that people throughout the world became acquainted with INTELSAT. INTELSAT's commercial approach was evident even in those early days when the television tariff was set. A "market development rate" was established in order to "fill up" the occasional use capacity. Thus, radio, submarine cable restoration, and seasonal peak telephony quickly joined television as significant occasional use services.

Another important service, but of a much different type, emerged in the mid-1970's in the form of a demand-assignment

---

Invited paper received October 11, 1983 (revised May 22, 1984).

*Director, Business Planning.

system, known as SPADE. That new system, which provided services from an available "pool" of circuits upon demand, was used for handling very small traffic streams where full-time circuits and the allied system costs could not be economically sustained. The SPADE demand-assignment technique gave rise to a new charging concept of a per-minute rate as an extension of the method widely practiced in network telephony. In the beginning, the basis for the charge had been the number of satellite channel accesses to an operating terminal, but that was soon changed to the number of minutes for which the satellite channel was assigned to the terminal making the call.

The next service milestone occurred in 1975, when bulk lease of capacity was initiated for domestic service with Algeria, Brazil, and Norway, and when United States Mainland-Hawaii traffic was transferred from the regular international network to the lease of a full transponder. The next important development occurred in 1978, when television service and domestic lease services became very significant revenue factors. This development grew into a global demand for television, eventually resulting in the establishment of a separate category of full-time leases for television channels.

Service development in INTELSAT is not confined to the major categories of services shown in Fig. 1 of Chap. 9. There are several other types of services that are established in certain circumstances and that may be used sporadically until regular service requirements emerge, e.g., program transmissions and one-way data transmissions on single channel per carrier (SCPC) 64-kbits/s data circuits.

The introduction of the time-division multiple-access (TDMA) system on a systemwide basis was fueled by two long-term trends: increasing demand for capacity for full-time services and the need to enhance the capacity of individual satellites. The establishment of a tariff for services carried on the TDMA system required new equations in resource utilization based on technical as well as other operational considerations impacting on INTELSAT's overall viability.

The TDMA tariff recognizes the benefit to INTELSAT of the substantial enhancement of the effective capacity on transponders dedicated to TDMA operation and also takes into account the benefit to users in reducing the need for additional Earth stations. This rationale led to the unit charge per standard 4-kHz voice-activated telephone channel being fixed at 0.875 times the charge for a telephone circuit derived on the FDM system. The use of digital speech interpolation (DSI) in conjunction with TDMA, of course, adds an even further boost to the telephonic capacity of INTELSAT transponders.

The next phase in service development arose when the International Maritime Satellite Organization (INMARSAT) began to lease capacity on INTELSAT V satellites that had been specially equipped for maritime services. INTELSAT, in a sense, had anticipated INMARSAT's requirements and had acted expeditiously to modify its satellites, then in construction, to carry a Maritime Communications System (MCS) package. The agreement with INMARSAT for a phased introduction of maritime services in all three ocean regions, starting in the Indian Ocean region late in 1982 and expanding to the Atlantic Ocean region in August 1983, demonstrated INTELSAT's flexibility and operational confidence in extending the benefits of economy of scale to maritime services.

Public perception of the value of telecommunications has sharpened in recent years, following the dramatic reduction in digital equipment costs and the emergence of the microprocessor. The potential capabilities of digital systems are increasingly attractive to a wide range of users because of their versatile applications (telephony, facsimile, video conferencing, remote interactive computation, and data base retrieval). In anticipation of international demand, INTELSAT has developed INTELSAT business service (IBS) for an integrated digital service with transmission rates ranging from 64 kbit/s to a full transponder that ultimately will be able to transmit up to 140 Mbit/s.

A. Leases

In a sense, all circuits provided through INTELSAT satellites operate on capacity leased from INTELSAT. However, the concept of bulk leases grew out of the need for large capacity trunks, or uniquely defined services, in domestic networks requiring a degree of flexibility in design and operation that distinguished them from normal international circuits and services. Although the leasing concept initially provided for standard transmission parameters, it was soon apparent that system integrity could be preserved by defining boundary conditions of operation for domestic services, leaving details of service exploitation and end-use quality to the lessee.

Since those services did not require global interconnection, a unified set of transmission standards was not necessary. INTELSAT thus introduced an important operational concept by developing "lease definitions" that declared the satellite resources available for each type of lease and defined the boundary conditions within which the lease system should operate so that other services carried on INTELSAT satellites would not be affected.

Lease services found a wider acceptance with the development of the "pre-emptible lease" concept. A large amount of usable in-orbit spare capacity was available on the INTELSAT system and if a lessee would accept the risk, admittedly small, of possible preemption in favor of international services in the event of an operational contingency, then low-cost domestic capacity was suddenly available. Given the exceptionally high reliability of INTELSAT services, pre-emptible lease service grew rapidly and, by 1983, over 40 such leases were in effect in some 30 countries. Although the pre-emptible lease service initially found ready acceptance in developing countries that were able to modernize their domestic networks with minimum lead time and capital investment, a sizable demand has arisen in recent years in developed countries in order to augment their television distribution systems.

The emergence of cable television and other distribution systems gave great impetus to the universal demand for INTELSAT television services. Thus, a corresponding demand for the flow of video material was to be expected. As a consequence, the demand for exclusive use of international television channels soon emerged, and INTELSAT responded by extending the lease concept to encompass this service. Although it was thought that the operational possibilities should be exploited to meet the emerging demand, it had to be ensured that revenue from the occasional use television service was not unduly siphoned off. A number international television channel leases have now started operation, and the demand is growing rapidly. Futhermore, there is no evidence that the rate of growth in occasional use television has been stunted by this development.

II. INTELSAT Business Service

INTELSAT business service (IBS) was introduced in 1984 and may be used to provide a variety of digital services including voice, facsimile, electronic mail, video-conferencing, computer data conferencing, computer data transfer, electronic fund transfer, remote printing, electronic document distribution, digital stereo audio, digital television, and other affiliated services. IBS may be provided directly to headquarters and branches of large corporations and other public and private agencies and organizations using small, low-cost ground terminals [3.3-m (11-ft) aperture size] located on or near the users' premises, or obtained through terrestrial networks connected to 5.5- to 8-m (18- to 26-ft) urban feeder Earth stations in major cities. Eventually, all communication services can be provided by INTELSAT within the framework of the global Integrated Services Digital Network (ISDN) and with provision of global connectivity at the earliest possible time.

Of special interest in such services is the global video-teleconferencing capability, typically requiring about 2- to 3-Mbit/s data channels. An International Teleconferencing Symposium (ITS), co-sponsored by INTELSAT and co-hosted by the INTELSAT signatories of Australia, Canada, Japan, the United Kingdom, and the United States, was held in Philadelphia on April 3-5, 1984, and was the most elaborate switched international videoconference ever staged.

A. Service Characteristics

Following introduction in the Atlantic Ocean region, IBS will be extended to other regions to provide global service as early as possible. The service will be provided using both Ku-band (14/11 GHz, with 12 GHz subsequently replacing the 11-GHz downlink frequency band) and C-band (6/4 GHZ) frequencies and will provide interactive point-to-point service as well as point-to-multipoint mode of operation. On-board cross-strapping will allow the interconnection of the Ku-band and C-band frequencies. A high degree of reliability and channel availability will be maintained, and the service will be offered on a non-pre-emptible basis. Encryption will be provided for privacy and security consistent with the transmission system of INTELSAT. Both full-time leases and occasional use will be provided. Provision will be made for information rates of 64, 128, and 256 kbit/s, and 1.544 and 2.048 Mbit/s, as well as higher data rates subject to a bit error rate of 10^{-6} or better for 99% of the year, assuming a three-quarter forward error correction (FEC) code rate. Exceptions for particular applications will granted.

The addition of other service features will be considered as the service develops, particularly with regard to videoconferencing and digital television.

B. Network Architecture

Three network arrangements are incorporated in the overall service concept for INTELSAT Business Service. This will permit a choice of the network architecture most suitable for individual needs. The INTELSAT space segment facilities will be designed to operate with any of these network architectures and to provide the necessary interconnection between different networks.

C. User Gateway Network Concept

The user gateway network concept is characterized by small Earth stations, 3.3- to 5.5-m (11 to 18 ft) antennas, in the Ku-band with very short and relatively inexpensive terrestrial interconnection to the business customers. Each of these small Earth stations could serve a single large business user, or it could be shared by a community of smaller users.

D. Urban Gateway Network Concept

Under an urban gateway network concept, a medium-sized Earth station, 8 m (26 ft) at Ku-band and 11 m (36 ft) at C-band, would be located in a city or its outskirts to minimize the terrestrial interference and interconnection problems. Signals from a customer would be routed via the local terrestrial digital network to the urban gateway Earth station in the metropolitan area, and then transmitted by the Earth station to the desired international destination.

E. Country Gateway Network Concept

A country gateway network concept envisions the use of the existing INTELSAT gateway Earth stations for international business communications. Therefore, the Earth stations are restricted to Standard A and B stations in the C-band and to Standard C stations in the Ku-band. This network concept will depend upon the availability and economy of a wideband national digital network.

F. INTELSAT Business Service Earth Terminals

In order to implement the user and urban gateway network concepts, small Earth terminals suitable for placement directly on customer premises or in congested urban areas are needed. INTELSAT has developed standards not only for a series of small Ku-band Earth terminals designed specifically for the new business services described above, but also for small C-band Earth stations for business services.

IBS--and especially video-teleconferencing and dedicated business voice and data--is likely to become a high-growth service, as it will reduce expensive "terrestrial tails" that tend to drive up the cost of international telecommunications. International businesses, governments, and other entities, for instance, can reduce their expensive and time-consuming international business travel, as well as reduce the cost of their conventional telecommunications services. The tradeoffs between air travel and videoconferencing are significant. One study has estimated that if only 7% of total overseas airline travel was replaced by videoconferences, total overseas telecommunications would be increased by a factor of up to 10. Telecommunications substituting for transportation of people is one effective answer to the world's energy crisis. When it is recognized that IBS data channels can be effectively utilized by businesses on a 24-hour a day basis, by combining videoconferencing, telephone, data, facsimile and overnight voice mail, and electronic mail, the appeal and cost-effectiveness of the service become increasingly clear. The Standard E Earth stations used for this service could, by the end of the decade, become far more numerous than conventional Standard A and B stations.

Thus, the new digital business service concept clearly has a wide range of potential applications. With this new service offering, INTELSAT is taking a significant step toward the realization of a truly global Integrated Services Digital Network (ISDN).

G. INTELNET

INTELNET is also exploring the introduction of a new type of digital broadcast network service that can work to very small Earth stations, known as microterminals. This service, which would operate in the 1.2 kbit/s to 56 kbit/s range would be quite effective for distributing news, as well as financial, weather, and scientific data to a larger number of distribution locations. The INTELNET service, which could be integrated into IBS, envisions microterminals as small as 0.65 m (25 in.).

H. Low Density Telephony Service

During 1982, a study by the South Pacific Bureau for Economic Cooperation (SPEC) of the telecommunications requirements of the South Pacific island nations concluded that INTELSAT would be an appropriate choice for providing space segment capacity for the satellite portion of the telecommunications services required to meet the needs of the Pacific Basin. INTELSAT has since introduced a new low-density telephony service--called VISTA--providing this capacity to the Pacific Basin and elsewhere. VISTA is based upon the use of small low-cost Earth terminals, with use of the space segment capacity shared by the island nations. The service could start as early as 1985 and expand in a phased manner over a number of years. Although the Pacific Basin region has telecommunications requirements that are unique to it, the VISTA service could also find application in many other areas of the world where basic telecommunication services are required for rural and remote areas.

I. General Characteristics of VISTA

As already mentioned, VISTA is primarily intended to provide basic and minimal communication for rural and remote communities currently having inadequate or no telecommunications. Thus, characteristically, VISTA will be provided to a large number of small user communities dispersed over a large geographical region, with each generating only a small amount of traffic. Small low-cost Earth stations with simple access and signaling are conceived to operate in a satellite telecommunications with automatic routing (STAR) or multi-modal (Mesh) network configuration (or with a mix of the two network configurations, as necessary). It should be mentioned that the term "telephony" is used in the context of VISTA to encompass the broad range of voiceband services including voice communication, telegraphy and telex, low-speed data, facsimile, etc.

J. Global Applicability of VISTA

Most developing countries have a considerable portion of their population scattered throughout villages and small communities with grossly inadequate, if any, means of modern telecommunications. For purposes of government development programs, local trade, health-care and extension services, education, as well as social and cultural exchange, the availability of simple telephony links to urban, metropolitan, and capital centers of administration, trade and commerce, health and educational facilities, etc., is urgently required.

The subject of rural satellite telecommunications has long been of interest to potential users as well as to telecommunications organizations, and the economy of the Earth terminals is obviously a paramount factor. In many cases, for reasons of Earth terminal economy, basic telecommunications could be provided by a simple "village telephone" utilized on a shared, 24-hour-a-day availability basis and typically served by one satellite circuit or perhaps a "party line" connected to as many as 10 telephones. Even a larger number of telephones (say, two or more sets for government and private use in a party-line or local reticulation mode) could be served by a single circuit. A larger number of circuits would be required to serve a larger population center. Additionally, higher volume telephony between larger urban and metropolitan centers would generally be required.

K. Network Configurations for VISTA

A STAR network configuration with a large size (Standard B), suitably located community user Earth station acting as the central hub would permit efficient utilization of the space segment capacity. This is based on the concept of a simple, small, low-cost basic user Earth station with one voice-circuit capability to serve the minimal communications requirements of the village and remote communities. Interconnections among the basic user Earth stations themselves are possible with the use of double hops via satellite and the central hub stations. Mesh network configurations could also be applied for direct interconnections among the basic user Earth stations. These examples should be considered only as typical configurations, since the service is designed to be flexible. It is assumed that the network configuration, Earth station design, and overall system planning will take into account the specific communication service needs and traffic requirements of the individual countries. Clearly, the foregoing general description would be equally applicable and suitable for low-density rural/remote communications anywhere in the world, assuming appropriate variations and modifications to accommodate the unique or specific requirements of the user countries.

L. Earth Stations for VISTA

It is obvious that an essential feature of VISTA is that the Earth stations to be employed for this service must be small and inexpensive. They should also be simple to install, operate and maintain.

Two standard specifications have been approved: Standard D-1, and Standard D-2. The basic user Earth station performance specification for VISTA (Standard D-1) is significantly different from the existing INTELSAT standard for Earth stations. Standard D-2, on the other hand, corresponds to an Earth station with rf performance more or less equivalent to the existing INTELSAT Standard B, and is designed to fulfill the role of the central node or hub station for a STAR or STAR/Mesh network. The G/T values specified for the Standard D-1 and Standard D-2 Earth stations are 21.7 and 31.7 dB/K, respectively; the corresponding typical antenna diameters being about 4.5-5.5 m (14.8-18 ft) and 10-11 m (32.8-36 ft), respectively. In particular, the Standard D-1 performance requirement has been simplified by allowing an axial ratio of up to 1.3, single polarization operation, and reduced bandwidth (72 MHz) operation for the frequency conversion equipment. These measures are expected to permit cost reductions for the basic user-type Earth station corresponding to a single voice-circuit capability, with the capability to expand to a larger number (approximately four) of circuits.

Interest on the part of manufacturers and the new possibilities that a Standard D Earth station provide are expected to lead to further improvements in the design and construction of small Earth stations that could be applied to VISTA with consequent cost reductions. Further consideration will be given to the possibility of the Earth stations being also equipped for simultaneous reception (only) of television, which would be a considerable asset. Moreover, efforts are underway to develop a digital version of the VISTA Earth station, in addition to the analog version now in effect.

In the Pacific Basin region, the coverage requirements essentially imply that a global beam is the only suitable choice. However, in other geographical regions of the world, alternative beam coverages--e.g., hemispheric or zone beams--may be considered, depending on the coverage, connectivity, and capacity requirements. The higher e.i.r.p. and G/T values for nonglobal transponders are expected to lead to improved performance and/or reduced Earth segment cost.

M. Future Potential

Given that a great segment of the world's population lives in rural and remote areas with inadequate telecommunications facilities, and the special advantages that a satellite offers for low-density rural/remote communications, the applicability and latent growth potential for VISTA are seen to be immense. Indeed, VISTA, in combination with IBS, represents a new era in satellite communications when increasing use is made of a large number of small Earth terminals for direct access to the satellite in contrast to a small number of large Earth stations being used as national gateways for conventional international satellite communications services.

III. Emergency Services

A. Cable Restoration Service

In the past, INTELSAT has provided emergency restoration service during failure of terrestrial systems and (transoceanic) submarine cables. The requirement for this service is expected to grow, and consideration is being given by INTELSAT to increasing its capability to provide this emergency service.

INTELSAT is a member of three regional groups responsible for developing and implementing restoration plans for submarine cables, terrestrial wideband facilities, and satellite Earth station antennas at the request of system owners. Circuit-routing configurations, which may involve the simultaneous use of multiple transponders in more than one satellite, are activated by notifying the INTELSAT Operations Center that a specific plan is to be implemented.

Currently, there are some 40 active restoration plans for international terrestrial systems utilizing INTELSAT space segment facilities. These plans encompass restoration requirements ranging from 24 to 2883 circuits. Over the past five years there have been 95 terrestrial system failures restored on INTELSAT satellites. The average requirement over this period was to restore 125 circuits for a period of nine days.

Beginning in 1988, optical fiber submarine cable technology will begin to supplant the existing analog design. This new technology promises substantial increases in capacity and system flexibility with a concomitantly enhanced restoration requirement.

Possible configurations under consideration for the proposed TAT-8 (transatlantic telephone) cable system, due to be placed in operation in 1988, utilize either two or three optical fiber pairs, each pair having an approximate capacity of 280 Mbit/s in each direction. Thus average capacity of the terrestrial systems requiring restoration is expected to increase over time. Several transponders are required to restore the larger analog cables being introduced prior to 1988, while a number of 72-MHz transponders are expected to be needed to restore the optical fiber cables in the post-1988 time frame. Consideration is being given to provision of the appropriate amount of space segment capacity and associated matters for providing the cable restoration service.

B. Other Emergency Services

INTELSAT is also considering provision of emergency communications service to potential users. This may include a few channels of telephony for emergency communication and disaster relief operation, etc.

IV. Audio-Conferencing Service

An audio-conferencing service was authorized by INTELSAT in 1983 to provide communications for such applications as "distant teaching" programs.

Campuses scattered over a number of remote locations can use the satellite capacity represented by only a few shared single channel per carrier (SCPC) channels. This approach can serve as an invaluable aid for communicating between different campuses or for audio broadcast of instructional material. This type of service may also find application in many other similar cases in the future. Distribution of classroom instruction as well as library or course material through audio broadcast of programs (including TVRO) may also prove feasible.

A. Aeronautical Services

Recently, INTELSAT reached agreement with the International Civil Aviation Organization (ICAO) to 1) determine the requirements of ICAO and its members for fixed and mobile communications; 2) consider the possibility of the requirements identified being satisfied utilizing INTELSAT space segment capacity starting in an early time frame; and 3) identify objectives for an early service demonstration over INTELSAT space segment if it was mutually agreed that this would be desirable.

It is also intended to attempt to identify the tangible benefits that the widespread utilization of satellites would have, such as improvements in transoceanic air traffic control and greater flexibility and efficiency in flight-path optimization.

Additionally, INTELSAT has been requested by the Agency for the Safety of Aerial Navigation for Africa and Madagascar (ASCENA), on behalf of the African Civil Aviation Commission (AFCAC), to assist in a feasibility study to determine whether it would be technically and economically desirable to use INTELSAT satellites in conjunction with mini-Earth stations (to be defined by INTELSAT) located in the immediate vicinity of aeronautical communication centers for the provision of ground-to-ground communications for the safety of flights over Africa.

B. Preliminary Service Description

Some of the projected service requirements for aeronautical communications via satellites are briefly outlined below.

Air navigation, air traffic control, and safety communications are normally provided by countries and are the primary interest of the ICAO. Aircraft and flight operations communications are normally provided by airlines or other aircraft operators and also are considered by the ICAO through the representative of the International Air Transport Association (IATA). Requirements for both of those categories can be considered with respect to fixed as well as mobile services.

C. Aeronautical Fixed Service

Requirements for aeronautical fixed services (AFS) include such services as ground-to-ground communications between airports and/or air traffic control centers and/or aeronautical meteorological centers. These would be generally described as follows:

1) Full duplex voice (controller-to-controller), slow-speed record (teletype and telex), as well as facsimile transmission between airports to improve flight safety.

2) A point-to-multipoint distribution of meteorological data from two World Area Forecast Centers to virtually all airports on a global basis.

3) An expansion of the one-way weather or meteorological distribution network to two-way operation, thereby providing both a collection and dissemination capability.

4) Additional communications in support of flight operations, e.g., aircraft status, maintenance reports, etc.

D. Aeronautical Mobile Services

Requirements for aeronautical mobile services (AMS) for communication between air and ground would include the following:

1) The use of slow-speed data circuits between aircraft in flight and aeronautical traffic control centers in order to more effectively use air space, i.e., reduce separation between adjacent aircraft.

2) The scheduling of crews and replacement crews; the reporting of in-flight emergencies; the re-routing of aircraft in flight and other activities in support of flight operations requiring communications between aircraft and ground-based installations.

3) Communications from passengers on business aircraft, interconnecting with the terrestrial telephone network or with corporate locations (headquarters, factories, warehouses, etc.) would be part of the future aircraft-to-ground and ground-to-aircraft requirements.

4) Potentially, there may be a public telecommunications service offered to passengers aboard commercial aircraft.

Consideration of the provision of the above types of services is currently underway and is expected to lead to their introduction. Extension of the Maritime Communications Subsystem (MCS) to include an aeronautical communications subsystem (ACS) is a likely approach due to the contiguous frequency band allocation for the two types of services.

V. INTELSAT Assistance and Development Program

An INTELSAT Assistance and Development Program (IADP) was established in 1978 to aid INTELSAT Signatories and users, and potential Signatories and users, to more effectively and efficiently utilize INTELSAT's facilities, including the space segment capacity. Under this program, an effort of up to two man-months is provided free of charge. More extensive programs may be considered on a case-by-case basis. In effect, this is a special support service that INTELSAT provides to help users avail themselves of various types of communications services, particularly the domestic lease services.

Since its introduction, the IADP has increasingly demonstrated its usefulness. While only 1 project was completed in the first year (1978), 3 projects were completed in 1979; 12 in 1980; 21 in 1981; and, in 1982, 33 projects were at various stages of consideration or implementation. Over 60 countries have benefitted directly from the program, but INTELSAT as a whole also benefits through more effective and efficient planning and utilization of the available resources.

The types of assistance being provided generally fall within three broad categories:

1) Planning and development of the international satellite services, including feasibility studies; Earth station developments; preparation of performance specifications and RFPs; evaluation of radio frequency interference; undertaking the evaluation of proposals, monitoring and implementation, verification and acceptance testing, upgrading and retrofitting.

2) Planning and development of the domestic satellite services, including feasiblity studies, system and network planning, frequency coordination, etc.

3) Training assistance, including seminars related to the planning, operation, and maintenance of Earth segment facilities intended to work with the INTELSAT space segment.

Assistance is also now being provided to groups of countries and telecommunications entities in the development of thin-route rural telecommunications in the Pacific Basin region, as well as in Africa and Asia.

In 1981, in response to an increasing number of requests, INTELSAT studied the need for training assistance, and concluded that, while there existed a large worldwide need for training in satellite communications and Earth station technology, there was only a very small number of institutions providing such training. A number of initiatives were undertaken under the IADP program to provide support and assistance in training activities, including presentation of seminars in various parts of the world.

In all IADP activities, special emphasis is placed on assisting the recipients in carrying out the work themselves. In this way the program aims toward a long-term transfer of skills, in addition to providing short-term assistance. There is already evidence that this approach is bearing fruit, as the requests for assistance are becoming increasingly more concerned with planning and development rather than "rescue and maintenance."

In summary, then, the INTELSAT system began by offering a very limited range of services in 1965. Today, the volume of traffic is more than 400 times larger than in the beginning, and there are scores of different services. As late as 1975 conventional telephone, telex, telegraph, and occasional use international television accounted for over 95% of INTELSAT's revenues. By 1984, however, 25% of all revenues were being obtained from a rich mixture of new services. By the year 2000, perhaps the bulk of revenues will derive from "new" rather than "conventional" services. On the long-range horizon, services such as mobile communications, personal communications, direct broadcast audio and video services, environmental monitoring or even communications to permanent colonies in outer space may become INTELSAT offerings, as the number of ways INTELSAT serves the world community continues to expand in the decades ahead.

11. INTELSAT and the ITU

David Withers*
British Telecom International, London, England
and
Hans J. Weiss†
Communications Satellite Corporation, Washington, D.C.

I. Introduction

INTELSAT's raison d'être and its principal activity--the provision of international telecommunications services by satellite--places the organization within the purview of the International Telecommunication Union (ITU).

The ITU, headquartered in Geneva, has its roots in the International Telegraph Union, which was established in 1865 and which it succeeded in 1934 as the result of a merger of the Telegraph and Radiotelegraph Conventions in 1932. In 1947, the ITU was incorporated into the United Nations as a specialized agency.

The ITU, as an international organization, draws upon as well as answers to its member administrations. Its purpose has been to facilitate action by its members, in some instances serving as their agent, toward the orderly use of the radio-frequency spectrum, the study of technical and operational matters, the development and promulgation of standards and procedures, the provision of technical assistance to countries in the development of their telecommunications systems, and, more recently, the equitable use of the geostationary satellite orbit for space telecommunications.

In its role of serving the administrations, the ITU maintains several permanent organizational elements: a general secretariat to deal with administrative and financial obligations; the International Frequency Registration Board (IFRB), which reviews and records frequency assignments made by the administrations and seeks to mediate and resolve possible radio-frequency interference conflicts, and the secretariats of the International

Invited paper received October 11, 1983.

*Chief Engineer, Holborn Centre.
†Senior Director, R&D Policy and ITU Matters.

Radio Consultative Committee (CCIR) and the International Telephone and Telegraph Consultative Committee (CCITT).

The deliberative and executive authority of the Union is vested in the national administrations and is exercised at various levels through different decision-making mechanisms at conventions of administrations' representatives: the Plenipotentiary Conference, the supreme organ of the Union, which meets on the order of every 5 years to set major policy; the Administrative Council, which convenes annually to implement the decisions of the Plenipotentiary Conference; World and Regional Administrative Radio Conferences, which deal with a variety of major issues at intervals determined by need; and the CCITT and CCIR, which address technical and operational issues at the working level and hold quadrennial plenary meetings, with more frequent interim meetings of the Study Groups comprising them.

In the broad range of documentation that the ITU publishes, of major importance are the Convention, in which are set forth the decisions of the Plenipotentiary Conference; the International Radio Regulations, which (comprising the table of frequency allocations, major planning arrangements, operating practices, calculation, coordination and notification procedures, and requirements for future activities) constitutes the principal code of interaction between administrations; and the Recommendations and Reports of the CCIR and the CCITT, which represent a substantial body of technical and operational knowledge relating to telecommunications and are the predominant source of information and the basis for new regulation as well as for day-to-day operation.

INTELSAT's operations are affected by those provisions of the Radio Regulations that pertain to space telecommunications as well as by relevant Recommendations of the CCIR and CCITT. Provisions for space telecommunications were first made in the Radio Regulations at an Extraordinary Administrative Radio Conference (EARC) convened in 1963 and have since been substantially expanded at subsequent Administrative Radio Conferences: at the 1971 World Administrative Radio Conference on Space Telecommunications (WARC/ST), which amended the space-related portions of the Radio Regulations; at two Administrative Radio Conferences for the planning of the broadcasting-satellite service in 1977 (ITU Regions 1 and 3); and 1983 (ITU Region 2 only); and at the 1979 Conference, at which the Radio Regulations were revised in toto. A similar expansion of relevant Recommendations and Reports has taken place in the texts of the Consultative Committees.

The common ground between INTELSAT's activities and the concerns of the ITU demanded and has resulted in a significant degree of interaction at the operational level in spite of the fact that INTELSAT, not being an administration, has had no direct access to the Union's proceedings that so vitally affect it. Such interaction as could and has taken place is marked by success.

Coordination of the INTELSAT system with other systems, as required by the Radio Regulations to avoid unacceptable mutual interference, is essentially a private undertaking in which INTELSAT deals directly with affected administrations. Notification and recording of INTELSAT frequency assignments, involving the services of the International Frequency Registration Board (IFRB), has been undertaken by the United States administration on behalf of INTELSAT. Participation in meetings of the Consultative Committees and at Administrative Radio Conferences has been possible through the observer status accorded to INTELSAT representatives, whose broad technical knowledge and wealth of operating experience have provided valuable commentary.

The following sections will examine in somewhat more detail INTELSAT's environment as it has emerged from ITU proceedings, the way in which it meets its own objectives within the Union's regulatory framework, and some aspects relating to its future role within the international telecommunications community.

II. Standards and Regulations

The INTELSAT global system, which includes satellites in geostationary orbit (the space segment) and the associated Earth stations (the Earth segment) is planned as a whole, and its operation coordinated, by the INTELSAT Organization through its various organs. Its satellites and Earth stations transmit to and receive from each other radio-frequency signals, and their operation is thus subject to the provisions of the Radio Regulations and guided by international standards as developed and documented by the CCITT and the CCIR.

National administrations retain ultimate authority over the use made of radio by stations within their territory, but uncoordinated spectrum use would lead to self-defeating chaos. Accordingly, the Radio Regulations (RR) of the ITU provide an internationally agreed-upon set of rules, designed to promote harmonious use of the spectrum by all nations. They have treaty status. The technical work of the CCIR and the CCITT is, among other objectives, intended to generate, through Recommendations, an international consensus for the further development of the RR and to facilitate the creation of a comprehensive international telecommunications network. These regulations and recommendations affect satellite communication in many ways, but the most significant aspects are as follows.

a) Frequency allocations: Article 8 of the current RR defines, for example, the frequency bands allocated to the fixed-satellite service (FSS) and lists any other services to which these bands are also allocated (shared allocations).

b) Circuit performance standards: CCITT Recommendations give agreed-upon basic performance standards for telecommunication circuits that are, or may form part of, international connections, regardless of the medium of transmission; and CCIR Recommendations interpret these standards in a form that is convenient for the satellite network designer.

c) Spectrum sharing constraints and coordination procedures: The RR and the CCIR Recommendations provide a basis for the simultaneous use of frequency bands by the FSS and and other services, in particular terrestrial radio services, without unacceptable interference.

d) Orbit/spectrum sharing constraints and coordination procedures: The RR and CCIR Recommendations provide a basis for the efficient simultaneous use of frequency bands by different networks of the FSS without unacceptable interference.

e) The International Frequency List (IFL): The RR establishes the conditions under which the IFRB registers frequency assignments made by national administrations to their Earth and space stations.

The first four of these aspects are reviewed in more detail below; the last aspect is discussed in Sec. III.

Frequency Allocations

The Radio Regulations prescribe the frequency bands that may be used by each of a fairly large number of "services," which are distinguished from one another by the technical characteristics of their transmit and receive terminals, by the type of information that is to be transmitted and received, by the use to which that information is ultimately put, or by a combination of these. A band of frequencies prescribed for use by a service is said to be allocated to that service. There are several degrees of allocation: primary and permitted allocations that are entitled to protection against interference from all transmissions in the band; secondary allocations that do not enjoy protected status; and "footnote" allocations that, for a variety of reasons, are treated as additions or alternatives to the normal, or "table," allocations and may be primary, permitted, secondary, or specifically constrained with regard to their use.

Prior to the EARC in 1963, there were no frequency allocations for the FSS. The whole radio spectrum from 10 kHz to 40 GHz was already allocated to other services, and the best frequency range for FSS--that between 1 and 10 GHz--was already in heavy use. Instead of withdrawing allocations from terrestrial services below 10 GHz, or subjecting the new FSS to the problems of operating above 10 GHz, the EARC decided to allocate bands near 4, 6, 7, and 8 GHz to both the FSS and terrestrial services, mainly the fixed and mobile services. (These latter services consist almost entirely of fixed or transportable radio-relay systems.)

The main FSS allocations were 3400-4200 MHz (space-to-Earth); 5725-6425 MHz (Earth-to-space); 7250-7750 MHz (space-to-Earth); and 7900-8400 MHz (Earth- to-space).

Some parts of the 4- and 6-GHz bands were already in use by radiolocation systems in various parts of the world. This usage continued, in the hope that excessive interference might be avoided by geographical separation. Still, the newly formed INTELSAT system used only the bands 3700-4200 MHz and 5925-6425 MHz; and the bands 3400-3700 MHz and 5725-5925 MHz were used for radiolocation in some NATO countries. Only the INTERSPUTNIK system used 3400-3900 MHz and 5725-6225 MHz. The 7250-7750 MHz and 7900-8400 MHz bands were taken into use for government FSS systems, mainly military.

By 1971, it was apparent that the 6- and 4-GHz bands allocated for commercial FSS services in 1963 would not be enough. The WARC/ST allocated several new bands for the FSS, in particular near 11, 14, 20, and 30 GHz, as follows: a) 10,950-11,200 MHz and 11,450-11,700 MHz (space-to-Earth); b) 14,000-14,500 MHz (Earth-to-space); c) 17,700-21,200 MHz (space-to-Earth); and d) 27,500-31,000 MHz (Earth-to-space).

Most of these bands were to be shared with terrestrial radio services, but 19,700-21,200 MHz and 29,500-31,000 MHz were allocated exclusively to the FSS, and the use of these latter bands is therefore free from the technical constraints that arise from sharing. The WARC/ST also allocated various other bands to the FSS near 2.5 GHz, near 12 GHz, and above 40 GHz, but these had not been put into use by 1979, when further changes were made by that year's WARC.

WARC-79 made yet more additions to the bandwidth allocated to the FSS, and the situation became very complicated, in part due to the fact that some of these allocations were not worldwide, but different in different ITU "Regions." The reader looking for exact and comprehensive information should refer to Article 8 of the RR. However, the FSS allocations that are of interest to INTELSAT now are broadly as follows:

a) 3400-4200 MHz and 4500-4800 MHz (space-to-Earth), all shared with other services. Note that the allocation of 3400-3600 MHz and 4500-4800 MHz to the FSS was not in accordance with the wishes of all countries represented at WARC-79. Some countries did not agree to withdraw the radiolocation service at 3400-3600 MHz, at least for the time being, and some countries indicated that frequencies between 4500-4800 MHz would not be assigned to FSS Earth stations in their territory. See Paragraphs 782-785 and 792 of the RR as revised by WARC-79.

b) 5850-7075 MHz (Earth-to-space), shared with other services.

c) 10,700-11,700 MHz (space-to-Earth), shared with other services.

d) 12,750-13,250 MHz and 14,000-14,500 MHz (Earth-to-space), all shared with other services, although sharing in the band 14,000-14,300 MHz is such as not to impose constraints on the FSS in many countries, and the band 14,300-14,400 MHz is unconstrained in Region 2 (the Americas).

e) Bands were allocated to the FSS (space-to-Earth) near 12.5 GHz. Some of these allocations offer the prospect of FSS operation without the technical constraints imposed by sharing. Precise frequencies and sharing conditions vary regionally and nationally. Others, applying to ITU Region 2 only, are associated with broadcasting-satellite allocations; see Fig. 1, which is a representative page from the Table of Frequency Allocations, dealing with these specific allocations. (This complicated situation has now been resolved at the Regional Administrative Radio Conference for the Planning of the Broadcasting-Satellite Service in Region 2, held in Geneva, June/July 1983, and awaits ratification by the Space Services Planning Conference ORB-85 (see Section VI).

f) 17,700-19,700 MHz (space-to-Earth), shared with other services.

g) 27,500-29,500 MHz (Earth-to-space), shared with other services.

h) 19,700-21,200 MHz (space-to-Earth), virtually unconstrained by sharing (but see RR 873).

i) 29,500-31,000 MHz (Earth-to-space), virtually unconstrained by sharing (but see RR 883).

j) In the bands 10,700-11,700 MHz (ITU Region 1 only, namely, Europe, Africa, USSR, and Asia west of Iran), 14,500-14,800 MHz and 17,300-18,100 MHz allocations were made for FSS (Earth-to-space), these allocations being reserved for feeder links serving broadcasting satellites.

k) Allocations for the intersatellite service are available at 23 and 32 GHz and at higher frequencies.

In addition to the foregoing, there are allocations for links between space and Earth at 2.5 GHz and at various frequencies above 40 GHz, but INTELSAT is not likely to be interested in the latter before the end of the century.

B. Circuit Performance Standards

Basic performance standards for circuits that are, or may form part of, international connections are defined in the following recommendations of the CCITT.

a) Noise in analog telephone channels: CCITT Rec. G.103.

b) Bit error ratio in digital telephone channels: CCITT Rec. G.821.

c) Noise in video channels: CCITT Rec. G.567.

d) Availability of telephone channels: CCITT Rec. G.104.

e) Acceptable transmission delay: CCITT Rec. G.114.

RR8-138

GHz
11.7 — 12.75

<table>
<tr><th colspan="3">Allocation to Services</th></tr>
<tr><th>Region 1</th><th>Region 2</th><th>Region 3</th></tr>
<tr>
<td rowspan="4">11.7 — 12.5
FIXED
BROADCASTING
BROADCASTING-SATELLITE
Mobile except aeronautical mobile

838 840</td>
<td>11.7 — 12.1
FIXED 837
FIXED-SATELLITE (space-to-Earth)
Mobile except aeronautical mobile

836 839 840</td>
<td rowspan="2">11.7 — 12.2
FIXED
MOBILE except aeronautical mobile
BROADCASTING
BROADCASTING-SATELLITE

838 840</td>
</tr>
<tr>
<td rowspan="2">12.1 — 12.3
FIXED 837
FIXED-SATELLITE (space-to-Earth)
MOBILE except aeronautical mobile
BROADCASTING
BROADCASTING-SATELLITE

839 840 841
842 843 844</td>
</tr>
<tr>
<td rowspan="2">12.2 — 12.5
FIXED
MOBILE except aeronautical mobile
BROADCASTING

838 840 845</td>
</tr>
<tr>
<td rowspan="2">12.3 — 12.7
FIXED
MOBILE except aeronautical mobile
BROADCASTING
BROADCASTING-SATELLITE

839 840 843 844 846</td>
</tr>
<tr>
<td rowspan="2">12.5 — 12.75
FIXED-SATELLITE (space-to-Earth) (Earth-to-space)

840 848 849 850</td>
<td rowspan="2">12.5 — 12.75
FIXED
FIXED-SATELLITE (space-to-Earth)
MOBILE except aeronautical mobile
BROADCASTING-SATELLITE 847

840</td>
</tr>
<tr>
<td>12.7 — 12.75
FIXED
FIXED-SATELLITE (Earth-to-space)
MOBILE except aeronautical mobile

840</td>
</tr>
</table>

Fig. 1 Excerpt from Article 8 of the Radio Regulations dealing with frequency allocations in the band 11.7–12.75 GHz.

(Figure 1 continued on next page)

RR8-139

836 In Region 2, in the band 11.7 — 12.1 GHz, transponders on space stations in the fixed-satellite service may be used additionally for transmissions in the broadcasting-satellite service, provided that such transmissions do not have a maximum e.i.r.p. greater than 53 dBW per television channel and do not cause greater interference or require more protection from interference than the coordinated fixed-satellite service frequency assignments. With respect to the space services, this band shall be used principally for the fixed-satellite service. The upper limit of this band shall be modified in accordance with the decisions of the 1983 regional administrative radio conference for Region 2 (see No. **841**).

837 *Different category of service:* in Canada, Mexico and the United States, the allocation of the band 11.7 — 12.2 GHz to the fixed service is on a secondary basis (see No. **424**).

838 In the band 11.7 — 12.5 GHz in Regions 1 and 3, the fixed, fixed-satellite, mobile, except aeronautical mobile, and broadcasting services, in accordance with their respective allocations, shall not cause harmful interference to broadcasting-satellite stations operating in accordance with the provisions of Appendix **30**.

839 The use of the band 11.7 — 12.7 GHz in Region 2 by the fixed-satellite and broadcasting-satellite services is limited to national and sub-regional systems and is subject to previous agreement between the administrations concerned and those having services, operating or planned to operate in accordance with the Table, which may be affected (see Articles **11**, **13** and **14** and Resolution **33**).

840 For the use of the band 11.7 — 12.75 GHz in Regions 1, 2 and 3, see Resolutions **31**, **34**, **504**, **700** and **701**.

841 The 1983 regional administrative radio conference for Region 2 will divide the band 12.1 — 12.3 GHz into two sub-bands. It will allocate the lower sub-band to the fixed-satellite service and the upper sub-band to the broadcasting-satellite, broadcasting, mobile except aeronautical mobile, and fixed services, all services being on a primary basis.

842 *Additional allocation:* the bands 12.1 — 12.3 GHz in Brazil and Peru, and 12.2 — 12.3 GHz in the United States, are also allocated to the fixed service on a primary basis.

843 In the band 12.1 — 12.7 GHz, the Region 2 space services, existing or planned before the 1983 regional administrative radio conference for Region 2, shall not impose restrictions on the elaboration of the plan for the broadcasting-satellite service in Region 2 and shall be operated under the conditions set forth by that conference.

844 In Region 2, in the band 12.1 — 12.7 GHz, existing and future terrestrial radiocommunication services shall not cause harmful interference to the space services operating in accordance with the broadcasting-satellite plan to be prepared at the 1983 regional administrative radio conference for Region 2, and shall not impose restrictions on the elaboration of such a plan. The lower limit of this band shall be modified in accordance with the decisions of that conference for Region 2 (see No. **841**).

RR8-140

845 In Region 3, the band 12.2 — 12.5 GHz is also allocated to the fixed-satellite (space-to-Earth) service limited to national and sub-regional systems. The power flux-density limits in No. **2574** shall apply to this frequency band. The introduction of the service in relation to the broadcasting-satellite service in Region 1 shall follow the procedures specified in Article 7 of Appendix **30**, with the applicable frequency band extended to cover 12.2 — 12.5 GHz.

846 In Region 2, in the band 12.3 — 12.7 GHz, assignments to stations of the broadcasting-satellite service made available in the plan to be established by the 1983 regional administrative radio conference for Region 2 may also be used for transmissions in the fixed-satellite service (space-to-Earth), provided that such transmissions do not cause more interference or require more protection from interference than the broadcasting-satellite service transmissions operating in accordance with that plan. With respect to the space services, this band shall be used principally for the broadcasting-satellite service. The lower limit of this band shall be modified in accordance with the decisions of that conference for Region 2 (see No. **841**).

847 The broadcasting-satellite service in the band 12.5 — 12.75 GHz in Region 3 is limited to community reception with a power flux-density not exceeding −111 dB(W/m²) as defined in Annex 8 of Appendix **30**.

848 *Additional allocation:* in Algeria, Angola, Saudi Arabia, Bahrain, Cameroon, the Central African Republic, the Congo, the Ivory Coast, Egypt, the United Arab Emirates, Ethiopia, Gabon, Ghana, Guinea, Iraq, Israel, Jordan, Kenya, Kuwait, the Lebanon, Libya, Madagascar, Mali, Morocco, Mongolia, Niger, Nigeria, Qatar, Syria, Senegal, Somalia, Sudan, Chad, Togo, Yemen (P.D.R. of) and Zaire, the band 12.5 — 12.75 GHz is also allocated to the fixed and mobile, except aeronautical mobile, services on a primary basis.

849 *Additional allocation:* in the Federal Republic of Germany, Belgium, Denmark, Spain, Finland, France, Greece, Liechtenstein, Luxembourg, Monaco, Norway, Uganda, the Netherlands, Portugal, Roumania, Sweden, Switzerland, Tanzania, Tunisia and Yugoslavia, the band 12.5 — 12.75 GHz is also allocated to the fixed and mobile, except aeronautical mobile, services on a secondary basis.

850 *Additional allocation:* in Austria, Bulgaria, Hungary, Poland, the German Democratic Republic, Czechoslovakia and the U.S.S.R., the band 12.5 — 12.75 GHz is also allocated to the fixed service and the mobile, except aeronautical mobile, service on a primary basis. However, stations in these services shall not cause harmful interference to fixed-satellite earth stations of countries in Region 1 other than those mentioned in this footnote. Coordination of these earth stations is not required with stations of the fixed and mobile services of the countries mentioned in this footnote. The power flux-density limit at the Earth's surface given in No. **2574** for the fixed-satellite service shall apply on the territory of the countries mentioned in this footnote.

Fig. 1 (cont.) Excerpt from Article 8 of the Radio Regulations dealing with frequency allocations in the band 11.7–12.75 GHz.

The work of the CCIR in defining circuit performance standards is based on these CCITT Recommendations.

CCIR Recommendation 353-4 (CCIR Rec. 353-4), in conjunction with CCIR Rec. 352-4, calls for the 1-min mean noise power added to an analog telephone channel by a single satellite hop not to exceed 10,000 pW0p for at least 80% of any month, and quotes relaxed limits for small percentages of the time under adverse conditions. Similarly CCIR Rec. 522-1, in conjunction with CCIR Rec. 521-1, calls for the 10-min mean bit error ratio added to a digital pulse code modulation (PCM) telephone channel by a single satellite hop not to exceed one part in a million for at least 80% of any month and quotes relaxed limits for small percentages of the time under adverse conditions. Much work remains to be done in CCIR Study Group 4 to make these recommendations fully comprehensive and to align them as completely as is feasible, having some regard to the economic facts of satellite communication, with CCITT Recommendations.

CCIR Rec. 354-2, in conjunction with CCIR Recs. 352-4 and 567-1, calls for the signal-to-weighted-noise ratio for continuous random noise in a video channel at the end of a single satellite hop to be at least 53 dB for 99% of any month and at least 45 dB for 99.9% of any month.

CCIR Rec. 579 recommends as a provisional planning objective that a one-hop satellite telephone circuit should be available for 99.8% of the scheduled time. In calculations of actual availability when designing to achieve this figure, predictable events such as solar eclipses and solar noise interference should be included, but short outages (less than 10 s) and the effects of adverse radio propagation conditions are disregarded, pending further study.

CCIR Rep. 383-4 deals with studies made of the transmission delay arising in satellite systems and on means of mitigating its subjective effects for telephone circuit users.

C. Sharing with Terrestrial Radio Services

Technical constraints and coordination procedures have been developed in the ITU to facilitate the shared use of a frequency band allocated to both the FSS and the terrestrial fixed and mobile services. Such sharing principles could be applied in other space/terrestrial frequency band sharing situations that are permitted by the ITU frequency allocation table, but in many cases they have not yet been elaborated in detail.

CCIR Rep. 209-4 reviews the general principles employed in the ITU to facilitate frequency band sharing. Two kinds of interference situations arise, namely:

a) Transmitters at terrestrial radio stations can interfere with satellite receivers, and satellite transmitters can interfere with receivers at terrestrial stations.

b) Earth station transmitters can interfere with receivers at terrestrial stations, and transmitters at terrestrial stations can interfere with Earth station receivers.

The first two situations are kept within acceptable bounds by applying constraints on the emissions of transmitters in systems. The second two situations are controlled by identifying all possible unacceptable international interference cases, in order that these interference cases can be examined in detail, and an acceptable solution agreed upon where a risk of unacceptable interference is confirmed, such agreement being a condition for the registering of frequency assignments in the International Frequency List (IFL) by the International Frequency Registration Board (IFRB).

Studies show that strong interference may reach a satellite receiver operating in a frequency band shared with the terrestrial fixed service either from a few fixed service transmitters with antennas illuminating the satellite with their main beams or from the aggregate sidelobe radiation from the generally large number of terrestrial stations, the antennas of which are not pointed at the satellite. In view of this:

a) RRs 2501-2504, CCIR Rec. 355-3, and CCIR Rep. 393-3 give general guidance on minimizing interference, in particular by configuring radio-relay chains so that terrestrial antennas do not point at satellites.

b) RRs 2505-2511 place limits on the e.i.r.p. and transmitter power of terrestrial stations operating in bands shared with FSS (Earth-to-space) with equal status. See also CCIR Rec. 406-5 and CCIR Rep. 790-1.

These constraints being observed, the interference level in any channel of any satellite system, due to up-path interference from terrestrial services, can be expected to be an acceptably small fraction of the total interference permissible from such services as recommended in CCIR Recs. 356-4 or 558-1.

In a similar way, the protection of receivers of the terrestrial fixed and mobile services is realized by a limitation of the power flux density (PFD) which is generated on the Earth's surface, within a specified sampling bandwidth, by any satellite system that transmits in a frequency band that is shared with equal status with those terrestrial services. Above 15 GHz, the specified sampling bandwidth is 1 MHz, it being assumed that most millimeter-wave terrestrial systems will use wide-band digital transmission; but at lower frequencies--where analog frequency-division-multiplex (FDM) systems are widely used--a 4-kHz sampling bandwidth is applied. For the actual limiting values of spectral PFD, see RRs 2552-2584; see also CCIR Rec. 358-3 and CCIR Reps. 876, 793, 877, and 387-4. With these constraints observed, the interference level in any terrestrial system, due to interference from FSS satellite transmitters, should be an acceptably small fraction of the total interference permissible from satellite systems as recommended in CCIR Rec. 357-3. However, RR 2585 provides for these constraints to be relaxed with the consent of all countries affected.

The radiation spectrum of many types of emission used in the FSS is highly nonuniform when the carrier modulation is light, and the spectral PFD constraints laid down in RRs 2552-2584 would be too severe for many satellite systems if no steps were taken to mitigate the problem. Carrier energy dispersal is commonly used to reduce the spectral nonuniformity (see CCIR Rec. 446-2 and Reps. 792-1 and 384-4).

Interference between FSS Earth stations and terrestrial fixed and mobile service stations using the same frequency bands is kept acceptably small through a process known as coordination. Thus, interference can be reduced by selecting Earth station sites which provide high radio propagation loss toward terrestrial station sites; see RR 2539 and CCIR Reps. 209-4, 385-1, and 876. A further reduction in interference can be achieved by limiting the power radiated from Earth stations at low angles of elevation (RRs 2540-2551 and CCIR Rep. 386-3). Another safeguard against unacceptable interference is the choice of compatible, noninterfering transmit and receive frequency assignments for Earth and terrestrial stations. This entire process of coordination is complicated. Its regulatory basis is set out in Articles 11-13 of the RR, and the technical basis can be outlined as follows:

a) A country intending to assign frequencies in shared bands to an Earth station publishes, through the IFRB, a map showing the area within which terrestrial stations operating in the same frequency band could sustain, or cause, interference that might be considered unacceptable according to the standards of CCIR Recs. 356-4, 357-3, or 558-1 under worst-case conditions. RR Appendix 28 gives an agreed basis for computing this "coordination area"; refer also to CCIR Rec. 359-5 and CCIR Rep. 382-4 (see Fig. 2 for an example of a coordination area).

b) It is presumed that the country making the proposal will ensure that unacceptable interference will not arise between radio stations within its own territory. If the coordination area includes territory of another country, that country must identify radio stations of its own lying within the coordination area, and more precise calculations are then made of the interference levels likely to arise. (See CCIR Reps. 388-4, 448-3, 792-1, and 449-1.) If unacceptable interference is found to be likely, means are sought for reducing it to an acceptable level. When foreseeable unacceptable interference problems have been solved, the IFRB is notified of the frequency assignments.

c) The process of coordination is repeated as required, for example, when a new terrestrial station is built within a published coordination area.

The regulations, standards, and procedures outlined in this section cover most of the situations where the ITU frequency allocation table allocates the same frequency bands to the FSS and a terrestrial radio service. One exception is sharing between FSS and the radio navigation service at 14 GHz; this case is covered by RR 856, CCIR Rec. 495, and CCIR Rep. 560-2.

D. Orbit/Spectrum Sharing Between Space Radio Systems

A basis is found for the operation of different geostationary satellites in the same FSS frequency bands without unacceptable interference by coordination, using procedures and technical standards agreed in the ITU. These procedures are reviewed in other sections of this chapter. The same procedures can also be used to coordinate geostationary satellites of other services, although the necessary technical standards are lacking in most cases at present. No procedures have been drawn up for coordinating nongeostationary systems, one with another, but interference from nongeostationary systems into geostationary systems is regulated by RRs 2613 and 2614.

These procedures, while they control the level of interference entering properly coordinated satellite systems, do little in themselves to increase the efficiency with which the geostationary orbit is utilized. CCIR Study Group 4 has been studying the problem of efficient orbit utilization since the mid-1960's and set up an Interim Working Party (IWP 4/1) in 1969 to assist in these studies. That IWP continues in existence, and considerable progress has been made in developing methodologies, standards, and equipment design options which, if used, will increase the number of geostationary satellite networks able to operate in a

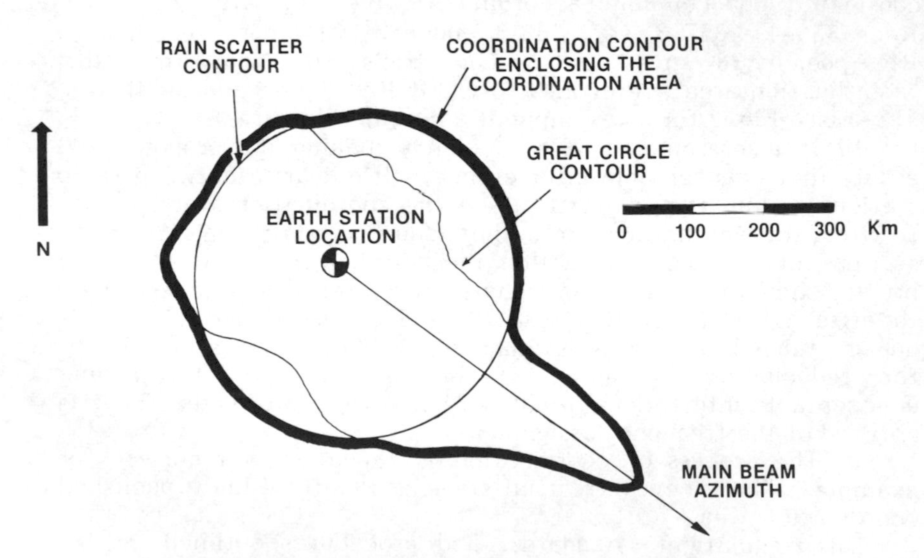

Fig. 2 Representative coordination area around an Earth station as contained within the composite rain scatter and great circle contours. One frequency band.

given arc of the orbit in the same frequency bands. See CCIR Rep. 453-3 for a general review of this matter. The more important issues are the following: satellite stationkeeping; Earth station antenna sidelobe suppression; use of efficient spot beam satellite antennas, accurately directed; permissible internetwork interference levels; use of frequency bands simultaneously in both upward and downward directions; use of bandwidth-efficient and interference-resistant modulation techniques; homogeneity of networks; and rational spectrum and orbit utilization.

CCIR studies on the use of frequency bands simultaneously in both upward and downward directions, reviewed in Rep. 557, led to the introduction of a number of bidirectional frequency band allocations at WARC-79. The other issues are reviewed in Sec. V.

III. INTELSAT and the IFRB

For all that INTELSAT is not an administration and therefore has no direct formal interaction with the International Frequency Registration Board, its frequency assignments and its system's technical and operational characteristics must comply with the ITU Convention and the Radio Regulations, and its networks must be coordinated with other networks through the regulatory process of the Radio Regulations, all of which involves the IFRB. Thus there is an intimate de facto interaction between INTELSAT and the IFRB that is conducted on the part of INTELSAT through an INTELSAT administration acting as intermediary; historically this has been the United States administration.

As discussed in Sec. II, the Radio Regulations, in addition to prescribing the use of frequencies by the various services (as set forth in the Table of Frequency Allocations, which is the substance of Article 8 of the RR) and setting certain technical and operational standards, provide also for the formal protection against interference of facilities and systems through the notification and recording of frequency assignments.

In the space services, notably the fixed-satellite service, which is the service of main current concern to INTELSAT, it is this category of provisions that has developed into an increasingly complex and comprehensive code of formal interaction between administrations owning or operating geostationary space networks.

The complexity of the procedures has its roots in a number of related facts. Foremost of these is that the 360 degrees of the geostationary satellite orbit constitute a finite natural resource that must accommodate all of the world's satellites using common frequencies. It has been shown that the number of times a given frequency can be "reused" within and among different geostationary networks, each reuse transporting different and independent information, is far smaller than the number of satellites that could physically be placed in the orbit. While several satellites can, in many cases, use the same orbit location when they serve different geographically separated areas on the Earth, generally satellites using the same frequency must occupy different locations in the orbit so as to ensure that mutual interference between networks does not exceed acceptable levels--the mean required "spacing" between satellites determining how many the orbit will accommodate.

Since the required intersatellite spacing or the conditions under which satellites may occupy the same orbit location are very sensitive to many technical and operational network characteristics, it became necessary to devise procedures by which administrations could be informed of such characteristics of each other's networks for the purpose of calculating interference, matching or negotiating changes to such characteristics, finding suitable orbit locations for their satellites within an ever-growing network population, and securing protection against unacceptable interference during their networks' operating lifetime. These procedures are set forth in a number of Articles in the Radio Regulations. To facilitate their application and to aid in the resolution of conflicts, the International Frequency Registration Board (IFRB) has been given key procedural and advisory responsibilities.

Historically, the IFRB has been the guardian of the Radio Regulations with, among others, the specific role of maintaining a record of frequency assignments for a variety of radio communications services. A frequency assignment is a set of data characterizing a transmitting or receiving facility. It includes not only the frequency of transmission or reception but also other information relevant to the assessment of interference that the facility may produce or sustain within an environment of other administrations' facilities using the same frequencies.

Frequency assignments are made by administrations but enjoy official recognition only after resolution of all potential or actual conflicts with other, prior assignments. Generally, the resolution of such conflicts, called "coordination" (see Sec. IIIA2), is a private undertaking among administrations; the ratification of an assignment, which formalizes its "existence" and thereby extends to it all due protection after successful coordination, takes the form of an entry of the assignment in the Master International Frequency Register (MIFR), which is maintained by the IFRB. The IFRB periodically publishes the information contained in the Master Register in a frequency list. The Master Register entry of an assignment follows the notification of the assignment by an administration to the IFRB and confers, upon the administration, in principle, a nonpermanent right to bring the assignment into use and to have it protected against unacceptable interference from subsequent assignments through the coordination process.

The relevant Articles of the Radio Regulations that provide the procedural framework of interaction between administrations and the IFRB for the recording of frequency assignments follow.

Article 11: Coordination of Frequency Assignments to Stations in a Space Radiocommunication Service Except Stations in the Broadcasting-Satellite Service and to Appropriate Terrestrial Stations.

Article 13: Notification and Recording in the Master International Frequency Register of Frequency Assignments to Radio Astronomy and Space Radiocommunication Stations Except Stations in the Broadcasting-Satellite Service.

Article 14: Supplementary Procedure to Be Applied in Cases Where a Footnote in the Table of Frequency Allocations Requires an Agreement with an Administration.

Relevant to the implementation of these Articles is also Article 10, which defines the functions and working methods of the IFRB. Fundamental to the IFRB's functions is the timely compilation and dissemination of all necessary information relating to the proposed or recorded frequency assignments that may be needed by other administrations to assess technical compatibility with their own proposed or recorded assignments. Although only the administrations themselves can determine such compatibility, the IFRB's involvement is necessary to ascertain that the assignments sought and the interactions among administrations comply with all pertinent technical, operational, procedural, and scheduling provisions of the Radio Regulations and of the ITU Convention.

Complex as the procedures are in practice, the principles underlying them are simple.

A. The Article 11 Procedure

The Article 11 procedure is composed of two major phases and must be completed before an administration can secure an entry of its frequency assignments in the Master Register.

1. The Advance Notification. Phase 1 is the preparation of an advance notification, which is sent to the IFRB no earlier than 5 and, preferably, no later than 2 years before the intended commencement of use of a frequency assignment (RR 1042). It gives preliminary information regarding the characteristics of a planned network for which an administration seeks an assignment. At this stage, the Earth stations that are to form part of any network are treated generically by making certain summary assumptions for their technical characteristics and, in the absence of knowledge regarding their specific locations, by bounding the geographical area within which they are expected to be located. For the space station, a specific orbit location needs to be given--although it is also necessary to indicate the total arc of the geostationary satellite orbit within which the satellite could be located (the service arc) to provide the desired service. This, as well as all other relevant information, is part of a mandatory advance notification information package, the contents of which

are prescribed in Appendix 4 of the Radio Regulations. This Appendix 4 information is necessary and sufficient to permit another administration to ascertain whether operation of any assignments it seeks or holds itself may be adversely affected by the network for which a more recent advance notification has been submitted to the IFRB. To make other administrations aware of the characteristics of a planned network, the IFRB publishes its advance notification, including the Appendix 4 information, in the next issue of weekly published circulars, notifying all administrations by telegram that such information is being published (RR 1044).

Administrations have four months to examine whether any assignments already held or in the process of being notified by them may be adversely affected and, if so, to comment accordingly to the administration having submitted the most recent advance notification, with copy of such comments to the IFRB (RR 1047). It is incumbent upon the notifying administration, in cooperation with any responding administration, to solve the problems that gave rise to the response comments (RR 1049). The IFRB may be called in to aid in resolving such problems (RR 1054), but other than that needs only to be informed, at six-month intervals, of the status of the problem. The IFRB, in turn, publishes this status information in one of its weekly circulars (RR 1056).

In the process, the notifying administration may well discover that its own planned network may be adversely affected by existing or previously notified networks of other administrations; it finds all information necessary to determine this either among the information published in the Master Register, or else in the IFRB's weekly circulars for networks for which no assignments have yet been recorded. This complex procedure tends to protract the interaction among affected administrations, and thus the Radio Regulations allow phase 2 (coordination, see below) to be initiated immediately with those administrations with which problems have been resolved or which responded favorably to the advance notification, and, in any case, to commence six months after publication of the advance notification in the IFRB circular (RR 1058), regardless of whether problems still remain.

2. The Formal Coordination Procedure. The second phase under Article 11--the coordination of frequency assignments--is, in principle, similar to phase 1, with four notable differences:

a) The coordination of space and Earth stations with respect to interference compatibility with other networks must take specific Earth station locations into consideration.

b) In frequency bands shared with terrestrial services, the specific Earth stations must also be coordinated with stations of these terrestrial services (and vice versa).

c) Actual transmission characteristics, including frequencies, must, where necessary, be taken into consideration, whereas phase 1 examination may have used only spectral emission power densities to characterize potentially interfering rf signals.

d) The basis for coordination is the technical information described in Appendix 3 of the Radio Regulations, which is considerably more comprehensive and specific than that of Appendix 4, which accompanies an advance notification.

The formal coordination procedure is, even more than the submission and examination of advance notifications, framed within carefully staged action sequences and deadlines. Again, the IFRB is crucially involved in the inspection and dissemination of information at every procedural step, but the onus of effecting coordination (i.e., the establishment of technical and operational compatibility between a network for which frequency assignments are to be recorded and those networks already recorded or at a more advanced stage in the Article 11 procedure) is the responsibility of the affected administrations, although the IFRB may be requested by an administration to provide assistance.

Figure 3 illustrates the major steps and actions to be taken to achieve coordination pursuant to Paragraph 1060 of the Radio Regulations (1060 coordination) in the form of a flow chart. An analogous procedure for the coordination of Earth stations with terrestrial radio-relay stations needs to be followed pursuant to Para. 1107 of the RR. Both procedures rely on technical information as set forth in Appendix 3 of the RR. The way in which this technical information is used is further discussed in the next section of this chapter (Sec. IIIA3).

The conclusion of 1060 and/or 1107 coordination sets the stage for recording of the frequency assignment in the Master Register, which is subject to the procedure of Article 13.

In the 1060 coordination process, there are four "clear" paths which lead to successful coordination and recording of a frequency assignment:

a) when no coordination is necessary (1066 applicable and unchallenged by 1080);

b) when coordination agreements are reached between the notifying and affected administrations (successful 1084 or 1087 completion);

c) when coordination agreements are reached between the notifying and affected administrations through assistance by the IFRB (successful 1097 completion); and

d) when affected administrations, having been assisted by the IFRB, either do not respond to a request for coordination (under 1096) or do not render a decision (under 1097), after a specified response time (completion under 1101 by acquiescence).

There is another path, characterized by "continuing disagreement" (after 1087) between the notifying and an affected administration, that could lead to the recording of a frequency assignment pursuant to Para. 1105, six months after first publication under Para. 1078.

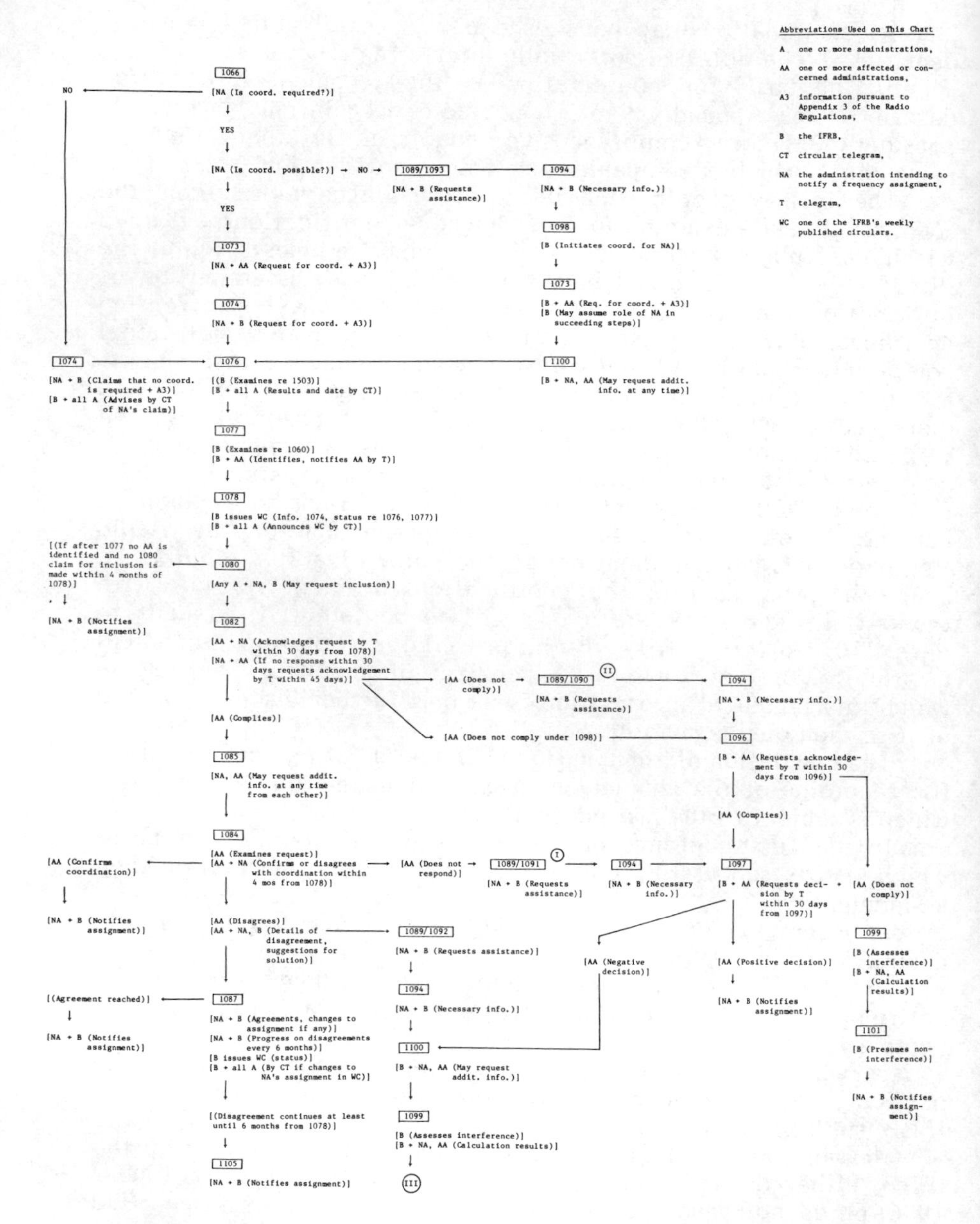

Fig. 3 Coordination of a geostationary space network under the provisions of RR 1060.

There are two paths, both involving IFRB assistance, for which Article 11 shows no clear completion of the coordination process: 1092 coordination and 1097 (from either 1090 or 1091 coordination), under continuing disagreement by an affected administration (marked as III in Fig. 3). Presumably these could terminate also in 1105 action, but at IFRB discretion.

At two further points in the procedure (marked I and II in Fig. 3), where IFRB assistance is shown to have been requested pursuant to 1091 and 1090, respectively, the notifying administration could, conceivably, attempt to continue without IFRB assistance, but since at these stages the affected administration has not responded to repeated requests, the only reasonable step would be to seek IFRB assistance.

3. INTELSAT Involvement. For the INTELSAT system, it is the INTELSAT Organization that complies with those provisions of Article 11 that define the administrations' responsibilities in the advance notification and coordination of their networks' frequency assignments. While INTELSAT will act on behalf of its members and the users of its space segment, it will, at the same time, confront those INTELSAT members with whose separate systems an INTELSAT network is being coordinated. To facilitate this process, INTELSAT has instituted an internal coordination procedure that precedes the ITU procedure with those ITU administrations that are parties to the INTELSAT Agreement, as discussed in Sec. IVD of this chapter.

To the extent that INTELSAT actions under Article 11 involve the IFRB, which deals only with administrations, the administration of the United States acts as the notifying administration for INTELSAT; however, the substance of the coordination process, the establishment of technical compatibility between the INTELSAT system and another system, is achieved by direct consultation between INTELSAT and the other system's notifying administration or, as the case may be, operating agency (see Sec. IV of this chapter).

For the specific coordination of Earth stations operating within the INTELSAT system, the operating administrations themselves are responsible. By and large, such responsibility extends only to the coordination of Earth stations with terrestrial services, but there have been instances where the blanket coordination that INTELSAT undertakes for its Earth stations in generic fashion (by assuming standard characteristics for them and considering them to be potentially located anywhere within the INTELSAT system's service areas) with respect to other networks has resulted in the need, for certain INTELSAT administrations, to notify changes to their Earth station characteristics relative to those they had notified on their own. This suggests that INTELSAT administrations carefully review INTELSAT's proposed coordination agreements for possible consequences on the notification status of their Earth stations with respect to other networks.

B. Article 14

Article 14 details a supplementary consultation procedure between administrations to be applied to frequency assignments which have footnotes in the Table of Frequency Allocations, and where such footnotes make reference to this Article.

Article 14 is invoked in cases where services with coequal primary allocation status have not adopted formal coordination procedures or mutually protective prior constraints on their emission characteristics such as have been adopted in some frequency bands (see RR Article 27, Sec. II, and Article 28, Secs. II and IV), and where, therefore, interference could result as these services are developed. To prevent this, Article 14 prescribes that an administration seeking to initiate the service which could give rise to unacceptable interference secure agreement on technical and operational characteristics from administrations whose systems or services may be affected by such interference.

For space services, the Article 14 procedure may be conducted concurrent with Article 11 action, even concurrent with the advance notification procedure of Article 11 (on the basis of Appendix 4 information).

An allocation of current interest that demands Article 14 consultation/coordination is that in the frequency band 11.7-12.3 GHz to the fixed-satellite service, with footnote 839.

C. Article 13

This Article governs the final step in the notification-coordination-recording chain of a frequency assignment and is intended to result in its final affirmation, conferring upon it all rights relating to operation and protection against unacceptable interference.

The procedure commences with the notification of the coordinated frequency assignment (i.e., coordinated as per Article 11 and, where appropriate, Article 14) together with all information identified in Appendix 3, including such information identified in Appendix 3 as requiring notification when it has been used to achieve coordination. This notification, as all official communications, must be submitted to the IFRB by an administration.

It is then the IFRB's responsibility to examine the notification for the following:

a) compliance with the ITU Convention and the Radio Regulations;

b) compliance with completed internetwork coordination under RR 1060, or with RRs 1066-1071 under which coordination is not required;

c) compliance with terrestrial coordination under RR 1107 (for Earth station notifications that, in the INTELSAT system, are the responsibility of the user administrations of the INTELSAT system);

d) the probability of harmful interference if coordination under either RR 1060 (internetwork coordination) or 1107 (terrestrial coordination) has not been successfully completed.

The provisions of Article 13 then give detailed instructions by which the IFRB is to find on the acceptability of an assignment for recording; ultimately the IFRB should reach a "favorable finding" on a frequency assignment if it is to proceed with its recording in the Master Frequency Register.

Among the detailed provisions of Article 13, two bear special mention. First, under RR 1513, an already recorded frequency assignment--the use of which has lapsed for 2 or more years--need not be considered in an ongoing coordination or recording procedure, and, in fact, such an assignment, to be reactivated, will require initiation of a new coordination process under RR 1060. Second, under RR 1553, an assignment notified prior to its being brought into actual use can be recorded, but only provisionally. Full validation of the recording requires a confirmation that the assignment is in use (RR 1554), otherwise it could be canceled (RR 1555).

The complexity of the Article 13 procedure may lead, in practice, to occasions where the IFRB is faced with having to rule on an issue without clear-cut instructions in the RR. It is most advisable that administrations as well as user organizations familiarize themselves thoroughly with the procedures and seek, in cases of any uncertainty regarding the status of their assignments, immediate clarification from the IFRB.

D. INTELSAT-IFRB History

The INTELSAT system has a successful history of making use of the Radio Regulations and availing itself of the services of the IFRB through the United States administration. Not surprisingly, many of the regulatory provisions of the Radio Regulations have their origin in concepts, technical considerations, and practical observations that grew out of the development of the INTELSAT system between 1965--the launch year of INTELSAT I--and the present.

INTELSAT has successfully recorded a fairly large number of assignments for its space stations in the three ocean regions, all having been made compatible with the assignments of other networks. In numerous coordination agreements, INTELSAT has accepted, for its assignments, constraints that are in fair balance with constraints accepted for others' assignments.

There is no question that, however complex, the provisions of the Radio Regulations and their application by administrations and the IFRB have worked well and that, in fact, the very complexity of the procedures has served as a solid guarantee for the interests and rights of the international space communications community.

IV. Tecnnical Aspects of Coordination

The formal framework of interplay between national administrations and the IFRB highlighted in the preceding section serves to safeguard the rights and interests of these administrations under the Radio Regulations in a qualitative fashion. It is the vehicle for the much less formal, but technically detailed, interaction between administrations or, as the case may be, operating agencies by which the administrations' interests and rights are quantified. This quantification is the heart of the coordination procedure discussed in Sec. III and is aimed at establishing, through calculations and consultation, whether the technical and operating characteristics of different space radio communications facilities that are capable of interfering with each other are compatible with one another and, where they are initially not compatible, at undertaking such action as is necessary to bring compatibility about--a prerequisite for the recording of a new frequency assignment.

Under the ITU technical coordination process, two steps need to be taken: the first is still formal, very approximate, and involves the administrations and the IFRB. Its purpose is to determine whether coordination of a frequency assignment that is to be brought into service is required at all, under Para. 1066 of Article 11 of the Radio Regulations. The second step, invoked if none of the conditions applies under which Para. 1066 provides for an exemption from the need to coordinate, is the more precise technical coordination process. That, while obligatory under the Radio Regulations, is an informal private undertaking between affected administrations or between operating companies and, once it results in agreement on the technical and operational compatibility between the facilities subject to coordination, its qualitative outcome (i.e., the fact that coordination has been effected) may be the only public and published remnant of the coordination process.

A. The Delta-T Concept/Appendix 29

As regards the first step, that of determining whether coordination is required at all, one of the provisions (a) of Para. 1066 is of particular interest. This provision refers to the fact that the occurrence of interference in a receiving system is causally related to the presence of unwanted radio-frequency energy at the input to a specific or typical receiver in the interfered-with system. Although the degree of interference and the amount of unwanted radio-frequency energy at a receiver input are related to each other in complex fashion, depending on the nature of the desired and the undesired signals involved, it has been possible to characterize the unwanted radio-frequency energy in such a way that it can be quantitatively related to the maximum interference caused by it. This is accomplished by considering the maximum unwanted radio-frequency energy in a reference bandwidth and relating it to the radio-frequency noise level that is inherent in the interfered-with system's receiver in the same bandwidth. A maximum limit is then imposed on this unwanted energy in such a way that any real transmission producing it would not cause unacceptable interference in any other real transmission.

Both (the unwanted radio-frequency energy and the internal radio-frequency noise, in the reference bandwidth) can be expressed in terms of an equivalent receiving system noise temperature, and the maximum acceptable amount of unwanted radio-frequency energy is given as a threshold percentage increase of the radio-frequency noise energy that would exist in the absence of the unwanted energy. Considering that there will usually be several sources of unwanted energy, corresponding to frequency reuses by a number of coexisting facilities (networks), the threshold percentage to be contributed by one source is chosen low enough to safeguard against an aggregation from several sources that would produce unacceptable interference as the interference environment grows more severe.

This threshold value is referred to in RR 1066(a) and is defined in Appendix 29 of the Radio Regulations as a 4% equivalent noise temperature increase in the interfered-with "link." The interfered-with link may be the total wanted link, made up of both the up- and downlink in a network, with the noise temperature increase measured at the receiving Earth station. It may also be only the up- or downlink (for example, where only an up- or downlink is affected, or where up- and downlinks are decoupled from each other by a change of modulation in the space station, with the noise temperature increase measured at only the space or only the Earth station receiver, or at both separately, as appropriate).

The equivalent link noise temperature percentage increase is calculated as $100\,\Delta T/T\%$, where ΔT (delta-T) is the actual measured or predicted noise temperature increase due to an unwanted emission in the reference radio-frequency bandwidth, and T is the total average receiving system noise temperature measured or calculated in the same bandwidth, both expressed in Kelvins. The reference radio-frequency bandwidth is 4 kHz at frequencies up to 15 GHz, and 1 MHz above 15 GHz, but calculations are based on parameter values that are normalized to 1 Hz.

The major purpose of Appendix 29 is to provide methods by which the noise temperature percentage increase in one network due to unwanted emissions from another may be calculated.

Special consideration is being given to the following interference configurations:

a) Networks using the same direction of transmission, having equal up- and/or downlink frequencies.

b) Networks using opposite directions of transmission, the uplink frequencies of one being the downlink frequencies of the other and vice versa.

c) Networks agreeing to consider additional polarization isolation.

d) Networks, one of which uses high-power density transmissions with the other using low-power narrowband transmissions--a situation for which the current delta-T threshold may be too high.

The delta-T calculations may involve the determination of the so-called link transmission gain, which bears the symbol Y. Administrations having difficulties with the quantitative determination of this parameter find additional information on the subject in CCIR Rep. 871. Fig. 4 illustrates one of the interference geometries used in ΔT calculations.

When a "notifying administration," that is, an administration applying for a new network (or even only a new frequency assignment), finds that delta-T calculations result in values of noise temperature increase in excess of 4% in any other network that must be recognized (either because it has its frequency assignments already recorded or because it is at a more advanced stage in the Article 11 procedures), it must seek coordination with that other network. This implies that the delta-T calculation must, in principle, be undertaken for all networks existing or having procedural priority. As the network population grows, this becomes an increasingly formidable effort, in spite of the simplicity of the delta-T concept relative to the detailed coordination procedure, and invites computer application.

B. Detailed Technical Coordination (ITU)

Detailed coordination, part of the overall coordination procedure discussed in Sec. IVA and a private undertaking between administrations or operating agencies, generally involves direct consultation between two coordinating parties; thus, multiple bilateral consultations to coordinate a new network or a new frequency assignment.

Its purpose is to bring about technical and operational compatibility between two networks by whatever means agreeable to the parties involved. The Radio Regulations provide little guidance regarding those means (except in certain limited applications, such as the protection of a frequency allotment plan: e.g., that for the satellite broadcasting plan of Appendix 30 of the RR). However, the CCIR literature provides substantial information on the calculation of interference and the ways in

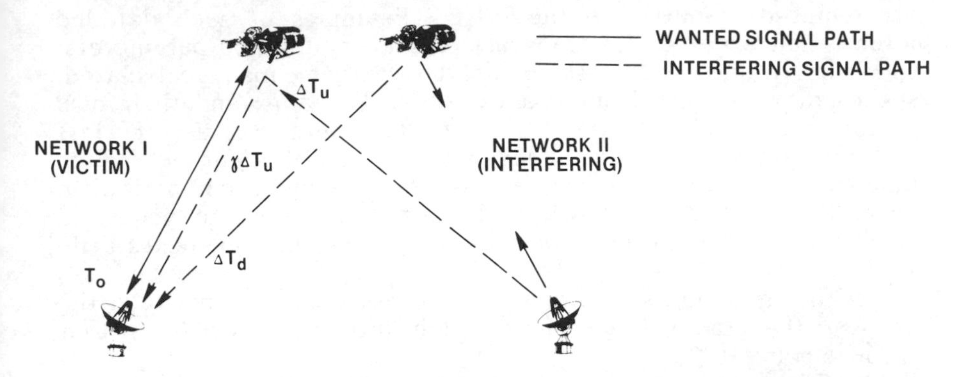

Fig. 4 Interference geometry.

which it can be reduced. For the fixed-satellite service, of primary concern to INTELSAT, Vol. 4, Part 1 of the Recommendations and Reports of the CCIR provides pertinent information (see also Sec. II of this chapter):

a) Relating to performance and interference criteria: CCIR Recs. 353-4, 354-4, 466-3, 483-1, 522-1, 523-1, and 579. CCIR Reps. 208-5, 455-3, 706-1, 710-1, 866, and 867.

b) Relating to the calculation of interference: CCIR Recs. 446-2, 465-2, and 580. CCIR Reps. 384-4, 388-4 (Vol. 4, Part 2), 391-4, 455-3, 555-2, 557-1, 558-2, and 870.

c) Additional relevant information: CCIR Reps. 453-3, 559, and 868.

Parties to coordination are not bound by what the CCIR documentation offers except that they should refrain from imposing on each other more stringent protection requirements than recommended by the CCIR. It would obviously be in the parties' interest, where they wish to depart from CCIR methodology, to remain in the realm of what is realistic and reflects current requirements. It is often found that yesterday's perception of what was unacceptable has given way to empirical and technological progress, offering new, more tolerant, and ultimately more spectrum-efficient ways to coordinate networks with each other. The fact that the CCIR's is not always the last word in matters of interference is exemplified by the adoption, in the United States, of intersatellite spacing for co-coverage networks that is much less than could be achieved with current CCIR methodology. Also, INTELSAT, in pioneering new system concepts and transmission modes, has repeatedly been faced with the need to deal with coordination requirements for which no CCIR precedent existed at the time (examples are the protection of SCPC transmissions and of digital transmissions).

Parties to coordination will generally take into consideration more detailed information regarding their networks than that they are required to notify to the IFRB. Examples of such detailed information are specific transmission and modulation parameters; specific frequency plans; measured rather than calculated, specified, or standard antenna radiation diagrams; and their own performance and interference criteria where these are less stringent than those recommended by the CCIR. Moreover, they may have to negotiate these parameters to achieve compatibility; generally, the new network will have more room to negotiate changes to its parameters than an existing one, but not necessarily so.

While the detailed coordination process is private between the parties, the Radio Regulations reach into the process at four specific points:

a) They set forth, in Para. 1085, the right of parties to coordination and, in Para. 1100, the right of the IFRB to seek additional information during the coordination process.

b) They indicate, in Para. 1052, the possibility of relocating the space station(s) of an administration with which coordination is sought. While this provision relates to the advance notification process of Article 11, it is invoked within a context of ongoing consultation between parties, assuming exhaustion of other means to achieve compatibility--this, for all practical purposes, is already the actual coordination process, whether or not at this stage formalized by a request for coordination under Para. 1060. Note that this relocation provision is the reason for requiring notification of the "service arc" (see Sec. IIIA1).

c) They make it mandatory that certain technical information, not subject to initial notification, must be notified when it has been used to achieve coordination. Appendix 3, which lists those characteristics and parameters of a frequency assignment that need to be notified to initiate coordination, also identifies, through footnotes, those that need to be notified when used in the coordination process.

d) They finally oblige, in Para. 1087, in the case of protracted efforts to achieve coordination, the notifying administration to provide progress reports on the coordination effort at six-month intervals for the purpose of publishing them in the IFRB's weekly circulars.

It is becoming increasingly unlikely, at least in the frequency bands around 4 and 6 GHz, that a new network can be coordinated successfully on the basis of its original Appendix 3 information alone. It is therefore generally necessary that, to achieve recording of a new assignment in the Master International Frequency Register, updated Appendix 3 information be notified, including specifically such information as has become subject to notification by virtue of its use in achieving coordination. There may be elements in the final agreement between parties that are not notifiable, such as declarations of intent regarding the use of certain transmission parameters and frequencies or regarding mutual notification of future actions; these may remain private to the parties, but may also be made available to the IFRB by either party as information.

An important option for administrations is set forth in Para. 1089-1093 of the Radio Regulations. These paragraphs provide for four conditions under which a notifying administration may seek assistance from the IFRB during the coordination process. When such assistance is sought, the IFRB may act on behalf of the notifying administration on procedural matters and in the calculation of delta-T. It is, however, not possible for the IFRB to calculate actual interference or to negotiate coordination agreements; the IFRB's role is limited to that of adviser, analyst, mediator, and, perhaps, intermediary, with the administration having requested IFRB assistance still responsible for decisions affecting its network or frequency assignments. This illustrates the degree of autonomy which administrations retain under the Radio Regulations, but it also underscores the importance of exercising this autonomy.

C. Appendices 3 and 4

Appendix 4 information, to be submitted with the advance notification, is little more than the minimum necessary to undertake the delta-T calculations of Appendix 29 for an entire network, including representative hypothetical Earth stations assumed to be located anywhere in a network's coverage area. Also, Appendix 4 information, submitted nominally as early as 5 years in advance of the actual network service date, will tend to give estimates for some of the parameters rather than known or specified values. However, delta-T calculations, comments, and even preliminary consultations based on Appendix 4 information are useful in the planning of a new network, in general, and in the formulation of the more detailed initial Appendix 3 information.

Appendix 3 information, while containing the necessary parameters to perform precoordination delta-T calculations by which to determine whether or not coordination with other networks is required, also contains much of the detailed technical information (separate for Earth and space stations) by which actual interference calculations can be undertaken. At the time submission of Appendix 3 information is required, the information available to perform the delta-T calculation and to initiate the coordination process can usually be expected to be based on progress in the network design and its planned utilization, perhaps even on the choice of an orbit location different from that of the advance notification.

Without going into details regarding the nature of the Appendix 3 information, it is noted (as mentioned previously) that, while it lists all characteristics that would be necessary to make precise coordination calculations, and while it gives fairly precise definitions for each characteristic (e.g., modulation properties of a transmitted or received signal), it makes some of these characteristics (specific carrier frequency, necessary bandwidth, description of a transmission, peak power, etc.) subject to notification only when they have in fact been used to achieve coordination. Hence, calculations of interference based on initial Appendix 3 information will often (and, at 4 and 6 GHz, usually) not result in successful coordination with some networks. On the other hand, if such still early information were already very specific, it would probably not survive the coordination effort either; the precise eventual characteristics are, therefore, generally and appropriately the result of the coordination process rather than the premise for it.

D. Technical Coordination Under the INTELSAT Agreement

In Article XIV of the INTELSAT Agreement it is stipulated that space segment separate from INTELSAT's to be established, acquired, or utilized by parties to the Agreement, persons under the jurisdiction of Parties, or Signatories shall be subject to a technical coordination process separate from that of the ITU. The Article distinguishes between space segment for the provision of domestic public [Subsection (c) of Article XIV], international public [Sub. (d)], and specialized [Sub. (e)] services. The objective of the technical Article XIV coordination procedure is the showing of technical compatibility of [the use of] such separate space segment with [the operation of] the INTELSAT space segment, prior to such use.

While the INTELSAT coordination procedure developed under Article XIV of the Agreement was, in its early stages, completely divorced from the ITU procedure, especially as regards criteria of acceptable interference, it has over the years been found to be impractical to maintain this separation. Objections were especially raised to INTELSAT's use of more stringent interference criteria than those of the ITU, thus discriminating against administrations to whom, as Parties to the INTELSAT Agreement, Article XIV would apply.

INTELSAT document BG-28-70 (Rev. 1; 12/14/82) outlines the INTELSAT coordination procedure. It is, like the ITU procedure, a two-stage process: a voluntary early informal consultation, roughly corresponding to the ITU's advance notification except for its voluntarism, and a formal technical consultation, corresponding to the ITU's actual coordination under RR 1060. The language of document BG-28-70 (Rev. 1) is considerably less comprehensive and detailed than the RR Article 11 language, since, unlike the ITU procedure that involves the IFRB, no third parties are brought in. The INTELSAT procedure does not supplant the ITU procedure; its main benefits derive from its being broader and more up-to-date than the ITU procedure in some areas, from being able to draw directly on the extensive experience and coordination aids available to INTELSAT, and from the fact that its successful completion anticipates the subsequent successful ITU coordination.

Part of document BG-28-70 (Rev. 1) is a list of characteristics to be made available at the outset of and during the INTELSAT coordination process: minimum information for the informal coordination step (roughly corresponding to Appendix 4 information of the Radio Regulations), and additional information that INTELSAT may require during formal consultation (analogous to RR Appendix 3 information).

The substance of the INTELSAT coordination procedure is compiled in INTELSAT document BG-43-71 (9/18/80) and its Addendum No. 1 (9/16/82). The document briefly describes the coordination process, comments extensively on relevant parameters and coordination methodology, and specifically addresses single-entry interference criteria (i.e., allowable interference from one interfering source). These criteria are equivalent to, in some cases identical with, those set forth in Recommendations of the CCIR. For interference assessments where the CCIR has not recommended a criterion or where the criterion recommended by the CCIR is ambiguous (interference in frequency-modulated television signals, digital time-division multiple-access transmissions, and single-channel-per-carrier transmissions), the INTELSAT document provides special guidance. Moreover, in Addendum No. 1 to BG-43-71, INTELSAT takes a first step in the direction of dealing with potential interference between transmissions in adjacently allocated frequency bands; in this case, potential interference from ITU Regions 1 and 3 broadcasting satellites transmitting above 11.7 GHz to INTELSAT Earth stations receiving below 11.7 GHz. The CCIR discusses this matter (Report 712-1) but has not yet formulated criteria. The Radio Regulations are silent on adjacent-band interference.

To facilitate the Article XIV coordination process, INTELSAT draws on its own technical documentation, in which it has set forth, and continuously updates, its networks' technical characteristics; such material is provided in document BG-28-7, Attachment No. 2, Annex 2, which lists INTELSAT's transmission parameters.

In its coordination process, INTELSAT attempts to stimulate practices that enhance the utilization efficiency of the geostationary satellite orbit, and, in fact, several of the coordination agreements INTELSAT has reached reflect this concern in a positive manner.

As part of the INTELSAT coordination procedure, successful coordination is generally documented in a coordination agreement that not only contains all the specific conditions that led to the establishment of technical compatibility, including necessary constraints on and amendments to the technical and operational characteristics of one or both affected networks, but also provides for recoordination in the event of necessary future changes to either of the affected networks. Coordination agreements are generally submitted to the IFRB in support of the ITU coordination process.

"Clearance" under Article XIV, i.e., the formal establishment of successful INTELSAT coordination, requires a favorable finding by the Board of Governors of INTELSAT, in the form of a recommendation. In the case of separate domestic public space segment, the Director General of INTELSAT conveys such favorable finding to the Party or Signatory affected. In the case of separate international public, or of separate specialized space segment, the INTELSAT Board's recommendation is addressed to the Assembly of Parties (to the INTELSAT Agreement), which renders its finding on the matter.

E. ITU Coordination by INTELSAT

INTELSAT, having in any case to comply with all relevant provisions of the ITU Convention and the Radio Regulations, is encumbered by the fact that its system is conjointly owned by many administrations. While the United States administration will act on INTELSAT's behalf in ITU proceedings, INTELSAT must internally secure consent from its member administrations to actions it intends to undertake on their behalf, and concurrence with decisions stemming from such actions.

This matter is the substance of Article X(a), Para. (xxiii), and of INTELSAT document BG-19-57, which details the internal INTELSAT procedures to secure such consent and concurrence.
In particular, when INTELSAT has concluded ITU coordination with respect to space segment separate from INTELSAT's, there is a one-month internal review cycle by means of which INTELSAT administrations which are Parties to the INTELSAT Agreement are given an opportunity to comment on the results of the coordination prior to the confirmation, to the IFRB, that coordination has been effected. It is noteworthy that under these provisions, INTELSAT administrations may pre-elect to be considered as giving their consent by silence or to be considered as having given consent only when having, in fact, expressly given it.

This complements the safeguards embodied within the ITU procedure for the protection of the interests and rights of sovereign nations.

V. The Finite Resource

The bandwidth allocated to the FSS is large but not limitless, and the usefulness of the higher-frequency bands is affected by severe radio propagation degradations in the lower atmosphere, especially due to rainfall. The geostationary satellite orbit, among Earth orbits, is much to be preferred, for operational and economical reasons, for most FSS applications, and it is obviously finite in length. By exploiting available techniques and technology to the full, it would be possible for very large numbers of geostationary satellites to operate simultaneously in this orbit/spectrum medium without unacceptable mutual interference. However, this very large total orbital capacity will not even be approached except by the conscious efforts of the users.

What are the main factors that must be considered if a satellite network is to carry out its mission without disproportionately impeding the use of the medium by other satellite networks? They fall into six groups: a) satellite stationkeeping; b) Earth station antenna sidelobe response; c) satellite antenna response outside the required service area; d) design of satellite transmission systems to facilitate orbit/spectrum sharing; e) homogeneity of systems; and f) rational frequency and orbit utilization practices.

A. Satellite Stationkeeping

The ITU Radio Regulations require satellites to keep in future to their nominal orbital stations within maximum longitudinal excursions of ±0.1 deg. This requirement applies to new satellites only and takes effect in 1987. Meanwhile, less stringent requirements apply; see RRs 2615–2627, also CCIR Rec. 484-2 and Rep. 556-2. The new standard is achievable now and INTELSAT has already maintained INTELSAT satellites within that tolerance for long periods. However, making the requirement still more stringent would have relatively little effect on the number of usable orbit locations.

B. Earth Station Antennas

Interference enters an Earth station receiver from a satellite that is outside the main beam of the Earth station antenna through the sidelobes of that antenna. Likewise, interference from an Earth station transmitter reaches the receiver of a satellite which is outside the main beam of the Earth station antenna also via its sidelobes. The sidelobe gain of an Earth station antenna tends to decrease with increasing angle off the beam axis. Therefore, if for a given separation between two satellites, the sidelobe gain level or the Earth station transmitter power level is so high that interference is unacceptable, it is possible to reduce this interference by increasing the separation between the satellites; however, this would reduce the number of such satellites that can use the orbit simultaneously.

A good present-day axisymmetrical Cassegrain antenna more than 150 wavelengths in diameter will have few sidelobe peaks exceeding a gain envelope described by 32-25 log Θ dB relative to isotropic gain (dBi) for values of Θ between 1 deg and 48 deg off the boresight axis (see Fig. 5). This expression has long been used as a template for sidelobe gain in the absence of specific data; see

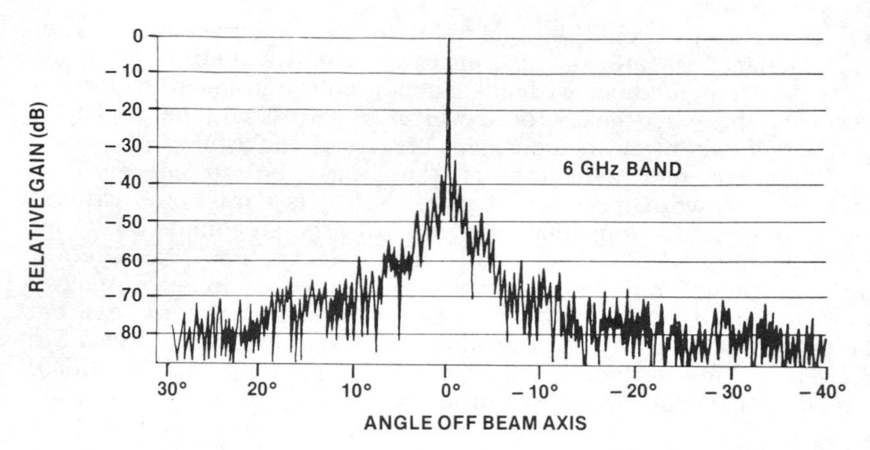

Fig. 5 Measured wide angle sidelobe pattern of a large aperture (D/λ~600) Earth station antenna.

CCIR Rec. 465-2. A poorly designed or constructed antenna may be significantly worse, due typically to subreflector spillover, diffraction at the edge of the subreflector, imperfections in the profile of the main reflector, and scattering at the subreflector supports. In order to achieve a long-term improvement in the sidelobe performance of antennas in service, the CCIR issued Rec. 580 in 1982, which provides for a provisional sidelobe gain design target 3 dB lower than that of Rec. 465-2 in the crucial directions for all large new antennas built after 1987. The determination of a design target for antennas with diameters smaller than 150 wavelengths has been delayed, pending further study. See also CCIR Reps. 390-4 and 391-4.

Smaller axisymmetrical Cassegrain antennas will usually have worse sidelobe characteristics than are called for in Rec. 580. However, offset-fed antennas can be designed so that sidelobes directed close to the geostationary orbit are much lower than 32-25 log Θ dBi. An offset configuration is rather costly to apply to large antennas but small offset-fed antennas are already in use on a small scale, and their cost is falling. A general use of offset-feed geometry for small Earth station antennas would substantially reduce the minimum satellite separation requirements for domestic systems.

Nevertheless, a universal achievement of low sidelobe gain will not in itself permit close satellite spacing. Another factor is high off-beam spectral power flux density that causes interference to other satellites, and this may arise for a number of reasons, such as low Earth station antenna main beam gain, low satellite receiving antenna gain, low satellite transponder gain, or insufficient dispersal of spectral energy, in addition to high Earth station antenna sidelobe gain.

To deal with this problem, CCIR Rec. 524-1 recommends a limit for the off-beam spectral e.i.r.p. density radiated in the direction of the orbit; for the present this Recommendation is confined to the 6-GHz band (see also RR 2636).

From early times in satellite communications, INTELSAT has taken a lead in setting high standards for the radiation patterns of Earth station antennas granted access to the system, and the concept of the limitation of off-beam spectral e.i.r.p. density also originated in INTELSAT. The reason for this is simple: INTELSAT has used several satellites in each of the three major ocean regions since the late 1960's. Long before the general satellite population began to build up, the narrow orbital arcs that are ideal for serving the INTELSAT system had to accommodate an increasing number of the Organization's own satellites, and self-discipline was essential. Even so, when global or regional INTELSAT satellite beams are accessed by small Earth station antennas as used, for example, in leased domestic networks, sidelobe radiation levels of these Earth station antennas are close to the maximum recommended by the CCIR.

C. Satellite Antenna Coverage

To the extent that beam coverages ("footprints") of satellite antennas could be limited to just the geographical area requiring service, especially when such service areas are geographically small, there would be much less interference between satellite networks serving different regions, and their satellites could be placed closer together in orbit; under certain conditions they can be collocated. Improved suppression of satellite antenna gain outside a beam that provides the desired coverage has already made it possible to serve different regions from the same satellite. The CCIR is studying the technical problems of matching footprints to service areas and containing antenna gain in other directions; CCIR Rep. 558-2 shows progress to date, and it is hoped that this will lead to a CCIR Recommendation in due course.

INTELSAT has done much to develop the technical means of matching footprints to service areas and of suppressing the gain of sidelobes. Its objectives have been to make feasible beam-division frequency reuse. Thus the 6- and 4-GHz "hemispheric beams" of INTELSAT IV-A and of INTELSAT V, made up of a multitude of small spot beams that, together, form the desired footprint, allowed twofold spatial frequency reuse to be used (see Fig. 6 for the generic design of a shaped-beam satellite antenna). Further refinements of this principle in INTELSAT VI provide fourfold spatial reuse. In INTELSATs V and VI, additional reuses have been made possible through orthogonally polarized satellite antenna beams. These same techniques can be used to enable substantial numbers of satellites serving different geographical areas to share a small orbital arc. For many countries' domestic service, much simpler designs, consisting of a single beam with good sidelobe suppression, would be sufficient.

It may be noted that WARC-BS that produced the Region 1 and Region 3 Plan for Satellite Broadcasting (1977) and RARC-83, especially the latter, imposed stringent constraints on satellite antenna radiation characteristics outside the service area; these may serve as a future model for the FSS.

D. Transmission System Design

Some types of radio transmissions are particularly liable to interfere with other emissions, for example, because their spectra have strong, undispersed "lines," typically at the carrier frequency. Others are particularly susceptible to interference. More satellite networks could share the frequency spectrum if:

a) The modulation techniques used were robust (that is, exhibited relatively small postdemodulation channel degradation for a given predemodulation carrier-to-interference ratio).

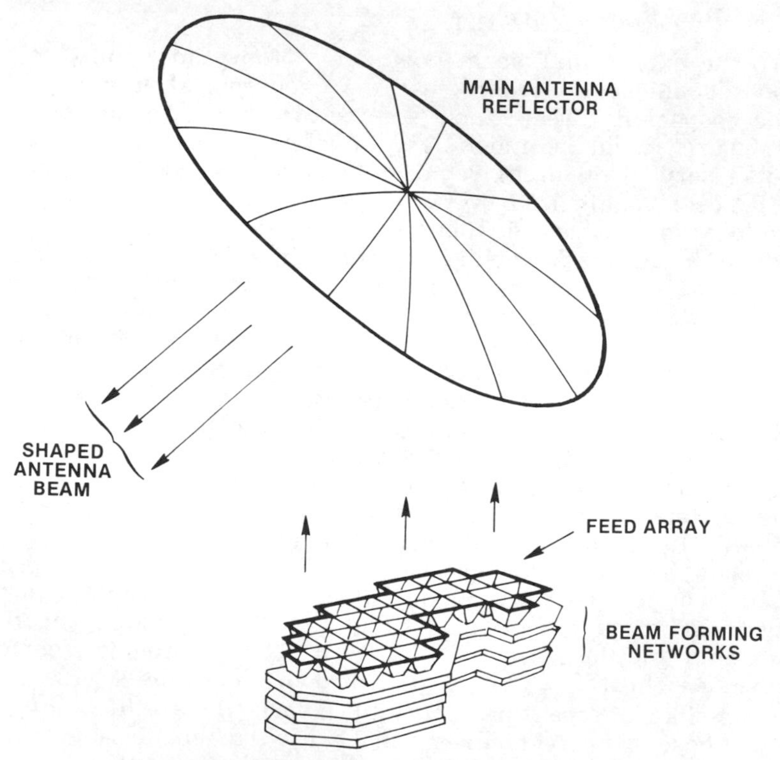

Fig. 6 Conceptual generation of a shaped satellite antenna beam. Feeds are arranged in a replica of the desired beam shape.

b) The transmission plans of satellite networks were arranged in an systematic manner with respect to each other, so that carriers highly susceptible to interference and carriers likely to cause severe interference were not assigned the same radio frequency in different networks.

c) Appropriate carrier energy dispersal techniques were employed wherever they would be likely to reduce internetwork interference.

d) A greater fraction of the total postdemodulation degradation budget (as represented by increases in channel noise or bit error ratio) than currently allowed were allocated to interference from other networks as "permissible interference."

The alignment of transmission plans is one of the objectives of the internetwork coordination process. The choice of modulation technique is fundamental for the design of a satellite network; the CCIR has not yet made any Recommendation on this point, although various CCIR Reports (386-3, 388-4, 448-3, 449-1, and 792-1) supply useful information on the basis for choice. CCIR Rep. 384-4 reviews carrier energy dispersal techniques, and Rec. 446-2 prepares the way for its use to facilitate the reduction of interference between satellite networks.

In the field of permissible interference, the CCIR has been able to provide a substantial initiative in recent years. In 1978 the percentage of the total channel noise, or the total radio-frequency noise responsible for bit errors in a digital telephone circuit, that should be allocated to permissible interference in new networks was raised from 10 to 20%. Exceptionally, where the wanted network employs frequency reuse, a limit of 15% has still been retained. The CCIR is giving consideration to yet further increases in the degradation budget allocated to permissible interference.

E. System Homogeneity

In an ideal situation, all the satellite networks operating in a given pair of frequency bands would be similar in their parameters. In the real world, the FSS has many quite different applications, and the optimum characteristics of different networks, and even of different user requirements in the same network, vary markedly. This inhomogeneity is a source of inefficient orbit utilization. (See CCIR Rep. 453-3.)

The CCIR has studied possible means of reducing inhomogeneity, either by reducing the range of values in the technical and operational characteristics that networks use, or by a division of the orbit or the spectrum between the various kinds of use. So far, none of these methods seems likely to achieve general acceptance. INTELSAT's networks are as inhomogeneous as any, a consequence of their versatility. Taking a long view, it can be foreseen that the parameters used in the different networks using a given pair of frequency bands will tend to converge, and there may be a better prospect of further improvement by CCIR Recommendation in 10 years' time.

F. Rational Frequency and Orbit Utilization

The processes employed by the ITU for regulating the use of the orbit and spectrum by the FSS depend primarily on coordination between users and potential users. Technical coordination, as discussed in Sec. IV, is a complex, difficult process, and if based on inefficient premises, the consequence will be inefficient orbit/spectrum utilization. There are a number of practices that, if adopted, would not only simplify the process of precise coordination but lead to more efficient orbit/spectrum utilization (see CCIR Rep. 453-3). Thus, in addition to good design and operating practices as already described:

a) Frequency band pairing conventions might be standardized; for example, if the 6-GHz band is used for uplinks, it might be agreed that the 4-GHz band will always be used for downlinks.

b) Satellites might generally be equipped for no more than one pair of frequency bands, decoupling the coordination processes of one band pair from those of another.

c) Satellites might generally be equipped for only one service, decoupling the orbital location requirements of one service from those of another.

d) Satellites might generally be designed and operated so that they could be stationed anywhere over a wide orbital arc and restationed elsewhere at any time during their working lifetime, where geographical considerations permit this.

e) Satellite antenna coverage areas might be minimized, in particular avoiding, where possible, beams serving adjacent service areas, to maximize the single-orbit location frequency reuse potential.

f) Combined with e above, satellite beams might be constrained to use only as much bandwidth as necessary to satisfy the demand within their associated service areas.

In practice, there are commercial situations in which it is economically attractive to disregard all of these practices. Perhaps as a result, no ITU endorsement for them has been agreed to date, and at present they are widely ignored. For the future, growth in the scale of operation of new services and further improvements in the technology of multibeam frequency reuse may lead toward single purpose, single band pair satellites. Some experts point to a reverse trend, in which the disposable communication satellites with which we are familiar would be succeeded by all-purpose repairable space platforms. The prospects are not clear in this respect.

VI. The "Space" WARC (ORB-85/88)

According to the ITU Radio Regulations, new satellites and associated Earth stations are registered in the International Frequency List only after frequency coordination has shown that unacceptable interference will not occur with "prior" networks, i.e., networks already registered or having procedural priority. Once registered, a satellite and similar replacement satellites have protection, possibly in perpetuity (the RR have not been clear on this point) against unacceptable interference from other satellites proposed later. This principle is often called "first-come, first-served." Under this principle there is concern to administrations that do not need or are not able to set up a satellite system of their own soon, and fear that there will be no room for them later.

This concern is not new. Thus, in the ITU Convention as revised in 1973, Para. 131 reads:

> In using frequency bands for space radio services Members shall bear in mind that radio frequencies and the geostationary satellite orbit are limited natural resources, that they must be used efficiently and economically so that countries or groups of countries may have equitable access to both ... according to their needs and the technical facilities at their disposal.

The 1982 Plenipotentiary Conference rewrote that paragraph, in particular deleting the reference to "... technical facilities at their disposal."

At WARC-79, many delegations pressed strongly for the RR to be changed, so as to establish not only the right but the assurance for every country to get a fair share of orbit and spectrum when it was needed. There was no disagreement with the justness of this principle, but it was recognized that it would be very difficult to devise a satisfactory way of achieving it by means of international regulations. In the event it was decided that the current procedure should be retained for the time being, with some clarifications and improvements, but more equitable procedures should be developed.

This decision is embodied in three Resolutions of WARC-79, numbered 2, 3, and 4. The key elements of Resolution 3 call for a World Administrative Radio Conference to be convened for the purpose of developing such more equitable procedures. Thus:

WARC-79 resolves:

(1) that a world space administrative radio conference shall be convened not later than 1984 to guarantee in practice for all countries equitable access to the geostationary satellite orbit and the frequency bands allocated to space services;

(2) that this conference shall be held in two sessions;
(3) that the first session shall
   3.1 decide which space services and frequency bands should be planned;
   3.2 establish the principles, technical parameters and criteria for the planning ... and provide guidelines for associated regulatory procedures;
   3.3 establish guidelines for regulatory procedures in respect of services and frequency bands not covered by paragraph 3.2;
   3.4 consider other possible approaches that could meet the objective of resolves 1.
(4) that the second session shall be held not sooner than twelve months and not later than eighteen months after the first session and implement the decisions taken at the first session.

Resolution 2 insists that the present procedures do not confer protected status on a registered satellite in perpetuity and calls on countries with registered assignments to take all practicable measures to help the entry of new satellites. Resolution 4 offers experimental rules, for use pending the outcome of the called-for Conference, which define the conditions under which existing registrations expire, or may be extended.

The ITU Plenipotentiary Conference in 1982 decided that the dates of the two sessions of the Conference should be somewhat delayed. The first session is now expected to start around the end of June 1985 and to last for six weeks, and the second session would run from the end of June 1988, also for six weeks. The two world administrative radio conference sessions thus called for now have the official designations ORB-85 and ORB-88. (The agenda for the first session was drawn up by the ITU Administrative Council Meeting in 1983. The CCIR will hold a Conference Preparatory Meeting in mid-1984 to prepare a technical brief for ORB-85.)

Resolution 3 applies to all space radio services, but the service that gave rise to most anxiety was the FSS, and it is that service to which the remainder of this section relates. Resolution 3 epitomizes the problem in the words "... guarantee in practice for all countries equitable access to ... orbit and the frequency bands ...." The word "equitable" implies natural justice; this cannot be well expressed in channels, orbital arc, and Megahertz, but it may be surmised that a process which provided the following would approach that ideal:

1) There should be access to spectrum and the geostationary satellite orbit to provide enough transmission capacity for every country

a) when required;

b) with equipment parameters appropriate to the application (i.e., at close to minimum cost, taking Earth and space segments together);

c) in a national system or in association with the international collaborators of the user-nation's choice; and

d) with secure tenure, as long as the need remains.

2) If sufficiency for all as in 1 is not achievable, then regardless of priority of entry into service there should be either equal proportionate cutbacks of orbit/spectrum use for all countries or a departure must be made from the use of minimum cost parameters, to allow the available resource capacity to be increased to meet the aggregate requirement. In the latter case, the additional cost of using nonoptimum parameters should be distributed in an equitable way between the countries involved.

There is no reason in principle why one or other of these objectives should not be achieved. The potential capacity of the geostationary satellite is very large, given good technology; and while many new demands will arise for satellite services, it can clearly be foreseen that new transmission media such as optical fiber underground and submarine cables will compete strongly with satellites for long distance services in the future. Nevertheless, the choice of rules and procedures, capable of being implemented with the rigor of an international treaty that will indeed permit these objectives to be achieved, presents many problems. In particular, difficulties arise because

a) forecasts of traffic requirements over 10 or 15 years are unreliable;

b) the service lifetime of Earth station equipment and satellites is long; and

c) the development and production of acceptably priced equipment with improved characteristics, such as will allow the number of satellites to be increased, is market-led.

Any solution that is to succeed in providing enough satellite capacity for all countries must be flexible enough to accommodate departures from traffic forecasts, despite the long life of equipment, and it must induce system designers and system owners to be progressive in the equipment they specify and buy.

Various possibilities have already been proposed. Three generic concepts have been identified and have been discussed in CCIR IWP 4/1, as follows.

1. Detailed, Long-Term, A Priori Allotment Planning. This process would probably involve identifying all foreseen service requirements up to 15 or 20 years ahead, including preferred key equipment characteristics and system parameters. A frequency/orbit allotment plan would be drawn up to accommodate these satellites, more severe mandatory characteristics being imposed where necessary to increase the aggregate capacity available. The allotment plan being agreed, changes to the

requirements during the plan's lifetime and the entry of unforeseen new satellites could be accommodated only to the extent that this could be done without causing unacceptable interference to other allotments.

2. Allotment Planning with Guaranteed Access. An initial a priori plan would be followed by the convening of conferences every few years with a commitment to adjust the prevailing plan to accommodate new satellite networks and changes in the requirements of satellite networks that are already in the plan. These adjustments might involve, for example, minor changes of orbital location or changes in Earth station or satellite characteristics after some appropriate period of notice.

3. Improved Coordination Procedures and Periodically Revised Mandatory Equipment Performance Standards. The present coordination procedures could be made more effective by reforms which

a) placed a positive obligation upon the owners of registered satellite networks to cooperate with a newcomer in solving coordination problems, including a requirement to modify the registered system where necessary;

b) gave formal recognition to the need for multilateral coordination negotiations where a new satellite network needs to be coordinated with several existing networks; and

c) imposed mandatory performance standards on satellites and Earth stations, designed to permit the orbital separation of satellites to be reduced without an unacceptable increase in interference; these standards would be made more stringent from time to time, as the pressure of requirements made necessary and the commercial availability of improved equipment made economically feasible.

Other regulatory possibilities will, no doubt, be developed. All will have strengths and weaknesses. All that are likely to be successful in meeting the long-term demand for service are likely to be extremely complicated to implement and to demand a large amount of expertise. Anything that will simplify the requirements that such a process must coordinate serves very usefully to reduce the complexity of the basic process.

At the present time, INTELSAT already provides the space segment on lease for a fairly large number of domestic satellite systems. For the owners of these systems, the complex, rigid, legalistic, and rather unpredictable processes that would be required for coordinating an independent system in accordance with the ITU Radio Regulations have been replaced by a simple commercial leasing agreement with INTELSAT, and the latter has taken responsibility for coordination with other systems. This is helpful to the lessees. It is also helpful to all the other organizations that operate systems of their own, who need, as a consequence, to negotiate coordination with fewer parties.

Where does INTELSAT stand in relation to ORB-85/88? Firstly, INTELSAT is a very large and expanding user of orbit and spectrum, and so it cannot be unmindful of the pressures for change which are now apparent; one consequence is an extensive program of studies of technical means whereby the use of the orbit might be made more efficient, so reducing pressure on the resource. Secondly, INTELSAT's system differs from most other systems in that its satellites need to provide coverage of virtually the whole Earth, some of its satellite antennas even serving the whole visible Earth with a single beam; these two factors cause INTELSAT satellites to require larger separations in orbital arc from other satellites than necessary for most spot beam satellites. The third point is that INTELSAT provides vital and valuable international services for almost every country on Earth, big and small; INTELSAT's orbit/spectrum needs should therefore probably receive widespread sympathy.

While, clearly, INTELSAT's needs will be an issue of some substance, INTELSAT's impact on ORB-85/88 could be of a much more fundamental nature. It has been suggested that the INTELSAT system might well become part of the solution to the problem of how to provide long-term guaranteed access to the geostationary satellite orbit for countries that perceive themselves as being in danger of seeing their chances to such access dwindle as the orbit is being used up.

Such countries view with suspicion assertions that technological progress and self-discipline will continue to provide room on the orbit for all comers and thus favor a planning philosophy that would reserve them both the right to later access and the protection for it ("a priori planning"). Arguments for such reservative planning are both political and economical. The ostensible political argument is, in addition to the concern with an exhaustion of access possibilities, the desire to do away with the increasingly cumbersome and discriminatory adversary process of coordination; and the economic argument is that, without reservative planning, in any case the cost of access would tend to increase with time, due to the required better technology which is perceived as a burden rather than as the key to the future orbit exploitation.

There is little doubt that INTELSAT, if given and accepting the mandate to provide for future domestic space segment to countries seeking it, would be in a unique position to do this under conditions which would be no less favorable to the individual countries than would be encountered by them under a regimen of reservative planning. In fact, the options for access to the geostationary satellite orbit through INTELSAT may well be more numerous and more flexible.

What INTELSAT would have to offer covers many areas associated with the establishment of a domestic satellite communication capability:

a) A highly competent and experienced technical staff versed in the design, evaluation, coordination, and operation of space and Earth facilities.

b) Facilities for the monitoring and control of system operation.

c) Established liaison arrangements for the long-term provision of launch services.

d) The capability to integrate domestic requirements into its overall system planning.

e) The capability to offer space segment capacity in increments according to scheduled demand growth.

f) A vigorous policy of advancing technology through research and development, and the capability, through economies of scale, to introduce technology improvements in a cost-effective manner.

g) The demonstrated capability, through centralized planning, to consolidate individual requirements on an apolitical basis.

The likelihood that INTELSAT would succeed in accommodating emerging domestic demands is enhanced not only by the vast resources and experience it has to offer, but also by the fact that such demands as they emerge would tend to place only relatively modest incremental requirements on the design and capabilities of the INTELSAT system, which already provides routinely for the handling of an enormous and continually increasing amount of communications traffic.

Acknowledgment

The authors wish to acknowledge that the reproduction in this chapter of material taken from the publications of the International Telecommunication Union (ITU), Place des Nations, CH-1211 Geneva 20, Switzerland, has been authorized by the ITU.

12. INTELSAT System Planning

William R. Schnicke*
Communications Satellite Corporation, Washington, D.C.

I. Planning as a Concept

An overview of INTELSAT system planning as a concept is provided in Fig. 1. There are three components to planning: the elements to be planned, the inputs to planning, and criteria on which to evaluate plans.

A. Elements to Be Planned

The elements to be planned are the INTELSAT space segment, including the number and type of satellites to be procured, the deployment and utilization of these satellites in orbit, and the launch vehicles to be used; the Earth segment to access the satellites, including the modulation techniques, which vary in their efficient use of available bandwidth; the utilization of facilities; and the schedule for each of the above activities.

In 1965, INTELSAT placed its first satellite--which was called the Primary--in service over the Atlantic region. Its second satellite--known as the Pacific Primary--entered service in 1967 over the Pacific region. The global system was finally completed in 1969 with the introduction of a satellite over the Indian Ocean region. Subsequently, additional satellite roles have been introduced in the form of a Major Path satellite in the Atlantic region in 1970 and a second Major Path satellite in the Atlantic region in 1978, as well as a Major Path satellite in the Indian Ocean region in 1983. The Primary satellite carries all users, generally speaking, with a single antenna operating in an ocean region; large, medium, and small streams are all accommodated on the Primary, which provides connectivity among all users. The Major Path

---

Invited paper received October 11, 1983. Revised June 1, 1984.

*Senior Director, International Systems Planning, COMSAT World Systems Division.

satellite carries the larger traffic streams of substantial system users who have provided several antennas to access separate satellites within an ocean region and provides a traffic diversity path to minimize the effects of any system failure on the grade of service.

Each step in the above evolution of the INTELSAT system was taken deliberately in order to meet expanding traffic requirements; thus one major element to be planned is how this space segment evolution should move forward over the remainder of this decade and into the 1990's and beyond. Are additional in-orbit satellite roles required to meet future demand? Can we accommodate future growth by simply replacing existing satellites with higher-capacity spacecraft when they reach end-of-life or are required for service in another role? What expansion of Earth segment facilities is required--both additional antennas and new modulation access encoding technologies (especially digital techniques) that may provide higher bandwidth use efficiencies? How will future traffic requirements be accommodated by the space segment, the Earth segment? In other words, how will facilities be utilized? What is the schedule for procuring and deploying additional space and Earth segment resources?

B. Planning Inputs

The inputs to planning define the mission to be served by the INTELSAT system. The mission consists of the traffic demand that is provided by the users themselves to INTELSAT at an annual Global Traffic Meeting. In the case of the United States, International Service Carriers bilaterally negotiate their traffic requirements with each foreign correspondent as input to INTELSAT's planning traffic data base. This forecast covers each of five years in the form of agreed circuit requirements on the INTELSAT system, with growth rates for such services provided for a subsequent 10-year period so that the traffic forecast covers a period of 15 years.

The international traffic forecast, INTELSAT's major requirement, is summarized in Fig. 2. By the end of 1983, INTELSAT system-wide was carrying some 42,000 two-way circuits. This is expected to double by 1988 and to double again by 1993, which represents an average annual compound growth rate of approximately 14%. Historically, INTELSAT system growth has averaged 25% over the last 10 years. Traffic forecasts are supplied by the carriers and their correspondents in full recognition of existing and planned complimentary submarine cable facilities throughout the world.

The geographic distribution of international service demand is illustrated in Fig. 3. Including the north-south traffic from Europe to Africa and from North America to South America, all possible

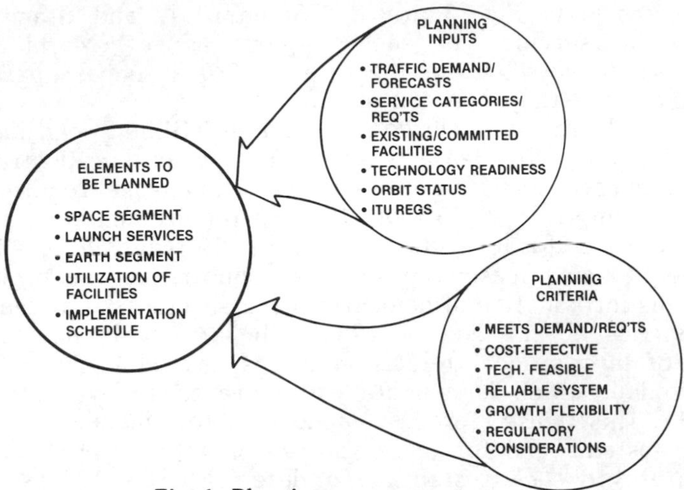

Fig. 1 Planning as a concept.

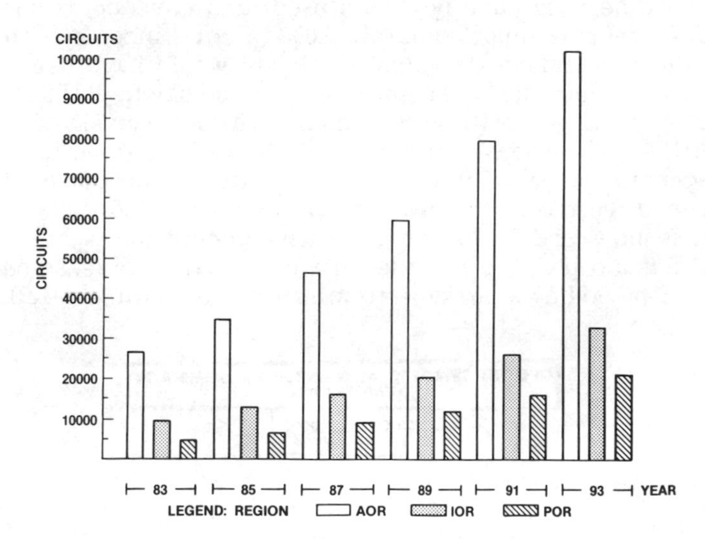

Fig. 2 International traffic forecast.

traffic that could be accommodated by satellites over the Atlantic basin amounts to 75% of the system's total. However, a significant share of this, namely, the Europe-Africa traffic, is handled on satellites over the Indian Ocean basin. The east-west trans-Atlantic traffic is approximately 40% of the system total. Forty-five percent of all traffic originates or terminates in the United States. Requirements handled by Indian Ocean satellites are approximately double those for the Pacific Ocean region.

Also shown in Fig. 3 are the Standard A and Standard C antennas in service in each region as of March 1983. Approximately two-thirds of the total of 170 antennas operate to satellites in the Atlantic Ocean region.

A second traffic requirement that INTELSAT considers in planning is transponder lease requirements, which are illustrated in Fig. 4. Domestic and international television lease requirements (the latter comprises 10-15% of all lease requirements) doubled in 1982 from its 1981 level to a total of 40 equivalent 40-MHz transponders being leased by some 23 countries. The projection includes the introduction of domestic lease services by at least ten new countries by the end of 1984. The relatively new service offering of international television leases has some six leases in service today, which is expected to increase to over 20 leases by 1988. This lease forecast accounts for planned domestic satellite systems and for leases currently within INTELSAT that will migrate to such systems. To date, INTELSAT leases have always been carried on spare or residual life spacecraft (a capacity that would otherwise have not been used) and have contributed to INTELSAT's income (approximately 10% of total income). This has reduced the operating cost, which otherwise would have been borne by mainstream international services. In addition, INTELSAT is currently providing maritime communications services capability to INMARSAT via a lease arrangement and will continue to provide a first-generation Maritime Communications Subsystem (MCS) package in three ocean regions through at least 1988. INMARSAT planning is now underway for a second-generation system that INTELSAT could possibly provide on a cost-effective lease basis to INMARSAT by adding a package to international service satellites.

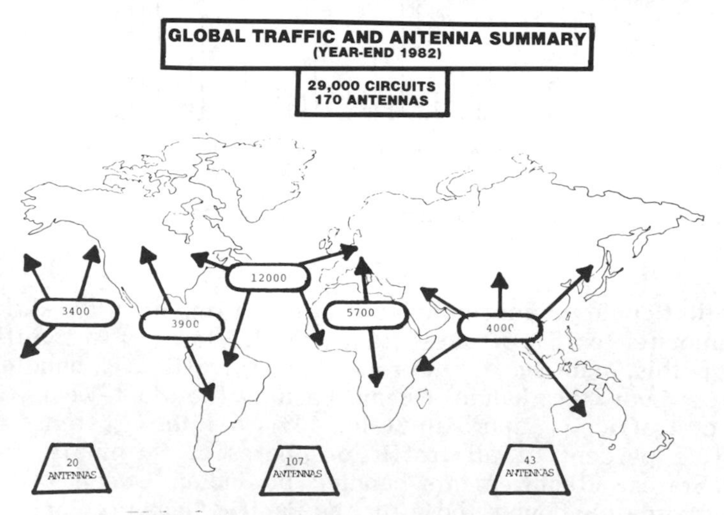

Fig. 3 Distribution of international service demand.

Finally, INTELSAT is planning to offer a small-dish 14/11 GHz–14/12 GHz digital business service commencing in 1984, and requirements for this service have been estimated to plan capacity needs. This is a relatively new service area, and confidence in the forecast for this type of service will grow only as time passes and more experience with the actual market for this service is gained.

The global system today includes three international service satellites operating over the Atlantic Ocean region (all of INTELSAT V design) plus an in-orbit spare. Over the Indian Ocean region, there is one operating INTELSAT V and an INTELSAT IV-A serving as a Major Path satellite plus an in-orbit spare. There are two IV-A's over the Pacific basin--one operating satellite and one in-orbit spare. There are 108 member nations of INTELSAT, 19 of which use domestic capacity also. There are 23 nonmember users of the INTELSAT system and approximately 15 remaining countries in the world that are neither members nor users of INTELSAT.

Among the approximately 170 international service Earth stations in the INTELSAT system today, the United States currently operates four Earth stations on the United States mainland, with a fifth under construction and due to commence service in 1984. There are two Earth stations on the East Coast and two on the West Coast, with a third East Coast station under construction. The United States also has an Earth station in Hawaii and Guam, as well as Standard B facilities in American Samoa, Saipan, and Palau, with approximately another half-dozen such smaller stations under development for Pacific Islands service.

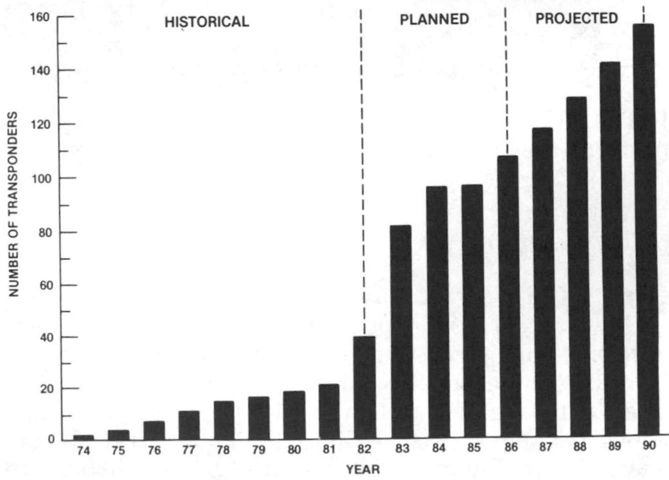

Fig. 4 Domestic lease requirements.

C. Planning Criteria

First and foremost, the system design must meet the traffic demand identified by the users, which means that it must provide for growth on a continuing basis. Second, the growth must be met in a cost-effective manner, which dictates that resources be used efficiently. The system design selected must not represent an undue technological risk, since INTELSAT is a commercial business organization and has to recognize its customer requirements. However, the future facilities should represent a logical extension in the use of evolving spacecraft and Earth terminal technology in order to continue to provide for services in an increasingly efficient, cost-effective fashion. INTELSAT must provide reliable service, which it has done over the years by incorporating adequate redundancy in the basic spacecraft design as well as in the Earth stations, and by providing an in-orbit spare to restore services in the event of a satellite failure. In terms of INTELSAT's projected traffic requirements some 10 years hence, it is virtually certain that the actual traffic that materializes 10 years from now will be somewhat different from that projected today--either in the level of or in the geographic distribution of demand. Hence the system design must be flexible enough to account for such changes and to accommodate reasonable variations in the pattern and level of demand while still achieving efficient use of space segment resources. Finally, INTELSAT must take into account orbit-use considerations in light of the growing demand for slots in the geostationary orbit and must, of course, satisfy all global and regional regulatory and ITU constraints, such as, for example, the downlink frequencies available in the three ITU regions to accommodate new business services as part of the fixed satellite service.

II. Current Satellite Programs

The INTELSAT spacecraft currently in-orbit or under procurement form the basis from which future space segment plans will evolve.

A. INTELSAT V Program

At present, INTELSAT has two active procurement programs underway. The first is for procurement of INTELSAT V series satellites from the Ford Aerospace & Communications Corporation (FACC). A total of nine baseline INTELSAT V satellites have been procured, the last five of which are equipped with maritime communications packages. As of July 1983, the first six of these satellites have been launched successfully by NASA on the Atlas/Centaur launch vehicle. Based on the launch success of this program, eight out of nine INTELSAT V's will be assumed, for planning purposes, to be successfully deployed. The first maritime package on the fifth flight unit of the INTELSAT V started service in the Indian Ocean region in December 1982, and the second package commenced service in the Atlantic Ocean region in mid-1983.

The INTELSAT V is the first body-stabilized satellite to be procured by INTELSAT, as opposed to the spin-stabilized designs of the first four generations of spacecraft. It is also the first satellite to introduce two new technologies simultaneously, namely, dual-polarization operation in the 6/4-GHz band and introduction of the 14/11-GHz band. According to the conservative approach of INTELSAT, the spacecraft includes redundancy to protect against reliability shortages of any of these new technologies.

In addition to the basic INTELSAT V satellites, INTELSAT has procured six modified INTELSAT V satellites, referred to as INTELSAT V-A spacecraft. These satellites expand the 6/4-GHz bandwidth of the baseline design by approximately 15% and will be used in both the Indian and Atlantic Ocean regions to provide for continued growth until the INTELSAT VI series is available. The V-A satellite is a straightforward derivative of the INTELSAT V design, and, for planning purposes, INTELSAT is assuming at least four successful launches out of the six INTELSAT V-A satellites.

The last several of the V-A satellites have been modified to provide for a business service capability by allowing in-orbit switching of the 11-GHz downlink (nominally 11.2-11.7 GHz) to higher frequencies in the 11- and 12-GHz bands. This will permit location of Earth stations near urban centers in both the United States and Europe. Use of the new frequencies, 11.7-11.9 GHz in the United States and 12.5-12.75 GHz in Europe, avoids interference from terrestrial microwave systems. The V-A satellites are scheduled for delivery in 1984 and 1985 and will be deployed for service shortly thereafter.

B. INTELSAT VI Program

The other major program INTELSAT has underway is the procurement of five INTELSAT VI series satellites from the Hughes Aircraft Company (HAC), with contract options for up to 11 additional INTELSAT VI spacecraft. A minimum of four are projected to be launched and deployed for the Atlantic and Indian Primaries, with spare service commencing in the 1986 time frame. The deployment of these spacecraft will release the INTELSAT V-A's from those roles for redeployment to Atlantic Major Path roles and for service in the Pacific Ocean region to replace INTELSAT V satellites that are nearing the end of their design lifetime. The first unit of the INTELSAT VI series is scheduled for delivery in December 1985 and for launch in early 1986 by NASA on the Space Transportation System (STS).

C. Launch Vehicles

Up until July 1983, INTELSAT launched all of its spacecraft on NASA launch systems. For the INTELSAT V program, INTELSAT has procured a total of ten Atlas/Centaur launch vehicles from General Dynamics and five ARIANE launch vehicles from the Arianespace organization. The seventh, eighth, and ninth flight units of INTELSAT V are scheduled for launch on the ARIANE as well as the 14th and 15th flight units.

For the INTELSAT VI series, a more sophisticated launcher is required, and the INTELSAT VI has been designed for launch on either NASA's Space Transportation System (STS) or on the ARIANE IV launcher, which is a planned derivative of the ARIANE I program. In addition, the INTELSAT VI satellites are compatible with the Titan 34D launcher, which may be offered on a commercial basis, and, potentially, with a derivative of the Atlas/Centaur, namely, the Atlas II/Centaur planned for development by General Dynamics. As of July 1983, INTELSAT had procured two STS launches for the first two flight units of the INTELSAT VI program.

III. Current Plan for Use of Satellites

Figure 5 illustrates the current operational plan for the use of INTELSAT satellites starting in 1986. This plan depicts the three operating plus one spare satellite in the Atlantic Ocean region, the two operating satellites plus one spare in the Indian Ocean region, and initially the one operating satellite evolving into a two operational satellite configuration in the Pacific, again with an in-orbit spare. Also depicted in the figure are several domestic lease service spacecraft deployed in the Atlantic and Indian Ocean regions, several of which also provide maritime service capability.

By 1987, the INTELSAT system will have in orbit four INTELSAT VI satellites, two in the Atlantic region and two in the Indian region; at least four INTELSAT V-A satellites, several of which are equipped with a business service package; two INTELSAT V baseline satellites without maritime capability; and an additional four INTELSAT V baseline satellites with maritime capability. Of the four maritime service packages, two will be in the Indian Ocean region, one in the Atlantic, and one in the Pacific. In terms of the need for additional spacecraft, the pacing requirement is saturation of the Major Path satellites in the Atlantic Ocean region, which should occur in the 1988-92 time frame depending upon the extent of use of bandwidth-efficient modulation techniques, such as time-division multiple-access/digital speech interpolation (TDMA/DSI) companded FDM (CFDM) and amplitude companded single sideband (ACSSB), and the extent of use of voice encoding, such as a 32-kbit voice. In addition, several satellites are reaching the end of their design life and need to be replaced, namely, the Indian Ocean region Major Path satellite in 1989 and the Pacific Ocean region Major Path satellite in 1991. The lease service spacecraft also reach end-of-life in the 1989-91 time frame, and consideration must be given to their replacement.

IV. Earth Segment Evolution

INTELSAT planning for the introduction of TDMA/DSI calls for first service on the Atlantic Major Path 2 in 1984, followed by introduction of the technique on the Indian Primary also in 1984, then on the Atlantic Primary satellite in 1985, followed by the Atlantic Major Path 1 satellite also in 1985. At present, traffic requirements do not dictate implementation of this technique in the Pacific Ocean region, although consideration is still being given to its use no earlier than 1986--motivated, in part, by a need to access the Pacific satellites digitally to interface efficiently with terrestrial feeder links to the Earth stations.

In 1984, the bulk of the traffic will be carried in the standard FM modulation technique along with a share in single channel per

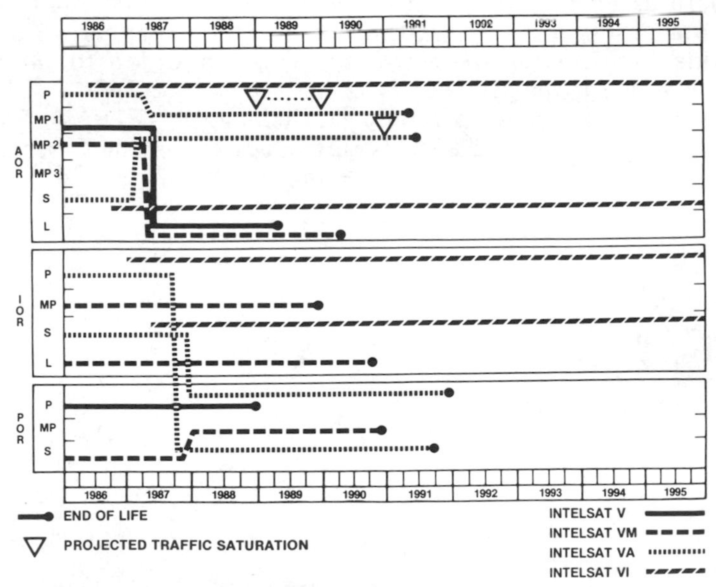

Fig. 5 Operational plan.

carrier (SCPC) and companded frequency-division multiplex (CFDM), with only a portion of the traffic carried in TDMA/DSI. From this point on, however, traffic growth will be accommodated in significant part by use of TDMA/DSI, with some 60% of total traffic carried in this mode by 1994.

One significant aspect of the introduction of TDMA is how INTELSAT plans to accommodate growth during the INTELSAT VI time frame that spans the 1986-93 and beyond period. During this period, INTELSAT expects at least 100% growth in international traffic requirements, however, INTELSAT VI provides only 40% more bandwidth than does the INTELSAT V-A spacecraft. Thus, to accommodate traffic growth requirements, the bandwidth available must be used more efficiently, and, to do this, INTELSAT will be relying heavily on the TDMA/DSI modulation access technique. TDMA/DSI is approximately two and a half to three times more efficient than standard FM communications and raises the 80-MHz channel transponder capacity of a satellite from some 1000 channels to approximately 2500-3000 channels, depending upon the number of users in a transponder and the efficiency of the TDMA operational plan, known as the burst time plan.

In contrast to the rapid growth in use of TDMA/DSI, INTELSAT expects relatively modest growth in multiple antennas to access the system beyond the mid-1985 time frame. One major factor driving the introduction of multiple antenna facilities within a country is to gain access to the several operating satellites serving in ocean regions. By the middle of the 1980's, adequate antennas will be in service to distribute the traffic over all currently foreseen operating satellites in the system. Additional antennas generally are not required, although a number of users will provide facilities as their traffic grows in order to provide diversified paths for the services.

A factor that may strongly influence INTELSAT Earth segment evolution is INTELSAT's introduction of new business

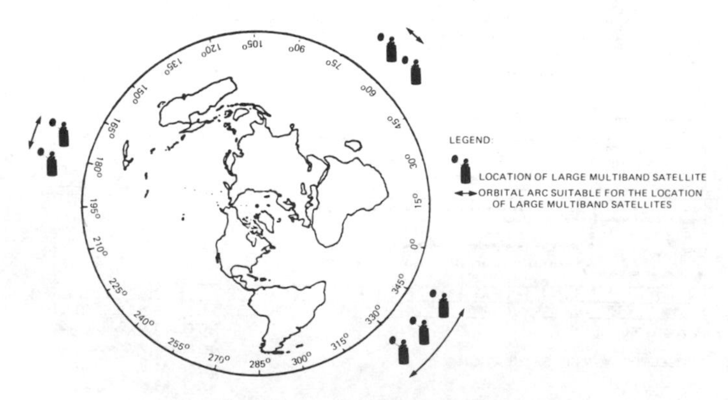

Fig. 6 Primary/Major Path concept.

services, with the attendant use of small urban gateway or customer premise stations to bring satellite services down closer to the user's actual source of demand. The eventual market for such a service remains to be seen, but service demand equivalent to several 40-MHz transponders is expected to evolve by the end of this decade in the North Atlantic region alone.

V. Post-1988 System Planning

The basic post-1988 INTELSAT system planning issue is the space segment facilities that will follow the first procurement of INTELSAT VI satellites. Only five INTELSAT VI satellites have been procured, and some 9 or 10 in-orbit roles (not including domestic lease service satellites) need to be filled eventually. The number and type of subsequent spacecraft depend upon the space segment architecture to be pursued.

A. Factors Influencing Planning Alternatives

Several alternative space segment architectures are under consideration that impact the design of the satellite. The current INTELSAT approach is referred to as the Primary/Major Path configuration; this configuration, in effect, uses the same satellite design in all system roles. Two alternative planning approaches under development are the trunk/connectivity and regional/transoceanic architectures, which are modular in nature and employ different satellite designs in different space segment roles to better match spacecraft capacity to requirements.

The concepts involving use of different satellite types are motivated by the imbalance of traffic levels between the Atlantic, with its high-density traffic routes, and the other ocean regions. This was illustrated in Fig. 3, which shows the geographic distribution of INTELSAT demand; the Atlantic/Indian/Pacific traffic levels scale roughly in a 4:2:1 ratio.

The cost of building even larger satellites, especially beyond the INTELSAT VI design, is a second major factor in system evolution studies. One unknown, however, is how large a satellite can be before it becomes unmanageable to design and manufacture.

Other services also impact the system design considerations, namely, the maritime, new business service, and domestic and international television lease requirements, which were reviewed earlier.

In the case of maritime service, INTELSAT V spacecraft currently carry a 30-channel system. The second-generation system, projected to be required in the 1988-89 time frame in the Atlantic by INMARSAT, requires a capability on the order of 125 channels. While INMARSAT has options other than leasing add-on packages from INTELSAT, the cost benefits of a multifunction spacecraft such as the INTELSAT VI with an add-on maritime communications package (which appears quite feasible) are very attractive. This is a major consideration in INTELSAT spacecraft design studies; furthermore, the operational implications in terms of INTELSAT's consequent flexibility to redeploy spacecraft between roles and ocean regions is an important consideration to both INTELSAT and INMARSAT.

In terms of the new business service capability, INTELSAT envisions a totally digital service operating in the upper 11- and 12-GHz portions of the K-band to permit Earth stations to be sited near urban centers in both North America and Europe. Such a service offering, where the services would be brought down closer to the user of small-dish antennas, should be attractive to private user networks. To this end, INTELSAT has adopted new standards for business service Earth stations, referred to as the Standard E antennas in the 14/11- and 14/12-GHz bands, which range from 3-1/2 to 8 m in diameter, and is developing analogous standards for small-dish 6/4-GHz antennas. Demand for this service is estimated to be between two and five equivalent 40-MHz transponders in the 1986-88 time frame.

B. Sources of Capacity Growth

There are basically three ways for INTELSAT to derive additional capacity, namely, the use of more orbital slots, the use of more bandwidth in each orbital slot, and more efficient use of bandwidth. INTELSAT is exploring all three alternatives to meet its requirements beyond 1988.

The use of more orbital slots, of course, requires additional satellites. The reverse is not true, however, inasmuch as satellites with complementary designs (for example, an all 6/4-GHz satellite and an all 14/11-GHz satellite) could be colocated in the same orbital slot.

The use of more bandwidth in a given orbital slot can be achieved in several ways: 1) through additional frequency reuse of the currently used bands employing a multibeam antenna; 2) through the use of dual-polarization in the bands not currently using this technology, namely, the 14/11-GHz band; 3) through the use of additional 6/4- and 14/11-GHz bands allocated to the fixed satellite service but not currently used by INTELSAT, or even the use of the higher-frequency 30/20-GHz bands; 4) or through the use of colocated satellites of complementary design to increase the bandwidth available in a given orbital slot.

The more efficient use of bandwidth can be achieved with several new modulation access techniques. In the digital area, the main technique is expected to be TDMA/DSI, perhaps coupled with voice encoding; 32-kbit voice encoding should be available this decade, and 16-kbit systems are under active development for the late 1980's/early 1990's. Also in the digital area would be PSK/FDMA/DSI, which achieves multiple transponder access in the frequency rather than the time domain, in contrast to TDMA/DSI. In the analog area, companded FDM is currently in service, and amplitude companded single sideband (ACSSB) is being reviewed by INTELSAT. This latter system is currently being used extensively in some American domestic systems.

C. Primary/Major Path Concept

The Primary/Major Path concept, illustrated in Fig. 6, represents an extension of the current system in that a single satellite design would be used for all roles in all ocean regions. This concept makes relatively good use of satellites in the Atlantic Ocean region, but somewhat less efficient use of satellites in the Indian and Pacific regions. This latter effect has been minimized by the time-phased introduction of spacecraft in the different regions. In the time-phased system, spacecraft are introduced first in the Atlantic, which has the greatest traffic demand, followed by introduction in the Indian and, subsequently (and this may be several years later), in the Pacific region. By that time, the demand in the Pacific has grown closer to the levels in the other two regions at the time in which the same satellite design was used there. Inefficiencies in use are further minimized by the use of different numbers of operating satellites in the three regions--currently three in the Atlantic, two in the Indian, and one in the Pacific. This approach would maintain our current orbital slot requirements and would provide system connectivity in each ocean region on the Primary satellite, with major traffic streams accommodated on the Major Path satellites. Under this design concept, the satellites continue to be located over the midocean portion of the orbital arc, which minimizes potential conflicts with domestic systems.

D. Trunk/Connectivity Concept

The trunk/connectivity concept, illustrated in Fig. 7, involves the use of two different satellite designs. The first would be a large satellite providing connectivity among the majority of system users, which, in effect, is much like the Primary satellites in use today. More than one connectivity satellite might be provided in the Atlantic Ocean region to cater for such services. These large satellites are complemented by smaller, more cost-effective satellites known as trunk satellites, which are designed and used in such a way as to carry larger traffic streams very efficiently. In the ideal case, the trunk satellites would have one traffic stream

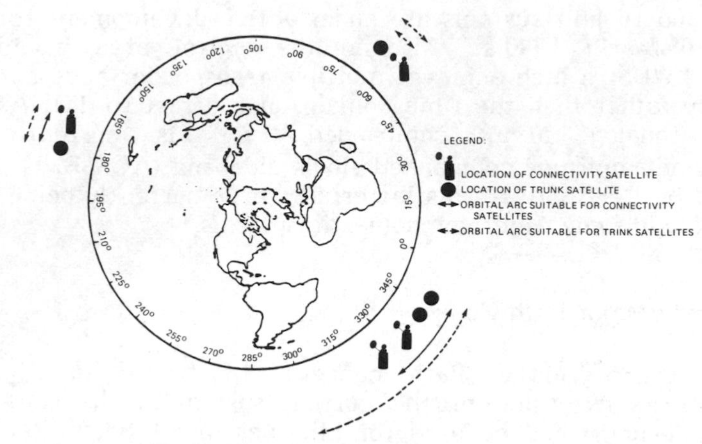

Fig. 7 Trunk/connectivity concept.

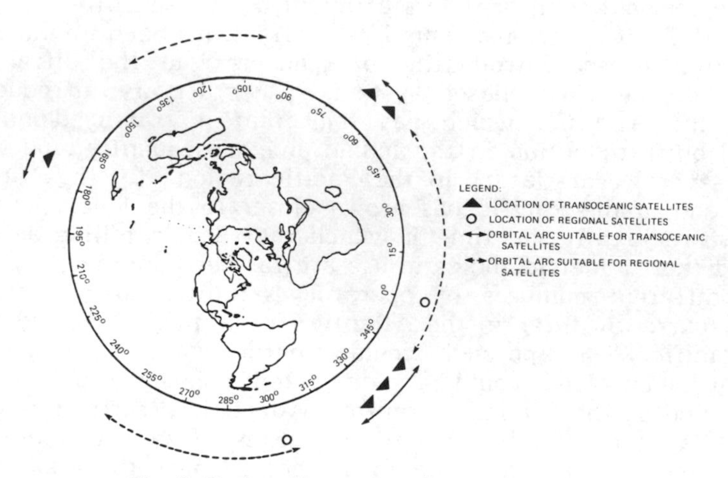

Fig. 8 Regional/transoceanic concept.

completely occupy a transponder on the satellite to avoid multiple-carrier operation, which maximizes achievable capacity. Implementation of this concept should provide a relatively good match of space segment capacity to traffic in all three ocean regions, and will require more orbital slots in the Atlantic and Indian Ocean regions due to the greater number of satellites that may be needed in the 1990's. While traffic would be partitioned according to the satellite role in the Atlantic and Indian regions, consideration is being given to a trunk satellite design sufficiently robust to provide Primary service in the Pacific Ocean region as well. In this case, the Pacific region will certainly require a Major Path satellite to meet foreseen demand. To ease transition requirements, the concept of trunk satellite needs must be fully developed early on to permit system users to plan their Earth segment facilities accordingly. Furthermore, the use of trunk satellites, if they are to be successful in achieving the space segment economy sought, will likely require the expedited introduction of voice encoding techniques.

E. Regional/Transoceanic Concept

The regional/transoceanic concept, illustrated in Fig. 8, like the trunk/connectivity concept, involves two distinct satellite designs--one large and one relatively small. The large satellites would provide for transoceanic service requirements. This could include some north-south traffic in both the East and West Hemispheres, but primarily the transoceanic satellite is to carry east-west services over each of the three ocean regions. These would be supplemented by smaller satellites, called regional satellites, which would be located closer to, and perhaps over, the continental regions and would provide full hemispheric coverage to support domestic service requirements. The concept of regional satellites would permit INTELSAT to satisfy the domestic service requirements of developing countries that will rely on satellites to achieve their basic telecommunications infrastructure and, at the same time, permit extension of this network, using the antennas of the domestic network, into the international environment, thus avoiding double-hop satellite transmissions. This concept would require the use of orbital locations close to domestic systems, which may be one potential drawback, and would permit accommodation of intercontinental traffic on the regional satellites other than that from the domestic system served by the satellite. In terms of relative timing, it is expected that the regional satellite design would not be considered for deployment before the end of this decade, in contrast to trunk satellites, which may be deployed earlier to meet pacing in-orbit needs, which currently is as early as 1988.

These three architectures are compared in the summary chart in Fig. 9.

F. Alternative Trunk Satellite Designs

To date, a number of alternative trunk satellite concepts have been identified. These are summarized briefly in Table 1. Three classes of designs have been considered. The smallest is the so-called "DELTA class," which would be based on a spacecraft bus launchable on the DELTA launch vehicle or the STS/PAM-D. This is the smallest spacecraft design under consideration, and, while it should be the least expensive on a unit cost basis, the spacecraft

SYSTEM CONCEPT	PRIMARY/MAJOR PATH	TRUNK/ CONNECTIVITY	REGIONAL/ TRANSOCEANIC
SPACECRAFT	SINGLE LARGE DESIGN	LARGE & SMALL DESIGNS	LARGE & SMALL DESIGNS
PROS	CONTINUES CURRENT ARCHITECTURE	MATCHES TRAFFIC IN ALL OCEAN REGIONS	FULL LAND AREA COVERAGE SUPPORTS DOMESTIC SERVICES
CONS	LOW UTILIZATION IN IOR & POR	MAY NEED MORE ORBITAL SLOTS	MORE SLOTS COMPETE WITH DOMSATS FOR ORBIT DIFFICULT TO SPARE

Fig. 9 Summary comparison of alternative system architectures.

Table 1 Alternative trunk satellite concepts.

Spacecraft size	Payload capability	Comment
DELTA class or STS/PAM-D	600-900 W 100-150 kg	HS-376 or DOMSAT derivative
(1280-kg transfer orbit)		Supports trunk mission only
		7-year on-orbit life
Atlas/ARIANE II or equivalent STS plus PKM	1000-2000 W 200-300 kg	HS-393 (new class) or INTELSAT V/V-A derivative
(2000-2400 kg transfer orbit)		Supports trunk/new business payloads
		10-year on-orbit life
ARIANE-IV class	1500-2500 W 500-750 kg	INTELSAT VI derivative
(3740-4200 kg transfer orbit)		May support maritime/ new business payloads
		10-year on-orbit life

could not serve the Pacific Primary requirement and, further, would likely provide only seven years of on-orbit life.

The second class of satellites is the Atlas/ARIANE II class design, which is also the INTELSAT V class. It would be launchable on the STS with a suitable perigee stage as well as on the Atlas/Centaur and ARIANE II expendable vehicles. Several satellite buses of this size either exist or are under development. This concept has the added virtue of potentially serving the total

Pacific Ocean requirements and thus could be procured in a larger quantity than the DELTA class, which should achieve production economies of scale. Furthermore, a 10-year orbital life should be achievable with this class of spacecraft, and other service requirements could be considered in designing the satellite; for example, the new business service requirements or the maritime service requirements. Either the DELTA class or the Atlas/ARIANE II class spacecraft would also be suitable for providing domestic lease services.

The third class of satellite is the INTELSAT VI spacecraft. In this case, to avoid considerable development costs, consideration is being given only to a derivative of the present INTELSAT VI satellite to make it more suitable for meeting trunk service requirements. Conceptually, this spacecraft could also be modified to provide Southern Hemisphere domestic service requirements along with high-density Northern Hemisphere services. A 10-year orbital life is achievable, and the spacecraft has the full potential to support maritime and new business service payloads, especially if launch on the STS is envisioned.

VI. INTELSAT Decisions and Schedules

The various alternatives for post-1988 system evolution are depicted on the INTELSAT "decision tree" in Fig. 10. This figure is most easily understood if reviewed from the bottom up.

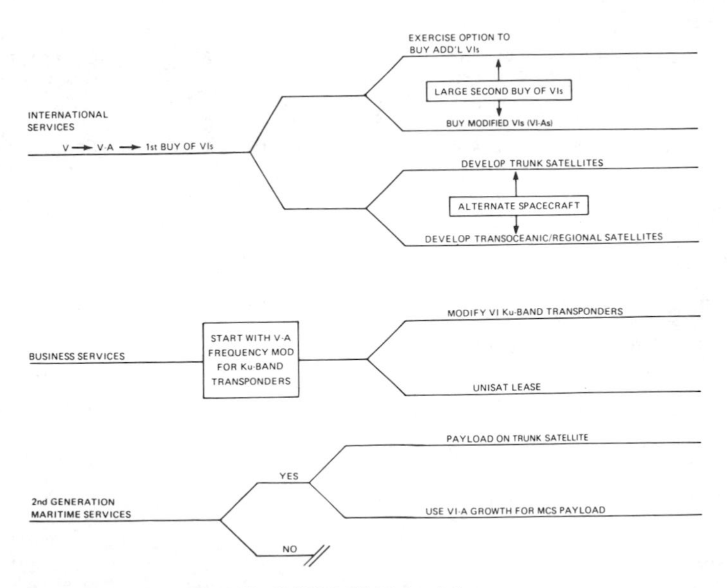

Fig. 10 INTELSAT decision tree.

The bottom line depicts second-generation maritime service. The basic decisions facing INTELSAT are whether or not to bid to INMARSAT to provide such services. If INTELSAT decides to bid and INMARSAT proceeds on its currently envisioned timetable for such satellites, INTELSAT's only alternative would be to accommodate a maritime package either on the INTELSAT VIs presently under procurement or on a growth derivative of the INTELSAT VI, which would be referred to as an INTELSAT VI-A satellite. This alternative is the only one that can be brought to fruition in time to provide a responsive bid to the INMARSAT Request for Proposals (RFP). If, however, INMARSAT eventually splits its procurement and delays spacecraft procurement decisions for service to ocean regions other than the Atlantic, then INTELSAT has the alternative of considering inclusion of a maritime payload on trunk satellites or on another new spacecraft design that may emerge. A complicating factor is the sizing of the maritime capacity required in the Indian and Pacific Ocean regions, and here the flexibility appears available to consider a package smaller than the 125-channel capacity package for use in the Atlantic. A decision to bid on second-generation maritime services is expected to be made by INTELSAT in early 1984, and a

Table 2 INTELSAT decision timetable.

Decision Date	Spacecraft	Deployment	Decision
Late 1984/early 1985	V-A or VI new business modifications	1987-89	Contract
Late 1984/early 1985	VI-A	1988-90	Requirements specification
Late 1985/early 1986	VI-A	1988-90	Contract award
Late 1985	Early trunk	1989-90	Release RFP
Late 1986	Early trunk	1989-90	Contract award
Early 1985/1986	Baseline VI	1989-90	Exercise contract options
Late 1985	Late trunk	1990-91	Release RFP
Late 1986	Late trunk	1990-91	Contract award
Early 1987	Trunk-variant or regional	1993	Release RFP
Early 1988	Trunk-variant or regional	1993	Contract award

vendor selection at least for the Atlantic is expected from INMARSAT by year-end 1985.

The second line from the bottom of Fig. 10 depicts business services. Here, there are three basic alternatives available to INTELSAT. INTELSAT has decided to modify three of the last INTELSAT V-A satellites to cater for business service demand. To extend this service beyond the Atlantic region, additional capability is required. Thus INTELSAT is considering either modifying additional INTELSAT V satellites, modifying satellites in the INTELSAT VI procurement, or procuring new satellites to provide a business service capability. Options available to provide a second-generation business service capability are much broader and less well defined at this time. Spacecraft decisions regarding extensions of first-generation business services beyond the Atlantic are expected in 1985, with decisions on a second-generation service due several years later.

The top line of the "decision tree" indicates the two basic options for evolution of the international service--involving continuation of the Primary/Major Path architecture or adoption of the trunk/connectivity architecture. The upper alternative to the right of the branch point envisions purchase of a relatively large number of INTELSAT VI class satellites. Such satellites could be of the baseline design currently under procurement, or they could be modified INTELSAT VIs that include a maritime capability, domestic lease capability, business services capability, or even enhanced capacity for international requirements. The second basic alternative is to proceed along the lower branch, which envisions the procurement of a new satellite design (either trunk or regional satellites), for deployment in this decade. This path does not preclude the purchase of additional INTELSAT VI satellites, but any such procurement would be in a limited quantity so that the total procurement of INTELSAT VI satellites does not exceed, for example, eight spacecraft.

Table 2 depicts the INTELSAT decision schedule associated with this decision tree. As can be seen, the earliest decisions are associated with the provision of business service. Subsequent decisions are motivated by the maritime service requirement and by the option terms of the INTELSAT VI contract, which provides economic incentives for early exercise of contract options. In the absence of a maritime requirement, INTELSAT need not take a major decision until late 1984 at the earliest.

As can be seen from the decision schedule, the regional satellite concept is not under active development for use in this decade. The exception to this would be if a trunk satellite design proved to be sufficiently flexible to also serve the regional role, in which case a hybrid spacecraft design, in effect, could be introduced during this decade to serve both the trunk and regional satellite roles.

VII. Conclusion

A very active decision planning process within INTELSAT is underway and will be continuing in response to a rapidly evolving planning environment. The driving force behind this planning process is the need to procure additional spacecraft for deployment in the 1988-90 time frame. The planning for such additional spacecraft must take into account the total system requirements, and, while the basic need for INTELSAT to provide mainstream international services is fully recognized and represents INTELSAT's major business, other service requirements are also key elements in planning.

Planning must maximize system efficiency and will likely require the use of new technologies in both the Earth and space segment; for example, a cost-effective trunk or other innovative design spacecraft to complement INTELSAT VI satellites expected to be in service through the mid-1990's. Furthermore, planning will probably rely on advanced spacecraft technologies to effect launch vehicles savings and to meet extensive system traffic requirements on a relatively modest payload. The use of digital modulation access techniques at Earth stations is envisioned along with the introduction of voice encoding to make efficient use of the spacecraft resource. Finally, major decisions for the follow-on system will be made in the 1984-85 time frame.

13. Research and Development

D.K. Sachdev*
INTELSAT, Washington, D.C.

I. Introduction

The launch of the first successful geosynchronous satellite in the early 1960's was a tribute to the precision rocket technology as well as the maturity of communications hardware. This advance was preceded by over two decades of terrestrial microwave system development. It is not surprising therefore that the payload configuration of that satellite had considerable commonality with a typical 6- or 4-GHz terrestrial microwave repeater. In fact, the early synchronous satellites were quite aptly described in popular literature as the "tallest microwave towers with the lowest cost per meter." However, as the satellite systems began to grow, the unique technology requirements of communications satellites began to surface, and the necessity of a dedicated and systematic research and development effort in all aspects of satellite communications systems became readily apparent.

Well before the first INTELSAT launch, several academic and industrial laboratories had devoted basic and applied research efforts to specific areas of satellite commmunications. Indeed, they have continued to do so ever since. However, INTELSAT was among the first to develop the whole spectrum of satellite technology in a planned manner.

Concurrent with the growth of the INTELSAT R&D Program, several industrial organizations and, even more noteworthy, national and international organizations--such as CNES, ESA, ISRO, NASA, and NASDA--have embarked upon extensive research programs in satellite technology as a whole. The primary focus of their efforts, however, is toward specific scientific missions or applications and toward building an industrial capability in the concerned organizations or regions. On the other hand, the INTELSAT R&D Program, as an integral part of the unique and

---

Invited paper received October 11, 1983.

*Manager, R&D Department.

largest satellite-based communications network, has continued to place major emphasis on advancing the state-of-the-art on a global basis.

This chapter is devoted to a brief overview of the past, current, and future INTELSAT R&D Programs. The introductory material is followed by brief summaries of major technological developments in specific areas. In each case, the description starts with past achievements, followed by a brief summary of current activities and those envisioned for the future. The final section pulls together a brief overall prognosis for satellite technology as a whole.

II. INTELSAT R&D Program

The R&D Program as it is known to us today, started in 1968 when the Interim Communications Satellite Committee (ICSC) established its basic objectives. After a few years of experience, in 1974 the Board of Governors formally adopted the INTELSAT R&D Policy and Procedures to provide an overall framework for R&D efforts and their interface with the INTELSAT network. The farsighted approach of this policy has been vindicated not only by substantial technological breakthroughs but also by the fact that the policy, when reviewed again in 1980, needed only marginal revisions in order to take into account the changing requirements and the modified INTELSAT management structure. Relevant extracts from this policy are set forth below.

A. Objectives

The INTELSAT R&D Program, through technological and scientific research and development, aims toward:

- the optimum design, development, construction and establishment of the INTELSAT system;
- the efficient maintenance, utilization and operation of the INTELSAT system;
- preparations for the longer-term future of the INTELSAT system, both for existing kinds of service and for appropriate new kinds of services.

The objectives go well beyond the development of specific technologies and place considerable emphasis on continuous interaction with, and benefit to, the system as a whole.

B. Program Elements

Before examining the specific elements of the INTELSAT R&D Program, it is well to understand that, unlike R&D efforts in corporations where products emanate directly from technological efforts, INTELSAT R&D has to play a background role in stimulating industry and providing in-house expertise in such a manner that future INTELSAT spacecraft and Earth stations benefit from the latest and most reliable technologies. This lack of directly identifiable R&D results is not unique to INTELSAT and is typical of research programs that are part of operating entities rather than industrial corporations. This so-called lack of direct visibility is, however, counterbalanced by the inherent flexibility available for undertaking relevant long-term research that is not necessarily justifiable from short- or medium-term commercial considerations.

In recognition of the above objectives and organization needs, the INTELSAT R&D Program has the following three broad elements:

1. Exploratory Research and Studies (ER&S). This part of the program includes research effort required to lay the basis for the long-term growth of communication satellite technology suitable for INTELSAT applications. The effort is directed toward research and advanced development on new and unexplored techniques, review and assessment of non-INTELSAT R&D, and technology studies directed toward the adaptation of new techniques to the needs of INTELSAT.

In addition, this effort is required to maintain a continuing staff expertise, utilizing external support wherever necessary to:

a) ascertain the technologies suitable for INTELSAT application;

b) determine the need for, and possible approaches to, INTELSAT development projects;

c) support the implementation of INTELSAT development projects; and

d) provide support for ongoing INTELSAT engineering implementation and operational systems.

2. Development Projects. This part of the program includes specific projects related to an INTELSAT need for technology development or for acquiring knowledge related to the special environmental and behavioral circumstances encountered in developing and operating a communications satellite system.

Development projects involve contracted-out effort as well as management effort related to the formulation of project objectives, postcontract evaluation, and reporting of results.

3. Management Effort. The overall management of INTELSAT R&D is the direct responsibility of the Director General. The Program provides for a continuing level of effort by the Executive Organ staff for this purpose. In addition, the

management effort provides for outside service contracts to assist the Director General in preparing specifications, contract monitoring, testing and integration of deliverables, etc.

C. Implementation

INTELSAT is not only a global communications network but also has a truly global ownership, which provides practically the entire relevant industrial capability in the world as a base for the INTELSAT R&D efforts. Accordingly, as a policy, a major part of exploratory research and studies (ER&S) and development projects is contracted-out through a competitive procurement process. While this global operation rides on the technological expertise throughout the world, it has a slight disadvantage in somewhat delaying the start of research effort. Superimposed on this, of course, is the need to interact with the diverse operating styles in different parts of the world. However, the benefits of this global and diversified operation have proved their worth, both in terms of providing the best available technology for the INTELSAT system and in making significant contributions toward advancing the state-of-the-art on a global basis, including a significant effort in developing countries.

The third element of the R&D Program, management effort, covers all efforts and expenditure by the in-house R&D team at INTELSAT. Specifically, it includes:

a) Program formulation and responsibility for its acceptance and approval by the Board of Governors;

b) implementation of the Program through contracted-out and in-house efforts;

c) systematic documentation of ongoing results and their assimilation and utilization by the system; and

d) support to future system plans, engineering management of spacecraft programs, and operation of the INTELSAT system.

D. Financial Aspects

Satellite communications technological advances, in spite of substantial gains in the last two decades, still show no signs of abating; therefore, a careful balance is necessary in arriving at the optimum financial investment in R&D. While a typical development project (and some ER&S contracts as well), extends beyond one or two calendar years, the new program authorizations are approved annually by the INTELSAT Board. Therefore, at any given time, the "dollars at work" are two to three times higher than the annual authorizations. To some extent, the annual authorizations have maintained a steady percentage relationship to

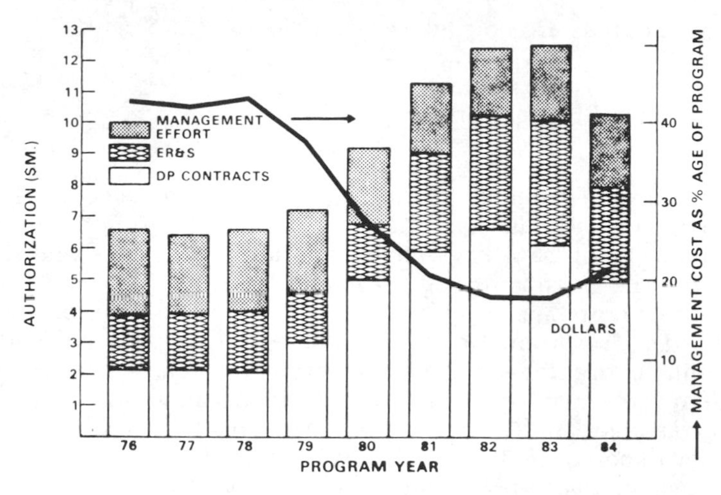

Fig. 1 R&D Program authorizations.

the annual INTELSAT revenues. However, as a policy, the major guiding principle in deciding the amount of effort to be expended is the future technological needs for the INTELSAT system over and above likely industrial developments, rather than a fixed percentage of revenues. Figure 1 summarizes the authorizations for the past few years. It also shows the relative authorizations for the three major elements of the program.

III. Major Technological Developments

Every R&D program, attempting to cover a broad range of system needs, adopts a certain classification that tends to be unique and to some extent related to the organizational structure. Of course, the structure itself needs to be modified from time to time to keep in step with the changing mix of R&D efforts.

For the past five years INTELSAT R&D has divided its effort into four different but flexible categories. These are: antennas and propagation; communications; materials and devices; and spacecraft technologies. The first two categories, and to some extent the third, are associated with what is described in the literature as the payload and overall communications link. The last category--spacecraft--covers all the disciplines associated with the spacecraft bus. Of course, in each of the categories, the overall program components described in Sec. IIB apply. Finally, new areas covering multiple disciplines, and frequently with the potential of leading to new system concepts, are addressed through a subcategory of ER&S, entitled Interdisciplinary Research (Sec. IIIE).

Individual areas are now addressed below:

A. Antennas and Propagation

1. Antennas. Antenna technology, from the very birth of satellite communications, has been crucial to its growth. In the early generation of satellites, due to relatively small bandwidth requirements and constraints on mass and launch vehicle envelope, the spacecraft antennas provided rather simple global or broad coverage. The resultant lower spacecraft e.i.r.p.s required a correspondingly greater emphasis on ground antennas. While the latter benefited considerably from radio astronomy techniques, a significant amount of R&D work was necessary before what is still the basic workhorse of the INTELSAT network, the Standard A antenna design, was perfected and standardized. The early emphasis was on maximizing Earth station receive gain-to-noise temperature ratio (G/T) performance. With the rapid increase in bandwidth requirements, dual-polarization capability was added to Earth station antennas. Increasing emphasis today is directed toward better sidelobe performance, lower costs, and smaller Earth station antennas.

Commercial spacecraft antenna technology development, due to the unique and very demanding requirements of INTELSAT, continues to be led by INTELSAT R&D. The long succession of efforts (both contracted-out and in-house) are best described via major INTELSAT spacecraft series.

INTELSAT IV was the first satellite to use the full C-band (i.e., the 3.7- to 4.2-GHz downlink and 5.925- to 6.425-GHz uplink). Even before the first such satellite was launched, however, it was realized that even greater bandwidth--through frequency reuse--would soon be required. Accordingly, R&D efforts in the 1970's addressed two major aspects: namely, frequency reuse through both spatial isolation and dual-polarization, the latter requiring corresponding R&D and implementation effort at Earth stations as well.

Frequency reuse by means of spatial isolation--utilized in INTELSAT IV-A for two reuses--required development of software and hardware techniques for shaped beams with low and controlled sidelobes. A series of efforts under INTELSAT R&D and within industry led to the successful realization of this technology. Offset reflector antennas were introduced to eliminate feed blockage and scattering. Multielement feed arrays (typically 40 feeds/antenna on IV-A) fed by beam-forming networks were utilized to synthesize the shaped-beams coverages and to provide sidelobe isolation into the adjacent beam coverage areas. New criteria for the far-field evaluation of such antennas were developed.

To avoid Faraday rotation-induced polarization tracking at Earth stations, circular polarization was chosen for C-band applications beginning with INTELSAT III [left-hand circular polarization (LHCP) for uplink, right-hand circular polarization (RHCP) for downlink]. An experimental dual-polarization global coverage antenna was flown on INTELSAT IV-A satellites (F-2 and F-3) to establish fully the use of this technique in subsequent satellites and to provide a test source for the calibration and retrofit of Earth stations prior to the introduction of dual-polarization reuse on the INTELSAT V satellites. Circularly polarized dual-polarization providing 30 dB of isolation required tremendous advances in polarizer technologies during the 1970's for both satellite and Earth station applications. INTELSAT R&D demonstrated several significant breakthroughs in polarizer capabilities during this period.

The tremendous growth in INTELSAT during the 1970's benefited substantially from the above two basic developments, as demonstrated by the spectacular success of INTELSAT IV-A and V satellites. INTELSAT V achieved higher capacity through four-times frequency reuse (two spatial, two polarization reuses) made possible through substantial advances in software analysis and synthesis, beam-forming technology, polarizer development, and feed array fabrication techniques. This development, of course, has since continued to grow through INTELSAT R&D and industry efforts and has several facets to it, as summarized in the following listing of several ongoing technology efforts.

1) Multiple-Beam Technology. From the mid-1970's onward, a major share of INTELSAT R&D antenna efforts has been devoted to this area, which is fundamental to higher reuse of all the frequency bands. One of the early developments was the adoption of offset geometry to avoid blockage and scattering inherent in the growing size of the feed structure relative to the reflector. As the number of beams is increased, it becomes increasingly difficult to use a fixed beam structure in all three of the INTELSAT primary ocean locations and for different coverage applications. This led to fundamental work on reconfigurable beam technology, a subject of substantial ongoing investigations. Through the use of low-loss variable power dividers (VPDs) and phase-shifters (VPSs), technology to realize a wide range of antenna footprints through a common feed and reflector for a multiport system has been brought to a high level of maturity. Current emphasis is toward refinement of software minimization of VPD and VPS losses; mass, size, and power requirements; quantification of the improvements in beam isolation and available gain.

2) Dual Reflector Systems. With increasing frequency reuse (six-, eight-, or tenfold), the single reflector size and the whole beam optics become large and lead to complex packaging and

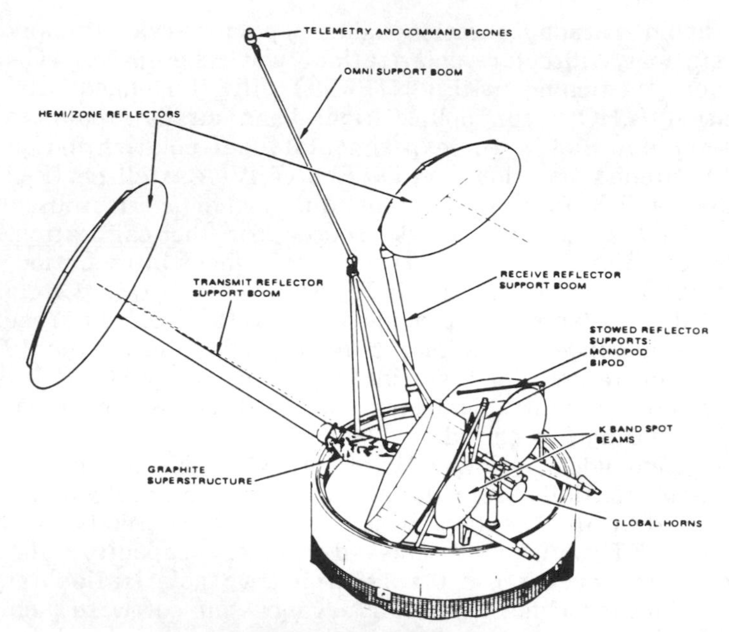

Fig. 2 INTELSAT VI antenna systems.

deployment and incur excessive losses due to scan aberrations, beam-forming networks, and long waveguides between the output amplifier and the feed systems. (Figure 2 shows INTELSAT VI antenna systems.) Use of a subreflector provides a technique for folding the optics and could lead to better packaging and also thermal control. However, the increased degrees of freedom, while they may substantially improve electrical performance, are associated with more difficult mechanical tolerances and alignments. Current R&D efforts are aimed at the development of new software and prototype systems for future applications.

3) Active Antennas. All spacecraft--both those of the past and most of those in production now--utilize discrete transmitters followed by a self-contained passive antenna system. This approach requires increasingly complex power splitting and phase-shifting to be performed after the output traveling wave tube amplifiers (TWTAs) or solid-state power amplifiers (SSPAs), resulting in losses at the highest power levels. To overcome this, development work is underway to integrate low-power solid-state power modules with the antenna elements themselves. Such an approach also shifts the beam steering function to a lower power level. Work is still necessary, though, to improve the linearity and efficiency of the low-power modules and the miniaturization of beam-forming networks before such a system could become

attractive for future applications. Ultimately, such an approach could lead to compact, directly radiating "triplate" systems, fully integrating the output power amplifiers (and receivers) with the antenna beam-forming network and feed radiator system.

4) Earth Station Antennas. The early work, as mentioned previously, was focused on maximizing receive G/T (primarily by adopting shaped Cassegrain reflectors) and developing reliable measurement methods for calibration purposes (radio star, satellite, and boresight). The introduction of circular dual-polarization required the development of new polarizer, orthomode junction, feed horn radiator, and tracking mode coupler technology. Significant effort under INTELSAT R&D was devoted to the development of measurement methods for calibrating dual-polarization Earth stations. INTELSAT sponsored two Earth station antenna technology seminars with worldwide participation of Earth station owners, contractors, and technology experts to consider all aspects of the transition to dual-polarization with the introduction of the INTELSAT V satellites.

Beam waveguide systems were introduced that permitted high-power amplifiers (HPAs) and receivers to be located at the ground level of the Earth stations, substantially improving the access for monitoring and maintenance as well as reducing waveguide losses and the overall cost of implementation.

INTELSAT R&D also sponsored early work on multifrequency (circular dual-polarization 6- and 4-GHz, single-linear polarization 14/11 GHz) Earth station antenna feed systems.

As a result of the 1979 World Administrative Radio Conference (WARC), additional bandwidth has been allocated to the Fixed Satellite Service (FSS) and INTELSAT R&D has demonstrated the feasibility of broader-band Earth station feed system technology.

With the introduction of dual-polarization, the uplink beam isolation and sidelobe performance from the Earth station have become the critical link parameters (rather than the earlier downlink beams G/T parameter). Significant efforts at sidelobe reduction for large- and small-aperture antennas are underway to improve the interference environment, satellite spacing, and frequency reuse capability consistent with cost effectiveness.

2. Propagation. During the first few years of INTELSAT operation, most of the Earth stations were in the temperate zone of the Earth. As more and more Earth stations came into the INTELSAT system, an increasing number reported rapid, and unexplained, signal fluctuations. It was necessary to find out what was causing these fluctuations, and thus the INTELSAT Propagation R&D Program was born!

That was in 1969. Since then dual-polarization has been implemented in the 6/4 GHz bands, and the 14/11 GHz bands have

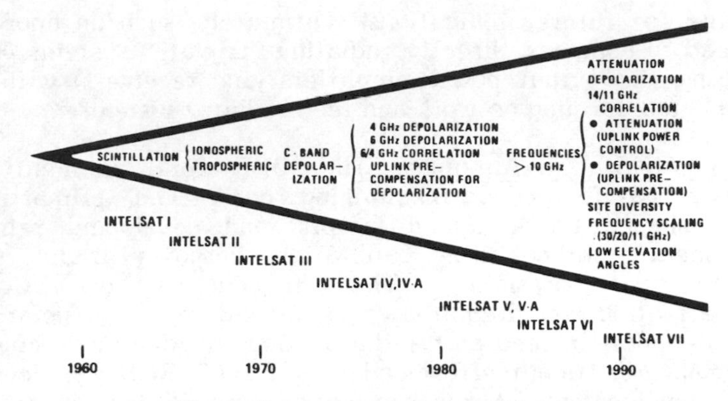

Fig. 3 Propagation-related activities.

begun to be utilized. As each new technique or band has been proposed, it has been necessary to investigate the impact these might have on the INTELSAT system and to predict the likely effect of the propagation medium.

Figure 3 shows schematically how the propagation topics requiring investigation have grown as the INTELSAT system has expanded. At frequencies of 6 and 4 GHz, the propagation medium is fairly benign. Even heavy thundershowers rarely cause a 3-dB fade. Rain, however, does cause significant depolarization and, to investigate this phenomenon, 6/4 GHz depolarization measurements have been carried out within the INTELSAT propagation R&D in Australia, Taiwan, Italy, Alaska, Hong Kong, Japan, and the United Kingdom. The latest experiment is seeking to define a depolarization precompensation system that could be used in the late 1980's within the INTELSAT system.

At frequencies of 14/11 GHz, rain not only depolarizes the radio signal, it can also severely attenuate it. To provide data for the worldwide INTELSAT propagation data base, extensive slant-path attenuation measurements using 11-GHz radiometers have been carried out within the INTELSAT Propagation R&D Program, yielding over 200,000 h of data. These data were instrumental in setting up the INTELSAT computer model that predicts propagation impairments.

Future propagation measurement programs aimed at acquiring data for uplink power control, site diversity, and depolarization compensation systems in the 14/11 GHz bands are underway or in their implementation phase on three continents. Much of the INTELSAT propagation measurement work has led to the direct involvement of many of the developing nations that are members of INTELSAT. These nations have benefited greatly through these, and other, R&D contracts. Future 30/20 GHz experiments should see these nations increasingly involved.

B. Communications

This area, covering a wide range of payload and overall system-related technologies, has accounted for a significant fraction of the effort under the R&D Program. Overall, practically all new modulation techniques and system concepts adopted to date by INTELSAT have emerged from R&D work in this area. Due to the wide range involved, only significant highlights are briefly described below:

1. Transponders. The early effort during the 1970's focused on the introduction of a range of advances in devices and packaging techniques. These included: 1) microstrip technology; 2) field effect transistor (FET) amplifier front ends and medium-power stages; 3) dual-mode waveguide filters; 4) filter fabrication technology, including graphite epoxy; and 5) transponder linearization (see below).

2. Echo Control. This has been one of the major areas of investigation for over a decade and has provided a series of advances, first through efficient echo suppression and more recently with echo cancelers. Current efforts are devoted toward the realization of miniaturized digital echo cancelers for the pulse code modulation (PCM) hierarchy at 30 channels.

3. SPADE System. This system was developed in recognition of the need for a cost-effective and efficient system to provide single channel access on demand. The system is now in extensive use throughout the network. Recent work has led to simplifications and cost reductions.

4. TDMA System. This is one of the major achievements of INTELSAT R&D and culminated recently in the adoption of such a system as the basic digital system in the near future for the INTELSAT network.

Work for this technology was initiated in the early 1960's, and a number of versions were developed and put on field trial. To some extent, the time-development multiple-access (TDMA) development paralleled the corresponding developments in digital large-scale integration (LSI) technology, and the standard 120-Mbps TDMA/digital speech interpolation (DSI) adopted by INTELSAT represents a sound culmination of efforts by both endeavors. A series of R&D contracts during the last 18 years has contributed to the industrial availability of cost-effective Earth station equipment on a production basis.

5. SS-TDMA. The use of TDMA over multiple-beam satellites requires efficient and dynamic interconnection techniques onboard

the spacecraft. INTELSAT R&D has been a pioneer in this area as well, and several outside contracts, in-house effort, and current INTELSAT VI related procurements have established this technology, which is being increasingly adopted by several other networks as well. Current R&D efforts are devoted toward improvements in onboard SS-TDMA capability and reliability through miniaturization and use of switching FETs.

The above is a very brief listing of past successful communication related endeavors under INTELSAT R&D. Current activities aimed at future generation spacecraft and systems are summarized below.

6. Solid-State Power Amplifiers (SSPAs). This has been an ongoing topic for several years; however, extensive exploitation by INTELSAT spacecraft is still mostly in the future. Both through device development (see Sec. IIIC) and amplifier development, INTELSAT R&D efforts have been a significant driving force in this field. Specifically, the first all-solid-state/power-amplifier (SSPA) satellite recently flown by RCA utilizes significant results from INTELSAT-funded effort. Current efforts--both at 4- and 11-GHz--are aimed at achieving reliability, efficiency, and linearity levels better than or equal to TWTAs.

7. Transponder Linearization. In recognition of the significant capacity/performance limitations due to nonlinear TWTs, for the past decade INTELSAT has funded a series of efforts to alleviate this deficiency. Earlier efforts included feed-forward techniques; however, recent major work has been on predistortion techniques. System level measurements have been carried out for alternative approaches for both FDMA and TDMA systems. Recently a flight-qualified unit was also built. Near-term applications are envisioned in several future payloads.

8. Monolithic Technology. As the satellite payloads become more complex, the overall reliability could deteriorate simply due to increased parts count. Monolithic technology, through miniaturization and substantial reduction in assembly and fabrication steps, provides an important tool for the future. Under INTELSAT R&D, an advanced 6-GHz receiver (used in rapidly increasing numbers in multibeam satellites) is currently under development. Several other efforts such as an SS-TDMA switching module and 11-GHz receiver are planned to exploit microwave-monolithic integrated circuits (MMIC) or hybrid technologies of the future.

9. Signal Processing. The mass and cost of practically every satellite are closely related to the total bandwidth requirement. The latter, in turn, is directly related to the radio-frequency (rf)

bandwidth used for each information or signal unit. Accordingly, efforts to reduce the bandwidth requirements without loss of quality have been an integral part of INTELSAT R&D as well as several other similar endeavors in the telecommunications field.

For voice, adoption of TDMA in its simple form requires 64 kbps for each voice channel. A series of R&D efforts worldwide, including INTELSAT's, has culminated in successful development of 32 kbps encoding without significant loss of quality and acceptable complexity. Current efforts by INTELSAT R&D will soon lead to field trials to be followed by adoption of 32-kbps/DSI as a standard. Concurrently, efforts are underway for 16 kbps encoding.

For video transmission, a full 36-MHz transponder has been a standard for the past 10 years. Half-transponder services have also been provided but with loss of quality. During the last 3 years, INTELSAT R&D has initiated development of efficient techniques for bandwidth reductions with digital techniques without loss of quality. A reduction by a factor of 6 (from 120 to 20 Mbps) is the target. Industrial development of codecs capable of interworking between all principal TV standards has also been started. Such codecs will join the growing family of lower bit rate codecs (down to 1.5 Mbps) for specific applications such as video-conferencing. Likely systemwide utilization of such a range of codecs, on demand, has also stimulated the development of a demand-assigned video TDMA system under INTELSAT R&D.

10. Onboard Processing. The "transparent" repeaters of INTELSAT I through V will be replaced in part by the SS-TDMA switch interconnected transponders of INTELSAT VI. This would represent the first step toward the new ongoing developments, grouped together under "onboard processing," which are judged to be for future payloads to provide the necessary interconnectivity, flexibility, and improved link performance. INTELSAT R&D and other efforts, notably by NASA and others, aim to achieve these objectives through several developments: 1) Onboard demodulation and modulation (regeneration) to substantially reduce the uplink impairments; 2) Onboard buffering and rate change to provide a range of transmission rates rather than one fixed rate as in TDMA and SS-TDMA; and 3) Onboard baseband switching with or without rf switching, an optimum combination of which can provide interconnectivity even at a single voice channel both within an rf band and between rf bands.

A series of efforts is envisioned in this area so as to provide all the tools for the future system designers, who may, in practice, choose to utilize only the most beneficial elements of onboard processing for specific system benefits.

11. High-Speed Digital Systems. The systems using 120-Mbps TDMA and SS-TDMA will efficiently provide interconnectivity and transmission during the 1980's. However, some of the very heavy

traffic streams will require a point-to-point capacity that is severalfold more than that of one 120-Mbps system. Such requirements are more efficiently served with a higher-capacity digital carrier such as 360 Mbps. Development of such a system is being evaluated from the point of view of the devices' availability and the necessary system architecture required and could be undertaken in the near future to allow more efficient competition with cable systems.

C. Materials and Devices

1. Travelling Wave Tubes (TWTs). From the beginning, INTELSAT R&D has played a lead role in the development of TWTs for space applications.

At 4 GHz, one major effort for an efficient TWT and its electronic power conditioner (EPC) was carried out during the 1974-78 timeframe. State-of-the-art 12 W TWTs were developed under this program, with several performance attributes not bettered even to this day; notably, high efficiency, low noise figure, low operating temperature of oxide cathode, etc. Unfortunately, the development work was performed too late to be a major influence on the INTELSAT V and V-A TWTs. However, the benchmark set by the TL4012A TWT spurred increased activity among other tube companies in the 1979-81 period--leading to major upgrading of performance between INTELSAT V and INTELSAT VI.

At 11 GHz, the INTELSAT-sponsored effort has been in two stages. The first on TWT proper, in the 1972-77 time period, and the second specifically for cathode technology.

The TWT development work sponsored in the industry made a major contribution to the INTELSAT V/V-A/VI technology base. Though the R&D work had been concentrated upon a high performance 20 W TWT (with EPC), the version selected for use initially on the INTELSAT V spacecraft was a very close, down-scaled, derivative. The tube manufacturers' indebtedness to INTELSAT as a sponsor of the development work associated with their commercial involvement in the INTELSAT V program is openly acknowledged in the literature. This tube was the only source of well characterized/well optimized devices available at that time.

Although early confidence in the INTELSAT V 11 GHz TWT design was high, the hidden intricacies of its compact configuration and the downscaling to 10.5 W output remained inconsequential until well into the INTELSAT V (and indeed INTELSAT VI) programs, when problems with cathode current decline, microdischarges (SSOs), and gain decay were noted. TWT hardware delivered under the two R&D contracts was readily at hand for use in diagnostic testing.

In summary, R&D efforts provided the impetus and technical guidance for subsequent INTELSAT spacecraft use of 11 GHz TWTs with well optimized basic RF and DC characteristics.

At 20/30 GHz, R&D work has been limited to TWT development for ISL application alone and, as such, this was originally part of the key technology base for INTELSAT VI. The technology, however, has ready application to Earth-space FSS requirements. After a series of seemingly insurmountable hurdles, this development is now approaching completion and represents a significant advance in the TWT design and fabrication technology.

2. Cathodes for TWTs. Although a subset of TWT efforts, the development of cathode technology is so crucial to the successful operation of TWTAs that INTELSAT has sponsored several projects in this area. Initially, the effort on cathodes dealt mostly with diagnostic testing of standard cathodes (oxide type, and simple dispenser emitters). More recently--and following the cathode performance shortfalls encountered in the INTELSAT V 11 GHz TWT program (see above)--attention has centered on characterization and reliability surveys of advanced dispenser cathodes. So it is convenient to subdivide attention in this area as follows:

1) Diagnostic Testing. One major benefit derived from this work has been the use of the Joule cooling effect as an in-orbit check on the operating margin remaining in cathode emission (via observation of the "hot-start" characteristics). Separately from this, some of the test methods for which correlation/justification were sought under the original contractual studies were also applied to INTELSAT V TWTs (both 4 GHz and 11 GHz devices). C-band TWT performance proved very consistent; the Ku-band cathode current decline has been referred to above and was characterized in detail.

2) Advanced Dispenser Cathodes. Upon realization that simple dispenser cathodes were--in essence--over-stressed parts within conventional satellite Ku-band TWTs, and (also) that new screening techniques alone could not restore acceptable reproducibility or longevity of performance, a deliberate shift of emphasis was made toward promotion of advanced dispenser-type cathodes for TWT application. This has been a very successful enterprise to date.

Initial work in this area (started in 1979) involved a survey of nine different advanced cathode types and some life-testing of two of those nine. That effort became a linchpin in the subsequent adoption of M-type dispenser cathodes in the 11 GHz TWTs for INTELSAT VI. Furthermore, some ancilliary work using the life-tested M-type cathodes afforded enhanced confidence in a whole range of 11-12 and 20 GHz TWTs for potential use on other spacecraft (possibly including later INTELSAT VI satellites).

More recently still (since 1982), contractual efforts have been concentrated upon actual incorporation of M-type cathode into high performance TWTs, but with emphasis on in-situ testing/screening of the cathode. The actual TWT used here is from a 20-40 W series of 11-12 GHz devices which may well see use on future INTELSAT spacecraft. Furthermore, INTELSAT participation in this work has promoted new approaches toward the quantitative estimation of cathode characteristics in operational life, new methods of temperature determination and initial emission screening.

In summary, INTELSAT R&D initiated a broadly based promotion of superior cathodes for use in Ku-band TWTs, for which reliability enhancement was sorely needed. This effort has been directly beneficial to the INTELSAT VI program and will have continued application and influence well beyond the current spacecraft series.

3. Power Transistors for SSPAs. At 4 GHz, prior to 1981, only limited INTELSAT R&D sponsorship had been applied in this specific area. Since SSPA performance/reliability is dominated by the individual characteristics of the component FETs, it was judged timely to expand the scope of the activity at that point. Work on the newest active project covering C-band gallium arsenide FETs is still in progress. Recently it has been established by in-house study that the final devices to be so developed should be compatible with applications in SSPAs up to 20 W output power. If this is indeed within grasp, then the way would be paved for fully solid-state amplifiers at C-band on INTELSAT spacecraft.

At 11 GHz, device (and technology) development for utilization in Ku-band SSPAs is altogether of a more exploratory nature than efforts at 4 GHz. Device performances attained under the more recent contract are state-of-the-art and are derived from a highly reliable FET.

4. Low-Noise Transistors for Receivers. It was concluded from the contracted-out effort under three earlier contracts that gallium arsenide FET preamplifiers provided optimum receiver performance. Certainly at that time (1978) the noise figure performance of gallium arsenide LNFET amplifiers was substantially better than provided for in the INTELSAT V 14 GHz specification. This conclusion prompted two subsequent activities. Firstly the introduction of FET preamplifiers into the late INTELSAT V and V-A spacecraft; secondly, it reinforced the requirement for high performance, high reliability FET components.

Within the timeframe reviewed above, work was already underway on basic reliability studies of low-noise 14 GHz FETs.

This was supplemented in 1977 by a contract aimed at advancing state-of-the-art RF performance. A number of innovations were introduced in this effort, notably electron beam lithographic definition of the gate electrode at 0.3 micron length. It was indisputable at the end of this effort that the specific device developed could find potential application in spacecraft LNAs. Some questions did remain, however, concerning ultimate reliability. This uncertainty--set principally by the metallizations and the expensive fabrication technology utilized--delayed its introduction as a commercial device for more than two years after the conclusion of the INTELSAT contract.

One aspect of the above FET that added to its expense was the difficulty of incorporating it into an actual circuit (even for purposes of testing). Thus, at the ISL 30/20 GHz frequencies, where the same requirement exists for ultra-low-noise FETs for the heterodyne receiver option, emphasis has been shifted towards producing a circuit module rather than simply a discrete device. This work is currently in progress.

5. Microprocessors and Memory Devices. Results of R&D efforts in characterizing digital devices to assess suitability for application at geosynchronous orbit have played a critical role in support of the INTELSAT VI program and have made essential contributions to other on-going R&D projects. In particular, improvements have been incorporated in the Distribution Control Unit (DCU) for SS-TDMA operation.

Several R&D programs required selection of suitable microprocessors for geosynchronous orbit applications where the radiation tolerance of the devices is critical. The knowledge gained from INTELSAT-sponsored and other related radiation tolerance testing allowed the selection of realistic microprocessors, RAMs and other circuitry to be used in these programs, which include: 1) The DCU for SS-TDMA; 2) IADES (Integrated Attitude Detection and Estimation System; 3) Development of Improved Earth Sensors; and 4) Earth Sensor for Attitude Control of Spacecraft.

6. Battery Cells/Batteries. INTELSAT R&D-sponsored activity, initiated in this area in 1971, has been limited almost solely to the nickel-hydrogen secondary system. Patents filed in 1972 covering the design of an aerospace-type nickel-hydrogen cell were issued in November 1974.

A parallel program of nickel-hydrogen R&D was established in 1973, sponsored by the United States Air Force. However, INTELSAT's long-term program, which was quite distinct and separate from this program, was 1) broadly-based technically (covering component development, individual cell design, cell testing and battery concepts and use); 2) broadly-based geographically (work was contracted out in France, the United States, and the United Kingdom); 3) centered on GEO application; and 4) cost-effective.

Progress on the initial R&D work through mid-1978 was very impressive, and generated an INTELSAT-led initiative to place nickel-hydrogen batteries on later INTELSAT V spacecraft (starting with F-5). At that point within the V program, it was clear that the baseline nickel-cadmium batteries were the principal life-limiting sub-system; indeed whether or not a full seven-year operation was attainable had become a serious concern. Furthermore, the later addition of the MCS package to the F-5 through F-9 spacecraft benefitted considerably from the added confidence of on-going nickel-hydrogen battery application engineering (also from the lower mass of the nickel-hydrogen sub-system compared to the nickel-cadmium).

Efforts from 1979 to the present time have been concentrated on optimized application of this technology to spacecraft larger than INTELSAT V (and V-A, which will use nickel-hydrogen cells of the INTELSAT design as baseline technology). Notably, 50 ampere-hour cells and a 1200 watt-hour battery are under consideration. In comparison, the INTELSAT V sub-system has 30 ampere-hour cells and delivers 650 watt-hours.

7. Opto-Electronic Technologies. From an initial effort to develop an electro-optical SS-TDMA switch, significant in-house expertise has been developed in electro-optics, optical systems and solid state lasers. This expertise has led to:

1) Development of initial concepts and definition of characteristics of an optical inter-satellite link which could provide a flexible medium to high capacity ISL link at greatly reduced mass, volume and power consumption compared to competing technologies; and

2) Developing a novel scheme for demultiplexing and demodulating several FDMA-BPSK channels to allow a FDMA uplink and TDM downlink. This would allow interconnectivity between low-cost Earth stations and be particularly beneficial for new business services and other thin route applications.

D. Spacecraft

1. Satellite Stabilization. Two driving factors have provided impetus for R&D on satellite stabilization: high pointing accuracy and long mission life. With the growth of spacecraft structures, the effects of structural flexibility on the stability of the spacecraft and on the pointing accuracy of the antenna beam were

studied through several research contracts. Even without anticipating future large antenna systems, the effects of thermal distortion and of mechanical alignment tolerances on the beam pointing accuracy are edging toward incompatibility with conventional spacecraft attitude control concepts based on the orientation control of the entire spacecraft.

Among different kinds of attitude control sensors, RF sensors, part of the antenna/feed system, permit better antenna beam pointing control than that obtainable with sensors (infrared Earth sensors, sun sensors, star sensors) which are mounted on the main body of the spacecraft and thus providing a reference orientation not at the antenna axis but at their mounting location. By virtue of their configuration, RF sensors permit significant reduction of the effects of structural alignment errors and of thermal deformation.

Another important area of research in the field of satellite stabilization is the interaction between the attitude control system and flexible appendages, such as solar arrays and antennas. A flexible inertia simulator for the air bearing platform was developed with the objective of extending the capability of the air bearing facility to simulate the effects of flexible appendages. The major challenge was to build an accurate simulator that could operate over wide parametric ranges of appendage natural frequency, modal damping ratio, and moment of inertia distribution.

More recently, a separate facility, in which the dynamics of flexible appendages are experimentally modeled by equivalent structural elements, has been designed, developed, and tested. The facility has been conceived as a single-axis system of three inertia disks, with a rotational degree of freedom about the vertical axis of suspension. The top disk represents the main body of the spacecraft, while the lower elements represent appendages and are mechanically coupled to the main body by torsion rods that are coaxial with the suspension axis. Each disk is independently suspended about the vertical axis by an air bearing. In this way, the torsion rods are only required to provide torsional flexibility and do not transmit any axial load. Different (interchangeable) torsion rods provide the capability of experimenting with different modal frequencies of the structure. Attitude sensors detect the torsional angular error of each disk with respect to a ground reference, and reaction wheels are used as control actuators.

2. Propulsion. After many years and several satellite programs (up to INTELSAT V) in which the monopropellant hydrazine technology dominated in the spacecraft propulsion area, bipropellants were selected for application to INTELSAT VI, to take advantage of the mass savings potential of a higher-energy propellant and of an integrated apogee insertion and on-orbit propulsion system (only practical with bipropellants). In parallel with this, R&D work was initiated to promote extension in the

range of bipropellant thrusters, toward lower thrust levels, and to investigate "dual-mode" propulsion using monopropellant hydrazine (N_2H_4) in combination with nitrogen tetroxide (NO_3), for operation in a bipropellant or monopropellant mode. The objective of this is to take advantage of the higher energy delivered in the bipropellant mode for applications requiring large velocity increments, such as the apogee insertion of the spacecraft into geosynchronous equatorial orbit (GEO), or the spacecraft latitude control (north-south stationkeeping), while maintaining the advantage of high efficiency in very small impulses, delivered in the monopropellant mode, for attitude control.

A current activity in the general area of propulsion is the renewed effort in electric propulsion. INTELSAT's R&D efforts in this area were initiated as early as 1969. A contract was granted to identify life-limiting mechanisms via an extensive endurance run of a small mercury bombardment thruster.

Current work aims at replacing mercury with the inert gas xenon, with minimum modifications to already developed thruster designs.

Ion thrusters have the following inherent advantages when run with the inert gas xenon at modest specific impulse (19,600 Ns/kg, 2000 s):

a) Operation with simplified power processors; potentially a sizeable improvement in overall reliability; and a desirable reduction in system mass;

b) Incisive start/stop transients using proven electromechanical gas valves. The start-up is also fast, 1 to 2 min cold, and 10 to 15 s warm. Direct flow measurement is also possible with gaseous propellants allowing precise start control without the flooding and shorting sometimes experienced with mercury;

c) Simplified start and cutoff control electronics, a further reliability benefit;

d) Reduced high-voltage isolation design problems, since xenon remains a gas at all conditions experienced in ground test and on-orbit operations; and

e) Reduced ground test time and expense because xenon does not condense in vacuum tanks, eliminating the need for periodic cleanup.

3. Mechanisms. The INTELSAT III and IV communications satellites were spin-stabilized with an Earth-pointing despun antenna platform. Their most important mechanical system, and that common to all spacecraft of this type, is the mechanical drive that counterrotates the communications antenna at the spin rate necessary to maintain its Earth orientation. In addition, spin-stabilized satellites having a despun platform require a means for transferring primary power and signals from the spinning portion of the spacecraft to the despun section. This is accomplished by a slip ring and brush assembly.

The performance requirements for these devices are rather stringent, since no unit redundancy is possible. The use of this device was anticipated in the INTELSAT R&D Program; since the expected life of the bearings and the slip ring and brush assembly was a significant concern, it was investigated under several projects.

4. Control Wheels. A variety of possible control system concepts may be used with body-stabilized spacecraft. The electromechanical devices required to implement these various concepts fall into two categories, reaction wheels and momentum wheels. Reaction wheels are small bidirectional momentum wheels that operate through zero speed. Momentum wheels operate over a relatively narrow speed range, but always with a significant angular momentum. In some types of systems this wheel assembly is suspended in a one- or two-axis gimbaled mechanism.

The one common feature of nearly all of the candidate stabilization system schemes is a flywheel that must rotate to provide the required angular momentum. Flywheels, particularly biased momentum wheels, operate at speeds much higher than those of despun antenna drives. At these higher speeds, the reliability of the bearing system is of greater concern than that of any other element in the device. As in despun antenna drive assemblies, the bearings constitute the single element that is not amenable to redundancy.

In addition to the momentum wheel, body-stabilized spacecraft typically require an electromechanical drive to orient the deployed solar arrays. The drive mechanism used to orient the deployed solar array and to cause the array to track the sun is a critical element in the spacecraft's reliability chain. A solar array drive mechanism using a stepper motor was designed, constructed, and tested under an INTELSAT R&D Program. This effort indicated the possible effects of the drive steps exciting the structure of the array. A periodic stepping was used to minimize these effects. The results of additional tests conducted recently to determine transient slip ring/brush performance were applied directly to INTELSAT V.

5. Bearing Technology. Fluid film bearings have been studied as alternatives to ball bearings for momentum wheels. The primary objective is the elimination of wear by eliminating metal parts in contact. Gas bearings, because of the chemical stability of the inert working fluid, are the most desirable in terms of lifetime. The fundamental difficulty is that the gaseous working fluid cannot be confined to the region of the bearing, but is also present within the momentum wheel housing. The high operating speed of the rotor results in prohibitive power consumption because of aerodynamic drag on the rotor.

The most viable operating fluid is a liquid. Windage losses can be made insignificant by selecting a low-vapor-pressure fluid. The lubricant can be contained within the bearing by suitable use of labyrinth seals. Because the optimum operating speed of the rotor is much lower than that of gas bearings, the rotor is heavier, but the bearing load capacity and power consumption are quite reasonable.

Magnetic bearings that use magnetic fields to support the rotor eliminate lubricants and their associated limitations and provide no contact between rotor and stator. The orbital environment of vacuum and zero gravity, which is a significant problem for liquid-lubricated bearing systems, is a natural advantage for magnetic bearings. While life and drag torque of conventional bearings are strongly related to operating speed, magnetic bearings have no speed/life limitation, and drag torque increase is much less dependent on speed. This tolerance to higher-speed operation can allow a specified angular momentum to be achieved with a lighter rotor than is possible with conventional bearings.

Magnetic bearing technology is relatively new, and there are several competitive suspension concepts for momentum wheel applications. Fully active systems use servocontrolled electromagnets on each axis, while hybrid systems use permanent magnets on certain axes, but always maintain one axis under active control.

A feasibility model of a magnetic bearing momentum wheel was constructed. The 4.34-kg rotor was operated at 12,000 rpm. Each of the two radial bearings consisted of two samarium cobalt annular magnets on the stator and one on the rotor. A magnetic bearing momentum wheel using fully active magnetic suspension has also been developed. A beryllium rotor operates at 16,000 rpm and provides 100 N-m-s of angular momentum. The fully active suspension permits the rotor to be gimbaled magnetically. The magnetic bearing momentum wheel work funded by INTELSAT now forms the technology base for high-reliability and lightweight momentum devices.

6. Spacecraft Structures. The INTELSAT R&D effort in spacecraft structures was initiated in the late 1960's. Four technical areas were identified that showed promise for providing more efficient spacecraft structures:

1) evaluation of environmental loads (handling, transportation, and launch);

2) development of methods of simulating the expected environment loading conditions;

3) development of computerized analysis to provide for reduced weight and to evaluate thermal stability; and

4) development of new structural concepts and utilization of advanced materials and fabrication techniques to advance the state-of-the-art.

The earliest work involved the development of a number of computer programs for evaluating spacecraft structural dynamics to provide a base for the continuing support of the development of INTELSAT spacecraft. Part of the R&D effort involved a continuing review of the technology in other structural areas, including development of advanced materials. An early contract indicated that significant weight savings could be achieved by utilizing beryllium and advanced composites for the spacecraft primary structure. An R&D project was authorized in 1974 to provide for the development of lightweight structures. As part of the contract, a beryllium thrust cone was developed and statically tested. The hardware delivered also included a lightweight boron aluminum strut. A significant advance in the cone design was the use of all-bonded technology and aluminum rings.

Current activity covers the application of advanced, lightweight materials, the development of deployable structures, and the testing of large structures.

7. Thermal Control. Research and development in this field has focused on potential thermal control coatings and radiation-resistant polymers; design, fabrication, and test of improved devices for spacecraft temperature control; and assembly of a large library of computer programs for analysis of satellite performance in the hostile environment of space.

The use of thermal control louvers on communications satellites achieves active temperature control by varying the effective radiating area of the component to be controlled in response to its temperature. Conventional louvers consist of movable blades covering the radiating area and rotating similarly to ordinary window louvers. INTELSAT recognized the need for an improved louver and awarded a contract for the development of a louver with counterrotating blades.

An active temperature control system, using variable conductance heat pipe, was developed later. Computer programs for thermal analysis were developed under INTELSAT sponsorship. These computer programs cover the full range of problems encountered in thermal analysis, from the satellite subsystem component level to the entire spacecraft. They compute such quantities as geometric view factors and radiant heat exchange factors for enclosures containing both diffuse and specularly reflecting surfaces and complete transient temperature histories for spacecraft components.

Current activity focuses on the design and development of active systems for large power dissipations, using a combination of deployable radiators and heat pipes.

8. Solar Cells and Arrays. Solar cells are the primary power source for INTELSAT spacecraft. Since the cells require sunlight to operate, they are also exposed to the hazards of space, such as particulate radiation (electrons and protons), ultraviolet (uv) light radiation, micrometeoroids, extreme temperature variations, and vacuum.

INTELSAT R&D work on solar arrays has been directed at two fundamental objectives: to increase the reliability of the solar arrays for 7- to 10-year missions and to improve the power density (W/kg) initially for spin-stabilized spacecraft and later for three-axis body-stabilized spacecraft, which have a much wider temperature range to withstand. Several INTELSAT R&D contracts have been directed toward developing high-reliability solar array modules for use with deployed solar arrays.

Different types of deployable solar arrays can be considered for use on three-axis-stabilized satellites: an array with a rigid substrate, an array with a flexible substrate, or a semirigid array, which utilizes a rigid substrate frame for support with a flexible substrate suspended within it. The rigid array generally employs some sort of spring activation with a pantograph or other controlling mechanism to deploy one or more panels of rigid substrates covered with solar cells. Until recently, the rigid substrates were typically made from aluminum honeycomb with aluminum faceskins. The newer substrates, used on INTELSAT V, are fabricated from aluminum honeycomb with carbon-fiber reinforced epoxy faceskins.

A solar array contract, jointly funded by INTELSAT and the German Research Agency for Aerospace Technology, was undertaken in the late 1970's. The array with a power goal of 30 W/kg developed in this program is semirigid with a carbon-fiber reinforced epoxy frame and a carbon-fiber reinforced Kapton solar cell blanket suspended in tension within the frame.

E. Interdisciplinary Research

As has been noted earlier, several apparently futuristic topics are by their very nature interdisciplinary. In both topics described here, these are related to likely new directions of system growth.

1. Platforms. In recent years, geostationary platforms have come to connote large space structures with or without assembly in space and frequently for more than one user or organization. INTELSAT satellites have so far led the industry in size and complexity, and thus the next-generation spacecraft could well be a large satellite or the so-called "platform" in general industry nomenclature. Specifically, it is conceivable that INTELSAT VII, if the present "distributed" network architecture is continued, could

be a full shuttle bay spacecraft or platform. Several technologies related to such a size are the subject of ongoing R&D projects within INTELSAT R&D.

Looking further ahead into the future and well into the next century, it is possible that cost-effective facilities for manned assembly in LEO may become available, along with remote capabilities for repair/assembly in GEO. Such a capability could be exploited to provide benefits such as:

1) Low-thrust transfer from LEO to GEO of a fully deployed satellite, thus eliminating the risk associated with complex deployment mechanisms.

2) The augmentation or replacement of space resources when required rather than being incorporated at the beginning-of-life of an expensive satellite of the future.

2. Intersatellite Links. With the largest network of satellites in orbit, INTELSAT could derive substantial system and operational benefits through direct interconnectivity in space. Such an interconnection, in specific situations, could also be with other networks. In recognition of this requirement, substantial resources have been devoted during the last 7 or 8 years toward developing an overall intersatellite link (ISL) concept as well as initiating hardware development. Initial efforts were directed toward microwave ISLs, with optical ISL investigation now underway.

During the last 5 years, practically all new hardware and components required for 23/32 GHz microwave ISLs have been either completed or are about to be completed (Fig. 4). Current in-house efforts are devoted toward development of an ISL test bed for simulation of such links in the ground.

Advances in optical devices have been so rapid in recent years that optical ISLs appear to be technically attractive for medium- (10-15 deg) to long-range (15-60 deg) high-capacity links. Selective hardware development is planned in the next few years following intensive ER&S efforts.

IV. Progress and Prognosis

The preceding sections have attempted to capture the broad objectives and achievements of INTELSAT R&D Programs. The topics described are only a few from a long and growing list of R&D contracts and in-house efforts. (Current active R&D contracts exceed 100.) In this section, a brief overview is presented of past progress and the prognosis for the future.

INTELSAT R&D is almost as old as INTELSAT itself and has grown in consonance with the system needs. Its prime role continues to be the development, on a global basis, of the technologies needed for more efficient and cost-effective

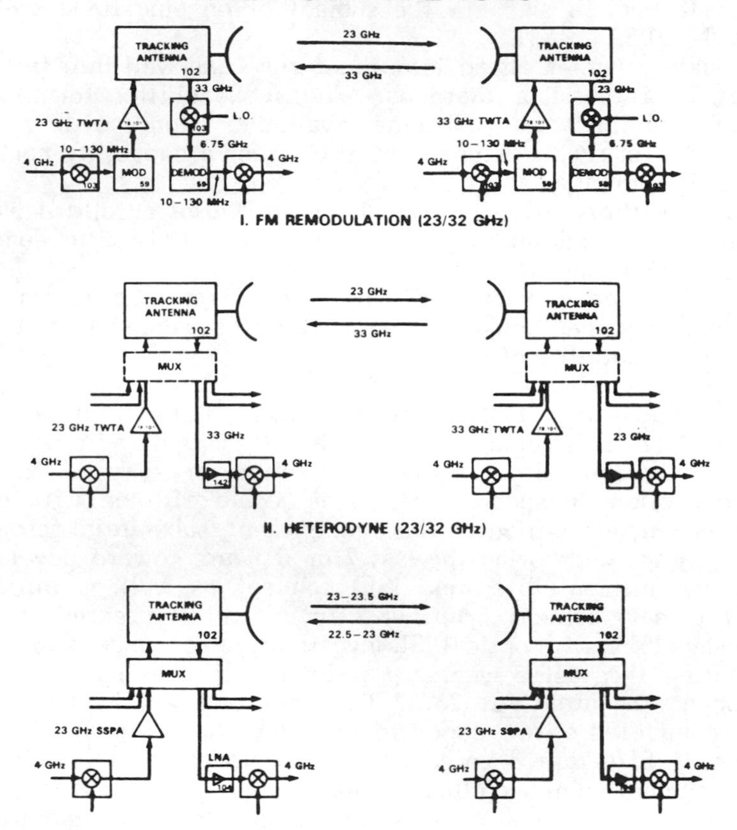

Fig. 4 ISL system configurations.

satellite-based services in the future. Due to the nature of the R&D effort, its direct traceability to actual utilization is not always straightforward. Nevertheless, it is possible to list some major successes here:

A. The 1970's

Four-times frequency reuse.
Dual-polarization at C-band (satellites and Earth stations).
Multiband Earth station antenna feed technology.
Dual-mode filters and fabrication technology.
SPADE system.
TDMA technology.
Nickel-hydrogen batteries.
Improved solar cells and arrays.

Echo control and cancellation.
Mechanical devices for spacecraft bus.
Wide-angle sun sensor.
Multiband satellites (6/4 and 14/11 GHz).
Major contribution to worldwide propagation data base in the 6/4 and 14/11 GHz bands.
Discovery of, and systematic research into, super-high-frequency (SHF) ionospheric scintillation.

B. The 1980's

Some are already realized; some are to come:
All solid-state transponders.
Six-times frequency reuse.
SS-TDMA.
First-generation ISLs.
Voice band coding and DSI.
Improved IBS services.
Digital TV distribution.
TWTA linearizers.
Near-field satellite antenna measurements.
Uplink depolarization precompensation system at C-band.
30/20 GHz propagation beacons and data collection.

C. Prognosis for the 1990's

Understandably, a forecast of technologies more than a decade away is always difficult. However, taking into account the lead time required from conception to maturity, a major part of current INTELSAT R&D Programs is indeed focused toward the 1990's. For maximum and timely benefits, all such work is carried out in close and iterative coupling with evolving future plans described in Chapter XII.

The technological goals for the 1990's are best described in terms of progress with respect to the present. This is summarized below through five principal technology driver areas:

1. Spacecraft Bus Technologies. So far each successive generation of INTELSAT satellites has utilized a bus larger than the previous one. This monotonic increase in size has motivated a variety of technological developments, which have been adequately reported in the literature. System plans currently under development for the 1990's are considering several alternative configurations and bus sizes, taking into account the traffic requirements, the expected launch capabilities and a host of individual technologies that are expected to mature in that time frame. While the system decisions for the 1990's are some years

away, several baseline trends have already emerged and form the basis of ongoing technology programs:

1) The INTELSAT VI spacecraft, including its transfer stage, is about a one-half shuttle-bay size, and it is expected that at least some of the spacecraft of the 1990's would require a bus twice that size. Therefore, as a broad technology driver, a full shuttle-bay spacecraft appears to be a good basis for scaling the performance objectives of relevant projects.

2) Spacecraft development will be heavily influenced by the corresponding developments in orbital transfer technology. Based on current efforts and trends, it would be reasonable to assume that liquid upper stages will be well proven in time for 1990's spacecraft. The available range could also include reusable orbital transfer vehicles (OTVs). If such stages also achieve a low enough thrust level to permit transportation of a fully deployed (and possibly checked-out) spacecraft from LEO to GEO, we could derive significant benefits in terms of spacecraft packaging and reduced risk inherent in deployment mechanisms.

3) Increasing investments in large long-life spacecraft point toward the alternative approach of modular step-by-step buildup of orbital resources. This could be achieved in several ways. A cost-effective approach would be the evolution of flexible

Table 1 Spacecraft Bus Technologies

	Current capability	Late 1980's	1990's Needs
S/C mass	500-1700 kg	700-2500 kg	1000-5000 kg
Configuration	Multiple and autonomous	Multiple and autonomous	Multi-user communications platforms; clusters and autonomous
Attitude control accuracy	±0.2 to 0.1 deg	±0.1 to 0.05 deg	±0.1 to 0.05 deg
Structures	Rigid deployable solar arrays (20-25 W/kg)	Flexible, ultralight deployable solar arrays (25-40 W/kg)	Large, flexible deployable solar arrays (40-70 W/kg)
Power generation range	2 kW	Up to 7 kW with some solar concentrators; thin solar cells	Up to 25 kW thin, high efficiency cells; solar concentrators
Power storage	NiCd: Some NiH_2 (30 W h /kg)	NiH_2: (35-45 W h/kg)	(40-50 W h/kg)
Propulsion	Mostly monopropellant: some AEHT; some bipropellant	Bipropellant: AEHT	Bipropellant: ion

medium-size spacecraft with capabilities to adapt transponders and antenna footprints to a variety of individual needs and to achieve the desired capacity via intersatellite links. Modular mechanical assembly, even in LEO, on the other hand, appears somewhat distant at this stage and ultimately may not be attractive or even cost-effective in the long run.

The foregoing broad scenarios form the basis of some of the individual technologies in current programs. Table 1 summarizes the broad technological goals as compared with more near-term objectives and capabilities.

2. Connectivity in Orbit. From the point of view of complete integration with other networks as well as more optimum utilization of orbital resources and investments, direct

Table 2 Intersatellite Link Technology (Full-Duplex Circuit)

A. MICROWAVE (23/32 GHz) - Full Duplex Operation

	Cluster	IBS Application (with sun outage) FM Remodulation	Heterodyne
Direction:	East and West	East or West	East or West
Frequency (GHz)	32/23	32/23*	32 or 23*
Antenna (m)	0.3	2	2
Operation	Full Band	Full Band*	Split band*
O/P power (W)	1	10	10
O/P device	SSPA	TWTA	TWTA
Range (km)	40	32,400	11,000
Range (degrees)	0.1	45	15
Capacity	3 x 72 MHz	1 x 72 MHz	1 X 72 MHz
Mass (kg)	41	80	67
Power (W)	143	76	56

*Due to diplexer isolation requirements, split-band operation is not feasible using FM remodulation.

B. OPTICAL - Full Duplex Operation Including Sun Conjunction

	1985 East or West	1990 East or West	1995 East or West	East and West
Frequency (THz)	200-600	200-600	200-600	200-600
On-board regeneration	No	No	Yes	Yes
O/P Power (mW)	15	100	300	300
Range (km)	40	11,000	42,300	42,300
Range (degrees)	0.1	15	60	60
Capacity	3 x 72 MHz	5 x 72 MHz+	5 x 72 MHz+	5 x 72 MHz+
Mass (kg)	40	70	55	95
Power (W)	150	160	250	500
Antenna (cm)	5	15	30	30

+2x FDMA; 3 x SS-TDMA

connectivity between satellites is the next major technological step.

During the 1990's, intersatellite links could play several roles, such as in:

1) global business communications network;
2) extending coverage areas and eliminating double hops;
3) modular build-up of orbital capacity;
4) enabling use of higher frequency bands in those places limited, at present, due to low elevation angle propagation impairments; and
5) direct spacecraft interconnection between previously separate space networks.

It is anticipated that both microwave (23/32 GHz or higher) and optical ISLs will find applications in the global networks. Microwave ISL technology is now mature and almost all the hardware required has been developed under INTELSAT R&D. Such ISLs could find applications in small-to-medium spacings (up to 30°) and low- to medium-traffic needs. One prime candidate could be the INTELSAT Business System (IBS) with its wide connectivity requirements. Optical ISL technology has the potential to provide really long (up to 60°) and high capacity links. However, substantial development work is still necessary to realize the space hardware. Table 2 summarizes the prognosis for these technologies.

3. Preparing for the Digital Era. Following the introduction of 120-Mbit TDMA and SS-TDMA, in INTELSAT VI, several additional steps are necessary to usher in an all-digital era as described in Sec. IIIB.

Table 3 summarizes the overall progress expected in digital satellite transmission as a prelude to a fully digital global network.

Table 3 Digital Satellite Communications

	Current capability	Late 1980's	1990's Needs
TDMA	60-120 Mbps, transparent	SS-TDMA with 120 Mbit/s	Higher-speed SS-TDMA and baseband switching
Voice processing	64 Kbps with DSI	32 Kbps with DSI	16 Kbps with DSI
Video processing	FM/18 MHz 1.5 Mbps teleconferencing	30 Mbps/18 MHz	20 Mbps extensive teleconferencing demand assignment
On-board processing	None	SS-TDMA	RF and baseband integrated processing

Table 4 Spacecraft Life and Reliability

	Current capability	Late 1980's	1990's Needs
Life Objectives years	7/10	10	10-20 with or without replenishment of consumables/modules
Reliability Contributors			
TWTAs	C-band 16 W K-band 20 W	Up to 250 W C-, Ku-, Ka-bands	Higher-efficiency linearity and life
SSPAs	C-band 8 W	C-band 20 W Ku-, Ka-bands 10 W	Ka-bands up to 30 W; higher-efficiency versions
Solar cells	Si 200 20 W/kg	Si 200 25 W/kg	Si 50 ; GaAs 35 W/kg
Radiation hardness	Mostly bipolar, some CMOS	Mostly CMOS	CMOS dominant
Deployment (antennas)	Mechanical	Mechanical	Retractable; electronic reconfiguration

Table 5 Spacecraft Antennas

	Current capability	Late 1980's	1990's Needs
Maximum aperture size (m)	2.4 offset (IS-V)	3.2 offset (IS-VI)	5 to 8 offset
Number of feeds per reflector	87 (IS-V)	150 (IS-VI)	300-500
Frequency band pairs/satellite	3 (IS-V)	4-5 (later IS-VI)	5-8
Frequency reuse a) polarization b) sidelobe	IS-V: C-band x 2 C-band x 2	IS-VI C-band x 2 C-band x 4	C-band x 8-10 Ku-band x 2-4 Ka-band x 2-4
Deployment	Rigid offset single articulation (IS-V)	Rigid offset double articulation (IS-VI)	Offset deployable with articulations
Beam pointing accuracy	± 0.2 deg (IS-V)	± 0.1 deg (IS-VI) ±0.05 deg with RF tracking	± 0.03 -0.05 deg with RF tracking
Mass (kg): a) beam forming network/feeds b) reflectors (percent of dry S/C)	IS-V: 18 (C-band Tx) 11 (C-band Tx) (9%)	IS-VI 141 (C-band Tx) 28 (C-band Tx) (18%)	500 - 1000 (25-35%)

4. Spacecraft Life and Reliability. This is, at once, a relatively simple and yet vast subject. If approached simply, all that is required to obtain, for example, 10-year-life satellites instead of 7, is to provide for extra fuel and solar cells and either accept lower end-of-life availability or increase payload redundancy. On the other hand, if increased life is closely coupled with increased reliability, there is a host of ongoing as well as future technology efforts that need to be addressed. For the 1990's, the latter should be the preferred approach consistent with the development resources. One aspect of direct relevance to this topic is the concept of refueling and repairing satellites in orbit, as discussed earlier.

Several subsystems in current spacecraft contribute toward upper bounds for life and reliability (obviously newer techniques would bring in their own attendant risks). Some of the major areas of effort have already been covered in Secs. IIIC and IIID.

Table 4 provides an overview of the progress expected.

5. Orbit Utilization. After two relatively "easy" decades in which relatively unrestricted growth was possible, satellite communications is already feeling the pressures of scarce orbital resources. The requirements of future decades will be even more demanding and can be met only by continuous efforts in practically the whole range of technologies, some of which already have been touched upon. However, one important technology will probably continue to provide maximum payoffs and has been deliberately singled out to be described under this heading. This relates to the antenna technology for both spacecraft and Earth stations.

For satellite antennas, INTELSAT R&D continues to play a lead role in increasing the frequency reuse, as described in Sec. IIIA.

Table 5 summarizes the objectives for the 1990's in this area along with the near-term capabilities.

For ground segment, the major effort currently is toward broadening the antenna bandwidth in order to fully exploit the expanded WARC-79 allocations at both C- and K-bands and to establish dynamic depolarization compensation systems. As the satellite e.i.r.p.s increase, smaller-diameter ground antennas will become viable and efforts will be necessary to improve the cost-effectiveness of dual-polarized feeds that meet the frequency reuse isolation requirements as well as ensuring adequate sidelobe performance.

Acknowledgment

The author wishes to express his deep gratitude to the following people for their generous and indispensable contributions to this chapter: Dr. W.J. English, Dr. P.P. Nuspl, Dr. G. Porcelli, and Dr. J.L. Stevenson.

14. The Next Hundred Years: INTELSAT and the Postinformation Society

Joseph N. Pelton*
INTELSAT, Washington, D.C.
and
Joel Alpert†
COMSAT, Washington, D.C.

I. The Past Is Prologue

The history of man, since the first appearance on Earth of the Australopithecus Man, covers a period of perhaps some five million years. Science historians Henry Adams, in the 19th century, and Gerard Piel, in the 20th, have discussed at length the motive force of technological development in man's evolution and the phenomenon they have called "the acceleration of history." Indeed, Gerard Piel has stated: "In the old taxonomy of primates it was supposed that man had made the first tools: toolmaking was the status symbol of membership in our species. Now, it would appear, tools made man. Certainly, toolmaking conferred a competitive advantage on the maker of better tools. But the meaning of this phase of history goes deeper. The truth is: Man made himself."[1]

If one reviews the history of mankind's use of energy, for instance, one finds a steady but ever-increasing access to power first with fire; then the domestication of draft animals; then the use of water and wind; and then chemical energy in the form of coal and oil (first to produce steam and then to fuel internal combustion and diesel engines). In the 20th century, the pace quickened, and man gained access to the power of the stars--first with the development of atomic fission energy and then, ultimately, nuclear fusion. The scale of the power accessible to man has thus increased by a factor of more than a billion times in the last five million years.

Invited paper received October 11, 1983.

*Director, Strategic Policy Formulation.
†President, World Systems Division.

There are, of course, many other ways to chart man's development--particularly in terms of his technological and scientific progress. Many would agree with Jacob Bronowski's basic division of history between the period before man mastered farming and city living and the time afterward. This epochal event, which occurred sometime between 8000 and 6000 B.C., clearly was a major turning point in man's history, not only in its own right but particularly in terms of its creating a process whereby man was increasingly able to innovate and to discover new technologies. With the innovations of farming and the city, man was, for the first time, able to release a small segment of humanity from the drudgery of simply surviving. The small gain that agriculture introduced in its initial stages, which allowed one family to be supported by four or more families who tilled the land, was, nevertheless, a significant breakthrough--because that 20% could find new and better ways of farming, transporting goods and materials, and, in effect, innovating.

The progress that ensued in the following centuries can be charted in various ways. One way is by calculating the amount of effort expended or the percentage of income spent to provide food. With the farm-based subsistence-type of agriculture, that figure was over 90%. Even with improved agricultural techniques during the 16th and 18th centuries, with seeds producing up to a tenfold yield per seed rather than the fourfold yield of previous centuries, those percentages were still shifted only marginally. In the 18th century, income devoted to producing food was still at about the 80% level. The development of farm machinery, fertilizers, chemical energy, and mass manufacturing techniques in the 19th and 20th centuries has dramatically shifted this balance. Typically, in today's industrialized countries, the percentage of income and activity devoted to food and agricultural production has fallen to the range of 25 to 30%. Perhaps even more significantly, the number of employees engaged in agriculture in the advanced industrial states has declined to the range of 3 to 10% of the entire labor force.

Although there are many significant aspects of the Industrial Revolution, there are only three that are particularly worthy of mention here. First, on a comparative basis, the agricultural revolution and the founding of the city released only about 20% of the population to perform services and to innovate. The Industrial Revolution, closely linked to the agricultural revolution of the 19th and 20th centuries, produced productivity gains sufficient to release more than 50% of the population in industrialized countries for "knowledge" and information activities. Second, while the agricultural society was based on agriculture and food products, the industrial society today is essentially based on industrial products. Yet, we are now in the midst of another transition--from an economy based on industrial products and goods to a postindustrial society based on information and services.

Within the Organization for Economic Cooperation and Development (OECD) countries, these sectors are already responsible for about 50% of all jobs and are the major producers of new economic wealth. The modern technologies of telecommunications, computers, genetic engineering, artificial intelligence, robotics, and advanced energy systems give promise of tremendous innovation that will significantly affect the future course of our world. Artificially intelligent robotic devices that can operate on a 24-hour-a-day basis can produce goods at lower and lower cost, utilizing less and less raw resources, resulting in the production of industrial goods by fewer and fewer employees. This portends tremendous change for 21st-century man, and signals major economic and social problems to be overcome.

We are indeed awash in the third wave of man's historical development. In the first wave, 20% of the labor force was released to think, innovate, and perform services. In the second wave--the Industrial Revolution--up to 50% of the people were released for these purposes. We are now, seemingly, entering an era where perhaps 90% or more of the people will be freed from the drudgery of repetitive or nonintellectual work. In the future, physical labor not related to recreation will largely be performed by robots or by artificially intelligent machines.

In returning to our historical time line, we see that man's first transition (from hunter/gatherer to farmer) required millions of years. The second transition (from farmer to industrial worker) required thousands of years, while the third transition (from industrial worker to information and knowledge provider) has occurred in a period of about a century and a half. If one accepts the thesis of the "acceleration of history" driven by scientific and technological innovation, this suggests that a fourth transition (from the information society to a postinformation, or cybernetic, society) could be expected to occur in a period of only a few decades and indeed could be anticipated in the early 21st century.

What might be the characteristics of this so-called postinformation society? The answer is, of course, no one knows. Nevertheless, certain aspects of the future can be expected to be played out within basic parameters. If we try to combine the best guesses of system analysts, economists, engineers, and research scientists from around the world, as well as demographers and social scientists, these trends and parameters suggest there is a great likelihood that the world population will stabilize somewhere in the range of 10 to 12 billion humans perhaps around 2025; that patterns of urbanization will likely continue, but at a less frantic pace; and that, ultimately, the patterns of human settlement will become more evenly distributed over the world's surface. Furthermore, adjustments of industrial processes to accommodate to the heat, energy, and environmental limits to the Earth's biosphere will become necessary. Likewise, adjustments to the finite limits of natural resources available on the planet Earth will also become critical in the 21st century.

Despite these limitations and adjustments, however, the remarkable thing about life in the postinformation (or cybernetic) society is that significant new options for mankind should open up as well. The potential for dramatic and epochal change could very well occur within the next 100 years. In the time frame of the year 2000 to the year 2020, we could very well see the evolution of so-called "thinking" machines or, as they are now commonly called, fifth- and sixth-generation computers. We can also foresee the

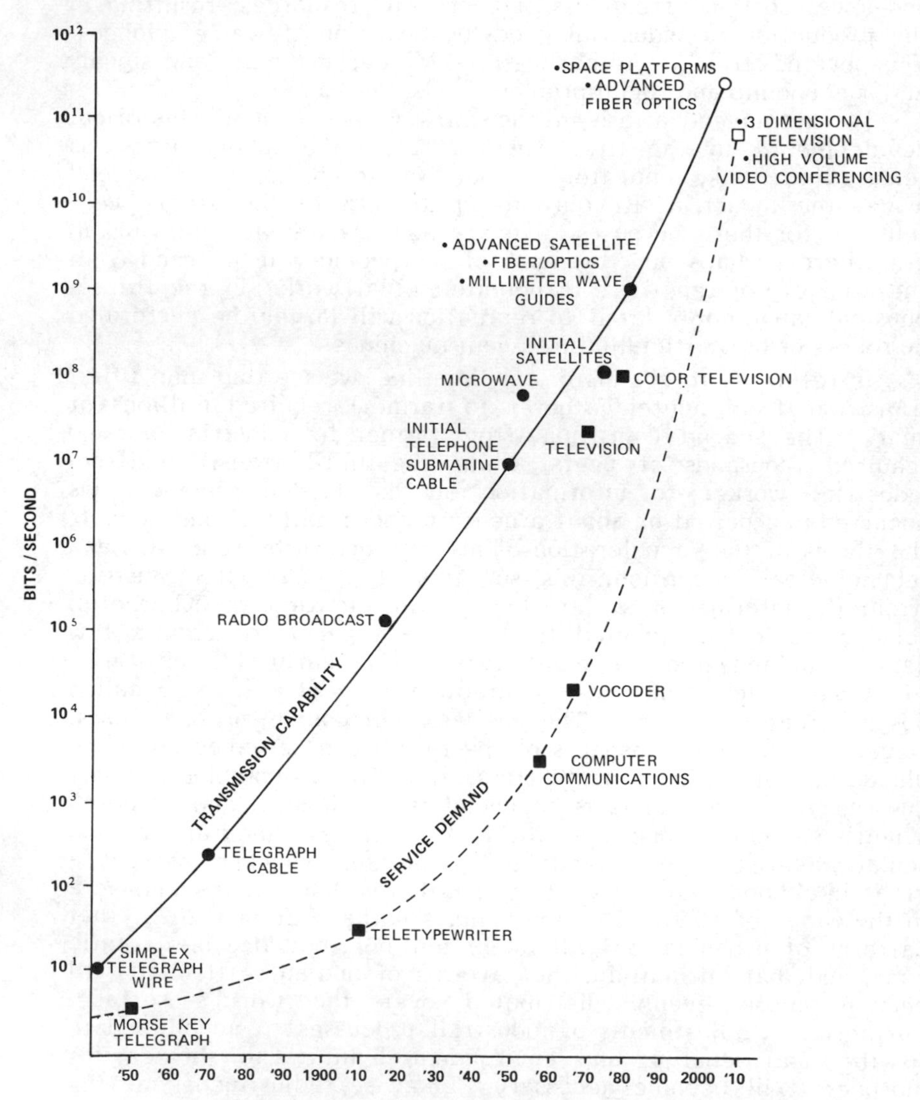

Fig. 1 150 year look at development of telecommunications (service demand vs. transmission capability).

evolution of space platforms, satellite clusters, and vast information-processing capabilities in space. We can foresee the return of man and machines to space in a much more serious, long-term, and economically important context--in effect, to exploit space to serve mankind's economic needs. Certainly, we can envision new employment and living patterns.

During the longer-term time period (from 2020 to 2085) the possibilities for the future soar way beyond the horizon of conventional thought and planning. Certainly, one can conceive of Von Newman probes to seek out life and intelligence throughout the universe, the possibility of star forming and bioforming within our solar system (a theme discussed by many science fiction writers and most recently and vividly by Arthur Clarke in his book 2010). Certainly, we can also anticipate the commercial exploitation of subterranean levels of planet Earth and undersea development. We can anticipate advanced robotic devices with sensors, locomotion, and cybernetic man/machine organisms--the "cyborgs" of science fiction writers' speculation. We can also envision a transformed world, in terms of population, human settlements, life styles, diets, and work and leisure options. We can anticipate low-cost, easily accessible, and worldwide communications--perhaps interplanetary as well. We can even hope for a refined and reformed set of human relationships, and perhaps even peace.

No matter how conservatively or how speculatively one views the future, however, it is hard to escape the proposition that computers and communications and the increasingly related technologies of energy and robotics will play a central role in the evolution of man, both on Earth and within the solar system, if not ultimately in star systems trillions of miles away.

As a point of departure in speculating about the future, we can certainly learn a great deal by analyzing trend lines of the past. In examining Fig. 1, which shows the symbiotic relationship between new communications services and new communications transmission capabilities over the last 150 years, it is possible not only to note the extent and force of this interrelationship, but also to perceive that on the horizon there are new service requirements and new types of communications applications that can lead us to continuing communications technology breakthroughs for at least 20 to 30 years into the future.

Figure 2 shows a similar set of curves with regard to the computing and cybernetics field, again with the expectation that the steady and rapid evolution of this technology will continue at least another 20 to 30 years.

Third, and finally, recent history has proven that computer and communications technologies are important tools that should make possible a continued flowering of scientific research, development, and innovation--in the life sciences, advanced energy generation systems, robotics, and data processing. Further, these tools will help us to enter new environments, in new ways, including the oceans, subterranean space, the solar system, and ultimately, the stars.

Before exploring this exciting future, however, one might try to convince the skeptics of how far we could progress merely by assuming a continuation of 20th century trends in terms of productivity, increasing innovation, and dramatic growth. One way to try to gage or calibrate the limits of the possible is to review INTELSAT's own achievements and accomplishments over the last 20 years. If one considers the development of INTELSAT from Early Bird, in 1965, to the INTELSAT VI satellite (to be launched in 1986), one finds a tremendous difference. In terms of satellite capacity, INTELSAT VI is 170 times larger; in terms of lifetime, it is expected to last some six times longer; in terms of international and national pathways to be serviced, the increase is several hundredfold.

Speaking in terms of satellite capacity and lifetime, if the same rate of growth were to be experienced for just another 21 years, then, by using straight-line extrapolation--a dangerous,

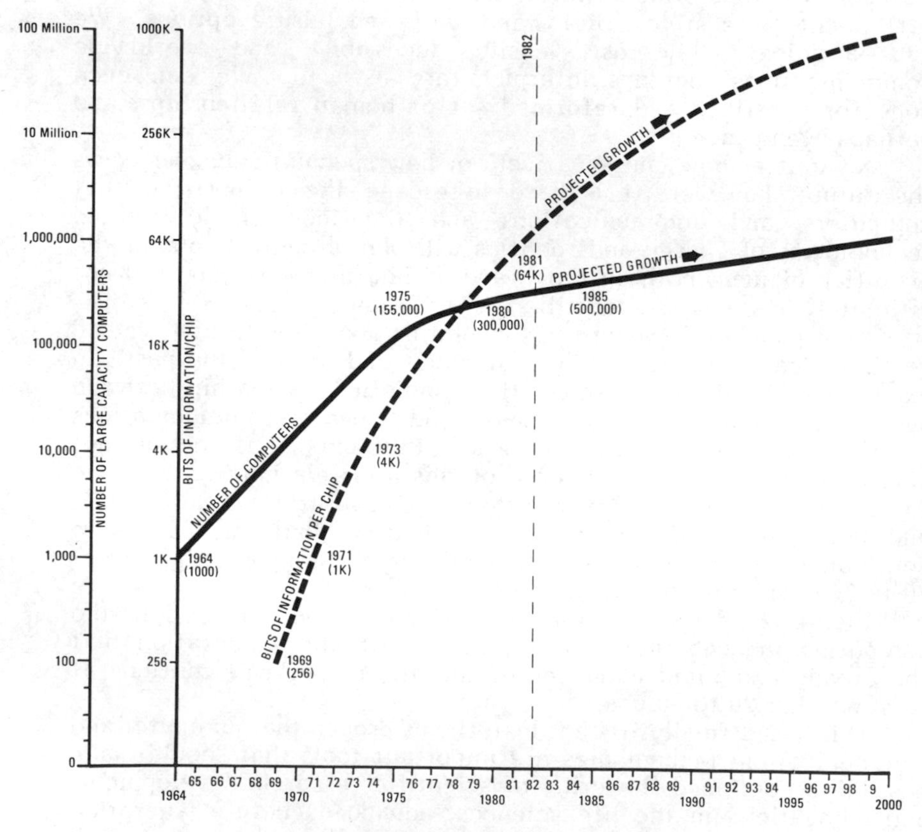

Fig. 2 Computer trends: 1964–2000.

but, nevertheless, interesting process--one could project the following results: a space platform of perhaps 50 metric tons, with a capacity equivalent to some 6.5 million voice circuits; a usable lifetime of some 60 years; and a capacity to serve millions of pathways. Engineers and scientists from INTELSAT and COMSAT Laboratories have conceived a profile of such a monster platform (see Fig. 3) based upon their conceptualization of the idea proposed by Arthur Clarke in his novel 2010.

Now, this projection of course assumes continuation of many trends that may not continue. For instance, it is possible that instead of one massive platform, a cluster of up to, say, twelve high-capacity satellites may be connected by intersatellite links, using a halo orbit around a single nodal point in the geostationary orbital arc. Furthermore, it seems unlikely that there would be a demand for this number of telephone circuits from a single point in a geostationary (Clarke) orbital arc. Yet, it is possible that equivalent capacity might be required to meet future new wide-band service requirements such as high-definition television; three-dimensional, multiple-raistered television systems; videophone; videoconferencing; etc. Furthermore, to achieve the number of pathways envisioned, one would have to think in terms of switching capabilities in the sky, with onboard signal processing and regeneration that would potentially allow the interconnection of millions of transceivers, such as might be used for mobile radio-wristwatch communicators or, more likely, mobile vehicular communications devices.

In performing the conceptual study of a "2010" space platform, INTELSAT scientists and engineers used a more or less straight-line projection of the future--something no one expects to happen. Thus such a future and such a space platform are certainly not being "predicted" by INTELSAT. Such an exercise merely probes one possible future to see what insights it gives up. Indeed, trend lines, such as certain governmental action to promote competition and stimulate multiple-services providers, would seem to be pushing us in the opposite direction--away from consolidation and centralization of facilities. Yet, technical and economic reasons related to the congestion of the geosynchronous orbital arc, the need for multiple frequency reuse, and economies of scale of a large system could have an opposite impact and push us toward very large space communications systems. Certainly service competition and facility consolidation are not necessarily mutually exclusive. Another key element to the future will be the relationship between space communications systems and the "wired city," based upon integrated services digital network (ISDN) standards and fiber optic laser communications systems. But that's another story.

The real purpose of this chapter is not to project what type of future might evolve for INTELSAT, fiber optics, and advanced communications systems, and their relationship to the world, but rather the reverse--to speculate about what the world might be

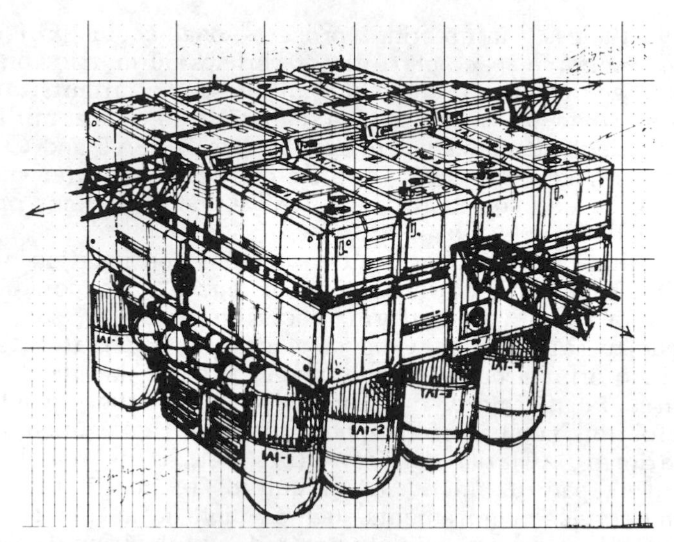

Fig. 3 "INTELSAT Atlantic-1 Antenna Farm" in the year 2010.

like in the next 100 years and how INTELSAT might fit into that world.

II. Cybernetics and the Postinformation Societies

Phase I: 2000 to 2020

As already mentioned, one of the more likely global aspects in the period 2000-2020 will be the evolution and ready availability of thinking machines. Today, literally billions of dollars are being invested in research, particularly in the United States and Japan, but also in Europe, to develop a so-called "fifth-generation computer." The exact technical characteristics and the performance levels of these fifth-generation computers and perhaps the sixth-generation computers that will follow are, of course, unknown. Yet the Japanese government and Japanese industry, in particular, have devoted an incredible amount of study to what they want to develop, as well as to why they want to develop such awesome new capabilities. A report by the Committee for Study and Research on Fifth-Generation Computers, of the Japan Information Processing Development Center--the results of a two-year investigation--gives us some insights into the "whys" and "wherefores" of thinking machines, as seen from the viewpoint of Japanese technical and political leaders. The report[2] states that the objective is "to develop an essential tool in all areas of social activity." The following excerpt from the report lists these objectives:

(1) To increase the level of computer "intelligence" and the affinity for "cooperation" with man.

(2) To enable computers to "act on behalf of all human beings," and assist in the development of unknown fields.

(3) To make a vastly broader range of information available when needed.

(4) To increase knowledge about unknown conditions by development of extremely precise, large-scale simulations and models.

This Japanese report goes on to note a wide range of possible applications for these super "thinking" machines, including: 1) an automated translation system, with over 90% accuracy and a 100,000-plus word capability in each language; 2) a question-and-answer capability, using some 5000 words and more than 10,000 inference rules; 3) applied speech understanding; 4) applied picture and image understanding; and 5) applied problem-solving capabilities (including a formula understanding system).

Furthermore, it is intended to develop a whole host of new knowledge and information-processing capabilities that represent a further evolution, well beyond the computer-assisted engineering, computer-assisted design, computer-assisted manufacturing, and computer-assisted instructional systems that are being applied in industry and in schools today. Finally, it is anticipated that fifth- and sixth-generation computers will prove valuable in extending the capabilities and flexibility of equipment, in terms of "intelligent" office automation equipment and artificial intelligent robotic devices, not only in the factory but in the "professional" office as well.

The ability of these advanced machines to "think," to "interact," and to become, in effect, a "partner" rather than just a tool of mankind, implies awesome changes for the future--not only in terms of the psychological adjustment to a man/machine interface and acceptance of a "thinking capability" within such machine partnerships, but also in terms of the potential impact upon employment. Thinking machines and robotic devices will undoubtedly redefine the meaning of jobs and work in the 21st century.

It has already been projected that the number of industrial jobs in the United States will decrease by the end of the century from 27 million to 12 million and that the number of jobs lost within the OECD as a whole will exceed 30 million. Indeed, this projection of a reduction of tens of millions of industrial jobs is considered by many to be conservative. If it should happen, this would mean that the number of people within OECD countries to be retrained for new jobs would be on the order of two million people per year, now through the end of the century. This, by way of comparison, would imply a nine times faster adjustment and retraining process than occurred over the last 50 years with respect to the agricultural revolution in the U.S. and Europe.

This, astonishingly enough, is only the tip of the iceberg. The real breakthrough that can be expected in the field of automation in the future will be with regard to professional services. Already computers play championship chess games, design new components, and perform a number of diagnostic tasks--from finding a fault in an internal combustion machine to accurately diagnosing a rare medical disorder. Thus, the future potential for automating professional and intellectual jobs poses staggering social dilemmas in terms of job retraining and impact on global employment.

A look to the third-world countries and their likely pattern of economic development during the 21st century does little to simplify the economic and employment issues for the future. Today, in the late 20th century, third-world countries, particularly industrializing developing countries, such as Singapore, the Republic of Korea, and Brazil, serve as a reservoir of so-called "cheap" labor for a variety of industrial manufacturing techniques, including assembly, product finishing, and testing. There has been some concern that many industrial jobs will be shifted away from industrialized nations (where the labor costs are high and trade unionism is strong) to the pool of cheap labor in countries where emphasis is being placed upon industrialization and where a reservoir of lower-salaried skilled and semiskilled industrial laborers is readily available. This concern, which is almost a preoccupation with the economic and scientific government officials of many industrialized countries, stems from seeing this issue in a short-term, as opposed to a longer-term, context.

If one takes a 20- to 30-year perspective of this issue, certain key facts and concerns come into focus. First of all, it is already true that reprogrammable, computer-assisted manufacturing devices and robotic assembly equipment are more cost-effective running on a 16- to 24-hour-a-day, two- or three-shift basis than are industrial workers in the United States. In broad terms, the fully amortized unit cost of production using programmable robotic devices in the United States runs in the range of $6 to $7 ($U.S.) an hour, while the fully allocated cost of an industrial worker is in the range of $15 to $18 per hour.

An industrial worker in a so-called "labor-cheap" country who earns in the range of $1 to $2 an hour is, of course, quite cost-effective by comparison. But, if one extrapolates some 10 to 20 years into the future, one finds that if current salary trends continue in industrializing countries there will be a problem here as well. This will be particularly so if the cost of producing robotic devices decreases while also achieving higher productivity at lower per-hour production costs. Thus, a crossover between robotic devices and industrial workers even in industrializing countries could well be reached as early as 7 to 10 years from now in some cases, and probably no more than 15 years in almost all industrializing countries.

This crossover point could very well be accelerated by the ready availability of low-cost global communications. Just as it is likely that the cost of robotic production will go down to as low as $2 to $3 per hour, it is also likely that international communications via satellite or fiber optics could also be reduced to $1 per hour or less for a high-quality, 16-kbit/s voice circuit. Again, this implies that the cost of international communications and certainly national communications will also be reduced to significantly less than the hourly wage of an industrial worker. This also implies that labor, whether it be an industrial worker, a professional, or even an artificially intelligent "being," can be electronically imported or exported from almost any place on Earth to any other place on Earth, with perhaps the main constraints being not of a technical nature but of an organizational, legal, or political bent--resulting from factors such as time differences for workers in diverse parts of the world, or restrictions on transborder data flow.

In such an environment, however, flexible industrial capabilities can be fostered in many places throughout the world and tied together electronically. Manufacturing capabilities can be located in "labor-cheap" areas, if desired. Fully automated robotic plants can be created well below the surface of the Earth to be near natural resources and to achieve low-cost accommodations. One conclusion seems clear. Namely, that the low cost of robotic production devices, as well as low-cost global communications, will have a significant impact on global employment patterns. It is certainly likely that, within the next 10 to 15 years, industrial plant locations will be less affected by proximity to a cheap labor force than they will be affected by such factors as access to natural resources, political stability, and ready access to low-cost communications and postmanufacturing, marketing, and distribution facilities and mechanisms.

One probable result of these changes is that a new multinational industrial consortium will emerge that will attempt to maximize the efficiencies that derive from the flexibilities created by low-cost global communications and widespread availability of very low-cost industrial and "professional services" robots. In order to tie these multinational complexes together, it is likely that wide-band video capabilities to achieve videophone and videoconference interconnection of these industrial networks will further stimulate the development of low-cost communications networks wherein even video applications are significantly reduced from today's services by a factor of at least 10 times or more.

The availability of low-cost, high-efficiency telecommunications, as well as high-efficiency, low-cost manufacturing, mining, and industrial robots, will also enable technologies to assist in man's large-scale return to space exploration and space exploitation. These technologies, combined with high-efficiency space vehicles that are perhaps 10 times more cost-effective than NASA's Space Transportation System (STS) and the European Space Agency's ARIANE 5 Prime Programs, should, sometime in the early 21st century, make possible a return to space for economic and commercial purposes.

After the last Apollo mission (in the 1970's), the noted space scientist and engineer, Wernher von Braun, was asked when man would return to the moon. Von Braun based his answer upon an analogy to man's exploration and settlement of the Antarctic region. He projected that, just as it had taken about 40 years from the time that Byrd, Perry, Amundsen, and Scott first reached the polar extremes of the Earth until the fully operational colonization by 5000 people could take place for scientific, industrial, and economic purposes, the same time span could be expected for settlement of the moon. Thus, he projected that, from the first moon landing in 1969, it would take about 40 years, i.e., perhaps sometime between 2005 and 2015, to achieve a truly permanent lunar presence for commercial purposes.

Von Braun suggested that the key to progressing from initial exploration of the Arctic regions by dogsled and tractor to practical economic development was, in large part, based upon a technological breakthrough--the development of an enabling technology. In the case of the Arctic regions, it was, of course, the development of sufficiently high-performance aircraft. To von Braun, the enabling technologies for the exploitation of the moon by a permanent lunar colony were the development of sufficiently high-performance, low-cost transport spacecraft, as well as a manned space platform in low Earth orbit to serve as a staging base between the Earth and the moon.

If one accepts von Braun's analogy and adds the further enabling technologies needed for lunar settlement (namely, low-cost wide-band telecommunications capability and low-cost, highly reliable self-replicating robotic devices), it is possible to roughly project that a lunar colony will make its appearance sometime between 2005 and 2015. The capabilities that such a lunar colony, particularly an industrial-based one, could make possible for the Earth are considerable. The Moon Colony--or perhaps the moon colonies--will likely serve not only as a scientific base for activities such as radiotelescope work from the "dark" side of the moon but for energy generation using the solar light and dark discriminator as a source of power and of natural resources. One study, commissioned by NASA and conducted by General Dynamics Corporation, has already conceptually demonstrated that it could be economically advantageous to do material processing on the moon for the purpose of manufacturing large-scale space platforms for communications and other purposes and then to "lower" these to geosynchronous orbit, as opposed to manufacturing such large structures on Earth and raising them up to geosynchronous orbit against the Earth's gravitational field. Thus, it is hard, when projecting the future, not to envision that in the first phase of the cybernetic society a permanent human presence in space will become technically feasible and also economically necessary.

Indeed, when the INTELSAT engineers designed the "2010" space platform with a 25-year projected lifetime, let's say the period 2007 through the year 2032, they decided to include a cislunar link to provide communications between the Earth and lunar space missions, as well as for space communications between the Earth and a permanent lunar colony with a capability of some ten full-time, dedicated telephone circuits. The question to the reader is, "Do you believe that this projected requirement was too optimistic or too pessimistic?"

Phase 2 of the cybernetic, or postinformation, society, has arbitrarily been set as the period 2020 through 2085 or, in effect, 100 years into the future. The only real justification for taking our projections so far into the future is simply the knowledge that one need not feel uncomfortable at the prospect of subsequently having to be held accountable for one's predictions. Again, it is perhaps easier to project this far into the future because you can envision new types of technologies as needed to accomplish almost any task. No one can quibble over an "uninvented" technology.

The range of technologies envisioned in a survey of science fiction, as well as the speculation of such farsighted scientists as Buckminster Fuller and Arthur Clarke, contains few boundaries as to what might occur within the next century. Technologies or technical capabilities that can be envisioned include: the development of advanced robotic devices or cybernetic organisms that could perform physical and mental tasks equivalent to the typical human being, with the intelligence of such devices or organisms being transmitted and received from a remotely sited, very-high-capability computer of the future.

One certainly can envision a wide variety of devices, equipment, and facilities suitable for the economic, industrial, and scientific exploitation of subterranean Earth and undersea colonies. Furthermore, one could also envision island factories and living quarters that float on the seas or perhaps, as suggested by Buckminster Fuller, drift in the Earth's atmosphere. One could also conceive of the ability (perhaps using the lunar colony as a base of operation) to undertake terra-forming of one or more planets of the solar system. Perhaps the most likely candidates would be Venus, Mars, or one of Jupiter's satellites.

It is also possible to envision Von Newman machines that are capable of sufficient intelligence not only to replicate themselves, but perhaps even, in doing so, to improve on their design to install new capabilities, intelligence, or flexibility for multiple replication. Such Von Newman devices, which perhaps represent the ultimate goals and objectives of the cybernetic process, could be used for a whole host of industrial applications on Earth, as well as in space. Even more so, they could be utilized as interstellar space probes that would allow, over vast periods of time, the systematic exploration of at least our galaxy. These devices might be designed in such a way as to exponentially increase the possibility of man's making contact with extraterrestrial intelligent beings.

Another intriguing technology that could be conceived is man's creation of an artificial star as a new energy source. As difficult and as farfetched as such a possibility seems, one only has to start from a scenario developed by Arthur Clarke in 2010, wherein he expounds the concept of Von Newman devices as instruments to generate the metamorphosis of Jupiter into a new stellar object in our solar system.

Certainly, there are few constraints that must be recognized as impenetrable barriers for man's intelligence, assuming that man's technologies are not used to destroy life and civilization on Earth in the coming decades. Of course some constraints must be respected, such as the likelihood that within 10 billion years the sun will cease to be a stable source of energy for the Earth's inhabitants. Presumably, with the capability to build Von Newman devices and to create new stellar energy sources, even these obstacles could ultimately be overcome. Clearly, however, in the much shorter time frame of perhaps the next few hundred years, man must be far more concerned about problems of environment, energy, and the physical limits of natural resources, described in the context of limits to growth debates that have emerged at the social and political levels in the last two decades.

Mathematical formulas show there are limitations to geometric growth. For instance, if it was assumed that the entire mass of Earth was composed only of petroleum, the entire supply source would be exhausted after only 342 years if the global rate of consumption (using the year 1972 as a basis) continued unabated. If one were also to assume that we had a one-million year supply of something--anything with a fixed supply--at a current rate of consumption, and assuming that the rate of consumption for that resource were to increase by just 2% per year (which approximates the current world population growth rate), then that fixed supply would be exhausted not in a million years, but in only 501 years. Finally, there is the example of how long it would take for the world's population to reach the absurdity of one person on each 1 m^2 of ice-free land at today's current growth rates. Again, the answer shows that this result would be achieved in only 600 years if growth continued unabated.

These conditions obviously will not occur, and indeed there are indications that an S-curve of growth--the type of growth that coincides with biological cycles--is already starting to take effect within the global economy. Even if all of the remarkable technologies just described, as well as others that seem beyond our grasp (such as mental telepathy as a communications device, or teleportation), are achieved within the next hundred years, it is clear that man will have to make some adjustments in his living.

He will need to adapt to new technological capabilities, to adjust to the finite physical resources of Earth in order to provide new types of employment and recreation, and to overcome the dangers that could evolve from a steady-state society with zero population growth and economic stabilization. Man's pioneering spirit, his will to innovate, and his other questing personality traits, however, suggest that change of one sort or another will remain with us, regardless of the technological innovations that emerge in the next hundred years.

Today, mankind is clustered on about 5% of the world's surface, with the remainder of the Earth being mainly unsettled. Almost 80% of the globe is ocean and Arctic regions. The rest of Earth's surface is largely covered by deserts, mountain ranges, and semi-arid space and jungles--all of which are sparsely inhabited. Advanced telecommunications and robotic capabilities can and should change this pattern. By 2085, it is therefore possible to envision men and women and children living beneath the Earth's surface; living under the oceans; living on floating islands, certainly at sea if not in the Earth's atmosphere. Yet, all of these people, regardless of where they are located--on, above, or below the Earth's surface--will be able to plug into a global telenet with amazing speed and at remarkably low cost. It seems likely that this more distributed global living pattern will have a strong impact on administrative and political jurisdictions and perhaps minimize the force of nationalism and, perhaps inevitably, diminish cultural traditions and distinctions between people of different genetic and linguistic backgrounds.

Diversity of life styles, diets, and customs will be enjoyed, aided by the many technological innovations of the future. It also seems probable that the Earth will become much more integrated as a single global planetary economy wherein the physical and ecological constraints become clear to all the Earth's citizens. A new "cosmic" understanding that our Earth is only a 6.6-sextillion ton mudball in an obscure solar system on the outer reaches of our quite ordinary galaxy seems almost inescapable in the century ahead. We hope that, with this new cosmic understanding, there will be better chance of establishing a global environment with broad prosperity and intellectual freedom, as well as physical and spiritual challenge for all mankind.

Unfortunately, despite this hope, there is nothing in man's social/political history, or even in the technological forecasts that have been made or implied in this chapter, that suggests that man will live in peace and that the ravages of war or economic deprivation could not be a part of the reality of the year 2085. Consequently, it is likely that the dangers of such a global social and political environment could be even greater than they are today.

As man has evolved rapidly from the 19th century, great stress has been placed on economic productivity, economic throughput, and cost efficiency, with little emphasis on the survival of the species, redundancy, and survivability. Perhaps one of the consequences of the development of artificial intelligence and highly sophisticated models and computerized simulation systems will be to show far more clearly the advantages to be gained by a socially and economically harmonious relationship for all peoples of the Earth.

Where will the INTELSAT of the future fit into this remote and, seemingly, strange world? Clearly, communications must be a vital universal service of this 21st century world. Not only will communications above, on, and below the surface of the Earth be an absolute necessity, but communications services within and perhaps outside the solar system will have become necessary as well. The economic efficiency, political cooperation, and technological capability of the INTELSAT system seem to be consistent with and, perhaps ultimately, essential to man's continued evolution toward the full realization of a cybernetic, or postinformation, civilization. As man spends more time at leisure, at sporting and athletic activities, and in putting aside the burdens of physical labors, the use of telecommunications can be expected to take on new roles and importance. Still, there is a danger that vicarious experience, rather than reality, could have a negative impact on man's evolutionary journey to make not only the Earth, but our solar system, a better home for humanity.

By the year 2085, INTELSAT can certainly be expected to have evolved and changed in many ways. It could become a more decentralized system, connecting different types of regional consortia and perhaps providing dedicated subsystem capabilities to corporations, national governmental agencies, and regional and international organizations. The key to INTELSAT's survival and growth over the next hundred years is surprisingly much the same as it has been for the last 20 years: Namely, to develop new, more cost-effective technologies; to develop new and innovative services while reducing the cost of conventional services; to remain as an effective global common denominator, responsive to the needs and demands of all countries and all peoples; and to constantly evolve new strategies and maintain institutional flexibility to respond to a rapidly changing economic, social, and political global environment. For INTELSAT, the sky can never be the limit!

References

[1]Piel, G., "The Acceleration of History," 50 Great Essays, edited by Elizabeth Huberman and Edward Huberman, Bantam Books, New York, 1964.

[2]Roth, A., "Japanese Fifth-Generation Initiative," Computer News, Vol. 2, No. 6, June 1983, p. 1.

15. INTELSAT and its Role in the Emerging World Telecommunications System

Joseph V. Charyk* and Irving Goldstein†
Communications Satellite Corporation, Washington, D.C.

Over the last two decades, international telecommunications have developed through a process of applying evolving technologies to meet an even faster evolving spectrum of service demands around the world. The challenge has been to provide increasing capabilities at decreasing prices with continuing reliability. The success of INTELSAT in this endeavor can be seen clearly by noting that the services now provided routinely by its satellites did not exist two decades ago. International satellite communications have evolved in those 20 years from a possibility, to a curiosity, to a necessity for governments, businesses, and individuals in widely diverse countries--diverse in their locations, systems of government, economies, languages, and levels of technological development.

INTELSAT was the catalyst and, increasingly, is the agent of this change. Its technical achievements are easy to enumerate, if startling in their rapidity: from no commercial satellites in 1964 to a global communications system in 1969; from a handful of ground stations in 1965 to nearly a thousand in 1984; from 240 telephone circuits on the INTELSAT I spacecraft to over 30,000 on each INTELSAT VI.

Today, INTELSAT carries some two-thirds of the world's international telephone traffic and virtually all international television transmissions on a system of more than 1000 pathways linking some 170 countries and territories. It takes an effort for us to remember that until 1956 there were no international telephone calls as we know them today; limited transatlantic calls were made via radio telephone. The first transatlantic undersea cable for voice services, installed in that year, made it somewhat easier to call between the United States and Europe, but provided no

---

Invited paper received June 25, 1984.

*Chairman and Chief Executive Officer.
†President.

communications links to South America, Asia, or Africa. It took almost ten more years before the entire Atlantic Basin was linked by the new technologies of communications satellites; within another three years, in 1969, the entire globe was spanned by three satellites providing INTELSAT's high-quality communications system for voice and television applications.

This global capability--the core of present international communications--is what INTELSAT satellites provide today. The ability to pick up a telephone and be almost instantly in contact with any corner of the Earth is a hard-won benefit of the changes made possible by INTELSAT's system. But while developing technology makes service improvements possible, the service applications, and not the technology, are the key to continued growth. INTELSAT seeks constantly to improve services, recognizing that all of us, as users of communications, don't care whether we derive our communications signals through hot water pipes or Earth station antennas and geosynchronous satellites--as long as they work inexpensively. While transoceanic television was non-existent in the early 1960's, today INTELSAT offers satellite capacity by the minute, month, or year, from one country to another or from one to many, in any geographical direction and under a range of prices; even broader applications are being discussed for the next few years that would allow integration of video capability with a basic telephone handset. The explosion in technologies and services has, in fact, made possible significant reductions in the cost of communications. The rate that COMSAT, as the Signatory for the United States in INTELSAT, charges United States carriers for a voice circuit has decreased since 1965, despite inflation, by some 73%--from $4200 per month to $1125. A television program transmitted between the United States and any point in the Atlantic or Pacific Ocean region in 1965 was $3000 for ten minutes; today, the rate is $126.50. These and other rates are being driven so low that we may someday be mailing bills whose postage is costlier than the communications services invoiced, and while working to achieve this we are also seeking ways for INTELSAT to lower the cost of international postage, such as electronic mail service.

Constant improvements in technology, expansion in the volume and scope of services, and reductions in price have occurred in many industries. The unique achievement of INTELSAT is that this has occurred through a cooperative commercial effort among many nations. Why is this a laudable achievement? Because each country on the globe has its own approach not only to government, language, and economics, but also to the preferred uses and goals for telecommunications and technological infrastructure. Why does this matter for telecommunications? Because each international telephone call, or data transfer, or television transmission requires agreement by entities in at least two countries on how such communications will be provided. And every communications satellite link is the result of the cooperative commercial agreements among 180 nations around the world.

As an international cooperative venture, INTELSAT has no precedent. INTELSAT's members have made it possible for the commercial cooperative to be just that--commercially oriented and cooperative--to establish an effective international system. The decisions reached by its governing bodies are implemented globally, regardless of the original views of the participants. As a corollary, the system has been designed to function at a common international level despite the wide variations in national systems and goals. There is an all too human tendency to assume that such cooperation, once achieved, will continue. To say that INTELSAT is a foundation of the world telecommunications system is accurate as a recognition of the current technological possibilities offered, but also implies an organizational solidity that will not continue without recurring international effort and commitment. INTELSAT's achievements, and the resulting benefits to communications users, are not based on the mechanical function of an autonomous machine that continues to operate effortlessly once established. Perhaps the most basic achievement of INTELSAT is that its members continue to search for--and find--the means to maintain and improve the international communications services the organization was created to provide.

The promise of such services is greater today than ever before. While the INTELSAT system is perhaps most commonly perceived as facilitating the basic international voice and television transmissions, which are the primary services of the system, the system does not stop there. INTELSAT is unique in providing international communications on both thin and dense routes. In each of the Atlantic, Pacific, and Indian Ocean regions, one satellite provides the primary means for interconnecting all countries within the third of the Earth visible from geosynchronous orbit -- hence the designation of that satellite as the region's "Primary". The telephone traffic carried on each Primary satellite ranges in volume from a handful to hundreds of circuits between any two countries in each regional network. This integrated network capability facilitates interconnection of a variety of applications for high- or low-volume services, with efficient use of orbital, frequency, and financial resources.

The achievement of INTELSAT over the past two decades has been the establishment of this international network. But its promise and challenge for the next decade is to combine with the basic network a significantly expanded range of services and connectivity at lower prices. Several current service applications are particularly noteworthy.

One example, which will be implemented over the next few years, is the expansion of high-quality communications to geographical areas so remote as to be unreachable in terms of communications by other than satellite systems. INTELSAT already links relatively high-volume inter-urban transnational users, and using existing resources, it will not be difficult to provide service to inexpensive Earth terminals in outlying areas such as the Pacific islands to meet requirements for one or two telephone and low-speed data circuits to urban population centers. Furthermore, the INTELSAT system can also be used to interconnect these remote areas with other rural areas, other adjacent national urban centers, or any part of the globe.

Another major service area is developing within the INTELSAT network even as we write. While both thin and dense route telephony have traditionally been provided primarily on a point-to-point basis between country pairs, INTELSAT satellites have the capability to offer point-to-multipoint services as well. This feature is often used for international television broadcasts, with the World Cup Soccer matches and the summer Olympic Games drawing very wide international audiences; the 1984 Olympics, for example, will be seen by nearly 2 billion people. Less dramatic but more frequent uses include weekly readings of the Koran, or daily news flashes from major cities around the world. Still, such multipoint interconnections for data and voice have become less common.

One new series of applications that will assist in expanding this interconnectivity is the digital business service now being implemented throughout the INTELSAT system. The technical parameters for this all-digital offering allow access by Earth stations that are either national access points, urban gateways, customer premise facilities, or any combination of the three. The service offers the means to communicate overseas more quickly, efficiently and economically, bypassing the cost and technical limitations of terrestrial facilities whenever desired. New capabilities offered by this service include a variety of digital transmission speeds from 64 kbits/s to multiple megabits and a flexibility to vary the types of communications provided. This latter feature is important as businesses consider, for example, how to use their leased capacity to integrate widely dispersed operating centers with, perhaps, voice communications among several nodes at one time of day, video-conferencing during others, and off-peak computer-to-computer transfer of information in a point-to-multipoint broadcast format, or on interactive or single point bases. The capacity is available on a full-time basis, 24 hours a day, for a certain number of hours each day, month after month; or in an occasional as needed call-up mode. In conjunction with this dramatic expansion in the range and flexibility of applications, expanded geographic coverage is also possible. This will enable direct communications from the west coast of North America to Europe, and from the easternmost portions of North America to India and Pakistan. INTELSAT's traditional division of the globe into Atlantic, Pacific and Indian Ocean sectors is being dissolved by the system's evolution at the same time the system's applications grow.

Efforts are also underway to improve specific existing services, such as data transfer. Communications administrations around the world are replacing their analog switching facilities with digital equipment, and their coaxial cable, microwave and ultimately local loop facilities with fiber optic cables. These remarkable changes will make it both feasible and necessary to shift the focus of innovation and cost reduction away from the provision of basic transmission facilities and toward the provision of enhanced services responsive to even more specific customer demands. Thus, the changing nature of the users' needs, and the network operators' plans for meeting those needs, are becoming increasingly important to satellite communicators.

For traditional analog communications services, residual errors present to some degree in any communications channel are tolerable and appear as background noise in telephony or as barely perceptible disturbance in a television picture. However, the significance of errors is radically different for digital data transmissions. Transmission errors may cause a particular application to fail, may take an entire computer center out of service, or may necessitate annoying retransmission of packet of data for error correction. A range of techniques for insuring error-free delivery of data has been developed and implemented over terrestrial media. As data users turn increasingly to satellites to take advantage of their low-error, wide-band, and multipoint characteristics, however, many more packets of information will be transmitted through the satellite system at a given moment than would be the case over a typical narrow-band, low-delay terrestrial link. New computer protocols are being developed to take account of these differences and to counteract the effects of the delay in transmission of a signal from a point on the Earth to the satellite 22,300 miles up and back down to the receiving point.

While such developments, which are most intimately related to developing computer services, may seem less important for voice and broadcast services, their key significance derives from their role in the growing international trend toward integrated services digital networks (ISDN). The essential proposition of the ISDN is a bold one: that the digitalization of national networks calls for advance planning of integrated voice, data, and image services and common end-to-end digital facilities to support such services. The International Telecommunication Union (ITU) has undertaken a massive standard-setting exercise to facilitate development of the ISDN, and INTELSAT's experience in interconnecting the diverse technologies of different national communications systems, including frequent national preferences for a local manufacturer or a particular type of equipment, provides a foundation and an approach to the development of ISDN. The challenge is to multiply this cooperation manyfold, and satellite operators will have to become much more integrated into service provision and network management than has been the case in the past.

The results of these and other developments underway will be a range of new, productivity-enhancing, cost-reducing services for telecommunications users worldwide. At the same time, the INTELSAT system must and will continue to interconnect countries that are not ready, or do not wish, to participate in such advanced digital systems with those that are on the forefront of such developments. It must offer both the advantages of highly specialized and technically innovative communications to those desiring them, and the continuing ability for all nations to communicate internationally using the traditional forms of service.

INTELSAT's achievements to date, and its promises for the future, are thus extremely dramatic as well as beneficial to all users of communications. This extraordinarily successful international cooperative has purposely fostered the flexible accommodation of a variety of national access methods, such as commercial or government control of communications, nationally subsidized or completely for-profit communications services. Thus, while communications in each country have evolved in accordance with domestic policies, it has been possible for all countries to jointly offer an expanding range of cost-effective services. Perhaps not surprisingly, INTELSAT's very success has recently led others to recognize that satellite communications can be not only cost-effective but also revenue-producing.

A large part of INTELSAT's achievement has been to foster international integration while retaining domestic independence in the control and use of communications. In recent years, the United States and some other developed countries have begun to seek increased competition in the provision of services to their national users. At the same time, these and other countries--often those with limited financial and other resources--have made substantial commitments to the international cooperative, based upon common agreements concerning INTELSAT's principles and goals. In 1983-84, however, several proposals have been made for separate trans-Atlantic commercial systems focusing on heavy route traffic. The economic burden that these would impose upon the INTELSAT cooperative is evident and has been a source of severe international concern. The challenge to INTELSAT's members will be to continue to promote cooperative action in the face of an apparent desire by one or two countries to transfer their domestic policies to the international arena.

Changes initiated within INTELSAT several years ago have begun to bear fruit as the number of service alternatives for users explodes, with significantly increased technical and price options. In fact, the vitality and viability of the INTELSAT system, and its value to the variety of users it serves in developed and developing countries, rest precisely on its ability to offer shared resources for a variety of applications. It is to be hoped that INTELSAT's members, while propelling the organization into the emerging global network, will similarly focus on the wider variety of services that the integrated international cooperative can make possible. Rather than looking for alternative, fragmented facilities structures to meet individual applications, it must be recognized that INTELSAT's mandate--to establish a global, commercial communications satellite network providing economical services to all users--is as important today as it was 20 years ago. While establishment of more networks may look attractive, unneeded duplication of facilities to provide limited services to a few users can only be more expensive, especially to the majority of users who can least afford it. It is new technologies and services, not networks, that are needed.

The technology of satellite communications was developed by INTELSAT to meet the uncertain service demands of 1964; the demands are being enunciated more clearly now, but INTELSAT continues to devise new technologies and new services to meet them.

Appendix

ABU = Asian Broadcasting Union
ac = alternating current
ACC = antenna control console
ACE = audio connecting equipment
ACI = adjacent channel interference
ACS = aeronautical communications subsystem
ACS = attitude control subsystem
ACSN = Appalachian Community Service Network
ACU = antenna control unit
A/D = analog/digital
ADCE = attitude determination and control electronics
ADCS = attitude determination and control subsystem
ADM = adaptive delta modulation
AEROSAT = Aeronautical Satellite System
af = audio frequency
AFC = automatic frequency control
AFCAC = African Civil Aviation Commission
AFETR = Air Force eastern test range (U.S.)
AFROSAT = African Domestic System (proposed ITU study)
AFS = aeronautical fixed service
AFSAT = African Satellite System (French, U.K., Italian proposed system)
AGC = automatic gain control
AIAA = American Institute of Aeronautics and Astronautics
AIEE = American Institute of Electrical Engineers
AKM = apogee kick motor
Alouette = international satellite for ionospheric studies (Canada)
AM = amplitude modulation
AMS = aeronautical mobile services
AMS = attitude measurement sensor
AMS = domestic satellite network of Israel (planned)
AMSAT = Radio Amateur Satellite Corporation (U.S.)
ANC = automatic nutation control
ANIK = domestic satellite network of Canada (owned by Telesat Canada)

ANS = astronomical Netherlands satellite (Netherlands/U.S.)
ANSI = American National Standards Institute
ANT = antenna
ANTEL = Administracion Nacional de Telecomunicaciones (El Salvador)
ANTELCO = Administracion Nacional de Telecomunicaciones (Paraguay)
AOR = Atlantic Ocean region
AP = Assembly of Parties (INTELSAT)
APE = antenna positioning electronics
APEX = satellite network of France
APKS = amplitude-phase-keyed system
apm = amplitude-and-phase modulation
APM = antenna positioning mechanism
APT = Asia/Pacific Telecommunity
APTU = African Postal and Telecommunications Union
ARA = angular rate assembly
ARABSAT = regional communications satellite for Arab countries
ARCOMSAT = Arab League communications satellite
ARENTO = Arab Republic of Egypt Telecommunications Organization
ARIANE = first commercial European launch vehicle system
ARIEL = ionospheric satellite (U.K.)
ARPA = Advanced Research Projects Agency
ARQ = automatic repeat request
ARRL = American Radio Relay League
ARTI = Arab Regional Telecommunications Institute
ASA = American Standards Association
ASBU = Arab States Broadcasting Union
ASCENA = Agency for the Safety of Aerial Navigation for Africa and Madagascar
ASEAN = Association of South East Asian Nations
ASETA = Asociacion de Empresas Estatales del Acuerdo Subregional Andino
ASIN = Action of National Information Systems
ASR = automatic send and receive
ASTRA = application of space techniques relating to aviation
ASU = automatic switching unit
AT = acceptance test
ATA = Aeronautical Telecommunications Agency
ATAA = Air Transport Association of America
ATB = all trunks busy

ATDA = augmented target docking adaptor
Atlas/Centaur = U.S. launcher (General Dynamics)
ATME = automatic transmission measuring equipment
ATS = application technology satellite
AT&T = American Telephone & Telegraph Company
ATU = Arab Telecommunication Union
AURA = astronomy research satellite (France)
AUSSAT = Satellite Network of Australia
AVD = alternate voice/data
AYAME = experimental telecommunication satellite (Japan)
Az = azimuth

BAPTA = bearing and power transfer assembly
BB = baseband
BBC = British Broadcasting Corporation
BBU = baseband unit
BCD = binary coded decimal
BCH = Bose-Chaudhuri Hocquenghen (SPADE)
BCW = burst codewords
BDE = baseband distribution equipment
BDF = baseband distribution frame
BDU = baseband distribution unit
BEF = band elimination filter
BER = bit error rate
BER CONT = bit error rate continuous
BFO = beat frequency oscillator
BG = Board of Governors (INTELSAT)
BG/BARC = BG's Budget and Accounts Review Committee (INTELSAT)
BG/PC = BG's Advisory Committee on Planning (INTELSAT)
BG/T = BG's Advisory Committee on Technical Matters (INTELSAT)
BHASKARA = satellite for Earth observation (India)
BID = Banco Interamericano de Desarrollo
Big Bird = U.S. Air Force reconnaissance (see SAMOS)
BIH = International Time Bureau
BINR = basic intrinsic noise ratio
Biosputnik = biological satellite (USSR)
BIPM = International Bureau of Weights and Measures
BIRF = Banco Internacional de Reconstruccion y Fomento
BIS = British Interplanetary Society
BLD = binary load dump (SPADE)

BLK = block
BOD = beneficial occupancy date
BOL = beginning of life
BONAC = Broadcasting Organization of the Nonaligned Countries
BP = bandpass
BPF = bandpass filter
BPI = bits per inch
BPO = British Post Office (British Telecommunications)
bit/s = bit per second
BFSK = binary frequency shift keying
BPSK = biphase shift keyed
BS = broadcasting satellite system of Japan
BSI = British Standards Institution
BSS = broadcasting satellite service
BST = block the SPADE terminal command
BSU = baseband switching unit
BSXF = burst sync failure (SPADE)
BTE = bench test equipment
BTI = British Telecommunications International
BTL = Bell Telephone Laboratories
BTR = bit time recovery
BTU = British thermal unit
BU = basic user
BU = baseband unit
BVU = brightness value unit
BW = bandwidth
BWR = bandwidth ratio

C&W = Cable & Wireless, Ltd.
C/I = carrier-to-interference ratio
C/N = carrier-to-noise ratio
C/T = carrier-to-noise temperature ratio
CANTV = Compania Anonima Nacional Telefonos de Venezuela
CAO = completed as ordered
CAR = channel assignment record
CAS = Commission for Atmospheric Sciences
CASSA = coarse analog sun sensor assembly
CASTOR = CNES technological satellite (France)
CATV = community antenna television
CBC = Canadian Broadcasting Corporation
CBS = Columbia Broadcasting System
CCC = Customs Cooperation Council
CCI = International Consultative Committee
CCIR = International Radio Consultative Committee (ITU)

CCIS = common channel interoffice signaling (U.S./AT&T)
CCITT = International Telephone and Telegraph Consultative Committee (ITU)
CCKF = continuity checktone failure (SPADE)
CCS = commercial communications satellite (Japan)
CCT = continuity check transceiver
CCTS = Coordinating Committee on Satellite Communications
CDE = control and display equipment
CDF = carrier distribution frame
CDMA = code-division multiple access
CEE = European Economic Community
CEEFAX = trade name for a videodata system in the U.K.
CELTIC = names of French digital compression equipment for telephony
CEMA = Canadian Electrical Manufacturers Association
CENTO = Central Treaty Organization
CEPT = European Conference on Postal and Telecommunications Administrations
CESA = Canadian Engineering Standards Association
CETS = European Conference on Satellite Communications
CETTEM = center of telecommunications for the Third World
CFM = companded frequency modulation
CFDM/FM = companded frequency-division-multiplex/ frequency modulation
cg = center of gravity
CG/MOI = center of gravity/moment of inertia
CGPM = general conference of weights and measures
Ch = channel
Chnl = channel
CHU = channel Unit (SPADE)
CI = critical item
CIC = common interface circuit
CIC = cover integrated cell
CIFAS = French-German industrial consortium for Symphonie program
CIGRE = international conference on large electric system
CIMS = Coordination and Interference Management System (INTELSAT computerized intersystem coordination project)

CIRM = International Marine Radio Association
CISPR = International Special Committee on Radio Interference
CITE = cargo integrated test equipment
CITEL = inter-American Telecommunications Committee
ckt = circuit
CLK = clock
CLR = clear
CMBD = joint committee on circuit noise and availability
cmd = command
CMI = joint international committee for tests relating to the protection of telecommunication lines and underground ducts
CMOS = complementary metal oxide semiconductor
CNES = Centre National d'Etudes Spatiales
CNET = Centre National d'Etudes des Telecommunications
CNET = Centro Nacional de Estudios de las Telecomunicaciones
CNIE = Comision Nacional de Investigacion del Espacio (Spain)
CNR = Consiglio Nazionale Delle Recerche (Italy)
CNR = carrier/thermal noise ratio
CNSS = Center for National Space Study
Co = circuit order
CO-POL = co-polarization
Coax = coaxial cable
Codec = coder-decoder
COM = computer output microfilm
COMDEV = Canadian aerospace firm
COMM = communications
Companding = compression and expansion
comsat = communications satellite
COMSAT = Communications Satellite Corporation (U.S.)
COMSTAR = U.S. domestic satellite (owned and operated by COMSAT General)
COMTELCA = Comision Tecnica de Telecomunicaciones de Centroamerica
CONT = controller
COPS = command operations
COPUOS = Committee on the Peaceful Uses of Outer Space
CORSA = cosmic radiation satellite (Japan)

COSMOS = European Multinational Industrial Consortium
COSMOS = USSR military satellite network
COSPAR = Comite Mondial de la Recherche Spatiale
COTC = Canadian Overseas Telecommunications Corporation (now known as TELEGLOBE)
COURIER = early low-medium orbit satellite (U.S.)
CPC = Committee for Programme and Coordination of ECOSOC
CPSK = coherent phase shift keying
CPU = central processing unit
CR = carrier recovery
CRA = Centro Ricerche Aerospaziali (Italy)
CRC = cyclic redundancy code
CRC = Communications Research Center (Canada)
CRPL = Central Radio Propagation Laboratory (U.S.)
CRT = cathode ray tube
CRYO = cryogenic
CS = communications satellite
CS = domestic satellite system--SAKURA (Japan)
CSC = common signaling channel
CSCD = common signaling channel demodulator
CSCE = communications subsystem checkout equipment
CSCM = common signaling channel modem
CSCS = common signaling channel synchronizer
CSELT = Centro Studi e Laboratori Telecomunicazione (Italy)
CSELT = Centro de Estudios y Laboratorio de Telecomunicaciones
CSG = Guiana Space Center (Center Spatial Guyanais)
CSM = communications system monitoring
CSM = Commission for Synoptic Meteorology
CSME = communications system monitoring equipment
CT = transit center
CTE = channel translating equipment
CTNE = Compania Telefonica Nacional de Espana
CTS = communications technology satellite (U.S./Canada experimental satellite)
CU = community user
CW = continuous wave
CW = clockwise
CXR = carrier

D/A = digital to analog
D/C = down-converter
DA = demand-assignment

DAMA = demand-assignment multiple access
DASS = demand-assignment signaling and switching unit
dB = decibel
dB/K = decibels per degree Kelvin
DBH = diameter at breast height
dBm = decibels relative to 1 mW
dBmOp = dBm referred to or measured at a point of zero transmission level, psophometrically weighted
DBP = Deutsche Bundespost
DBS = direct broadcast satellites
dBW = decibels relative to 1 W
dc = direct current
DCA = Defense Communications Agency (U.S.)
DCP = data collection platforms
DCPSK = differentially coherent phase shift keying
DCRT = data collection receive terminal
DCU = display and control unit
DD/C = dual down-converter
DDI = direct digital interface
DEC = declination
DELTA = U.S. launcher (McDonnell Douglas)
demod = demodulator
demux = demultiplexer
DET = detector
DEV = deviation
DFS = domestic satellite network of Germany (planning)
DG = director general
DGT = Directorate General of Telecommunication (France)
diam = diameter
DITEC = digital television communication system
DLF = direct through-connection filter
DM = delta modulation
DMA = direct memory access
DMSP = Defense Meteorological Satellite Program (U.S.)
DNI = digital noninterpolation
DNO = down
DOC = Department of Communications (Canada)
DOD = Department of Defense (U.S.)
domsat = domestic satellite system
DPA = destructive physical analysis
DPCM = differential pulse code modulation
DPSK = differential phase shift keyed
DR = dynamic range

DRSS = Data Relay Satellite System
DRT = data recording terminal
DSB-SC-AM = double sideband-suppressed carrier-amplitude modulation
DSB-SC-AM w/QM = double sideband-suppressed carrier-amplitude modulation with quadrature multiplexing
DSB-SC-ASK = double sideband-suppressed carrier-amplitude shift keyed
DSCS = Defense Satellite Communications System (U.S.)
DSI = digital speech interpolation
DSIF = deep space instrumentation facility
DSIR = Department of Scientific and Industrial Research (U.K.)
DTI = data transfer interface

EACSO = East Africa Common Services Organization
EAGE = electrical aerospace ground equipment
EARC = Extraordinary Administrative Radio Conference
EBR = electron beam recorder
EBU = European Broadcasting Union
ECA = Economic Commission for Africa
ECC = eccentricity
ECE = Economic Commission for Europe
ECHO = early low-medium orbit passive satellite (U.S.)
ECLA = Economic Commission for Latin America
ECMA = European Computer Manufacturers Association
ECOM = electronic computer-originated mail
ECS = European regional communications satellite system
EDD = Hughes Electron Dynamics Division
EDP = electronic data processing
EFI = error-free intervals
EFT = electronic fund transfer
EFTA = European Free Trade Association
EIA = Electronic Industries Association (U.S.)
e.i.r.p. = equivalent isotropically radiated power
EKRAN = Statsionar satellite (USSR)
EL = elevation
ELDO = European Launcher Development Organization
ELF = extremely low frequency
ELU = existing carrier lineup
ELV = expendable launch vehicles (U.S.)

EM = engineering model
EMBRATEL = Empresa Brasileira de Telecomunicacoes S.A.
EMF = electromotive force
EMS = electronic message service
ENTEL = Empresa Nacional de Telecomunicaciones de la Republica Argentina
ENTEL = Empresa Nacional de Telecomunicaciones S.A. (Bolivia)
ENTEL = Empresa Nacional de Telecomunicaciones S.A. (Chile)
ENTEL = Empresa Nacional de Telecomunicaciones del Peru
EO = executive organ
EOL = end-of-life
EOW = engineering order wire
EPC = electronic power conditioner
EPR = Environmental Planning and Research, Inc.
EPS = electrical power subsystem
EPTA = Expanded Program of Technical Assistance (United Nations)
EPTEL = Empresa Publica de Telecomunicacoes (Angola)
ER&S = exploratory research & study
ERDA = Energy and Development Administration (U.S.)
ERS = environmental research satellite
ERTS = Earth resources technology satellite
ES = Earth station
ESA = European Space Agency
ESC = European Space Conference
ESC = engineering service circuit
ESCAP = Economic and Social Commission for Asia and the Pacific
ESOC = European Space Operations Center
ESRO = European Space Research Organization
ESSA = Environmental Science Services Administration (U.S.)
ESTEC = European Space Research Technology Center
ESTL = European Space Tribology Center (U.K. Atomic Energy Authority)
ETR = eastern test range (NASA's Cape Canaveral)
ETS = engineering test satellite
ETT = electrothermal thrustors
EUROCONTROL = European Organization for Safety of Air Navigation
EURONET = European Public Network (data)

EUROSAT = Organization to implement a European system of telecommunications and TV by satellite
EUROSPACE = European Industrial Space Study Group
EUROVISION = European and Mediterranean TV services
EUTELSAT = European Telecommunications Satellite Organization
eV = electron volt

F = facsimile
FAA = Federal Aviation Administration (U.S.)
FACC = Ford Aerospace and Communications Corporation
FAO = Food and Agriculture Organization
Fc = sampling frequency
FCC = Federal Communications Commission (U.S.)
FDFM = frequency division/frequency modulation
FDM = frequency division multiplex
FDM/FM = frequency-division-multiplexed/frequency modulated
FDMA = frequency division multiple access
FDR = final design review
FDSS = fine digital sun sensor
FDSSA = fine digital sun sensor assembly
FEC = forward error correction
FET = field effect transistor
FF = flip flop
FFP = fixed fee procurement
FIEG = International Federation of Newspaper Publishers
FINTEL = Fiji International Telecommunications Limited
FLTR = filter
FLTSATCOM = U.S. Navy communications system
FLU = full lineup
FM = frequency modulation
FMEA = failure modes and effects analysis
FMWA = fixed momentum wheel assembly
FP = full period
FPA = final power amplifier
FRG = Federal Republic of Germany
FSK = frequency-shift keyed
FSS = fixed satellite service
FT = full time

G = gain
GA = group one A
GaAs = gallium arsenide

GAGR = group automatic gain regulator
GALAXY = U.S. domestic satellite owned by Hughes Communications
GARP = Global Atmospheric Research Program
GAS = special autonomous study group
GATT = General Agreement on Tariffs and Trade
GCA = ground control approach
GCE = ground control equipment
GCS = ground control system
GDF = group distribution frame
GDPS = global data processing system
GEO-IRS = geostationary orbit-infrared sensor
GFRP = graphite fiber reinforced plastic
GGSE = gravity-gradient stabilization experiment satellite
GGTS = gravity-gradient test satellite
GHIUD = global human information use per decade
GHz = gigahertz
GIP = group interface processor
GMS = geostationary meteorological satellite
GMT = Greenwich mean time
GND = ground
GOES = geostationary operational environmental satellite
Gorizont = communication satellite (USSR)
Gp = group
GRD = ground resolved distance
GRGS = space geodesy research group
GRP = group reference point
GSFC = Goddard Space Flight Center
GSTAR = U.S. domestic satellite network (owned by GTE Satellite Corporation)
G/T = gain-to-noise temperature ratio
GTE = General Telephone & Electronics
GTE = group translating equipment
GTS = global telecommunication system
GUATEL = Empresa Guatemalteca de Telecomunicaciones
GUY = French Guyana Space Center (ESA)

H-SAT = European heavy experimental satellite for broadcast TV
HAC = Hughes Aircraft Company
HAKUCHO = cosmic radiation satellite (same as CORSA)(Japan)
HCE = heater control electronics
HE = heat exchanger
HEAO = High-Energy Astronomy Laboratory
HELIOS = FRG experimental satellite
HELVESAT = broadcast satellite system of Switzerland

HEOS = highly eccentric orbit satellite
HF = high frequency (3000-30,000 kHz)
HONDUTEL = Empresa Hondurena de Telecomunicaciones
hp = horsepower
HPA = high-power amplifier
HPF = high-pass filter
HPM = hybrid phase modulation
Hu = high usage circuit
Hz = hertz (cycles per second)

I = international code to CCITT signaling format
IWP = Interim Working Party
I/O = input-output
IAAB = Inter-America Association of Broadcasting
IADES = Integrated Attitude Detection and Estimation System
IADP = INTELSAT Assistance and Development Program
IAEA = International Atomic Energy Agency
IAF = International Astronautical Federation
IAGA = International Association of Geomagnetism and Aeronomy
IALA = International Association of Lighthouse Authorities
IAM = initial address message
IAMAP = International Association of Meteorology and Atmospheric Physics
IARU = International Amateur Radio Union
IATA = International Air Transport Association
IAU = International Astronomical Union
IBA = International Broadcasting Authority (U.K.)
IBI = Intergovernmental Bureau for Informatics
IBI = International Broadcast Institute
IBM = International Business Machines Corp. (U.S.)
IBRD = International Bank for Reconstruction and Development
IBS = INTELSAT business service(s)
IBTO = International Broadcasting and Television Organization
IC = integrated circuit
ICAO = International Civil Aviation Organization
ICC = International Chamber of Commerce
ICES = International Council for the Exploration of the Sea
ICI = International Commission of Illumination

ICN = idle channel noise
ICRP = International Commission on Radiological Protection
ICSAB = International Civil Service Advisory Board
ICSC = Interim Communications Satellite Committee
ICSU = International Council of Scientific Unions
IDA = International Development Association
IDB = Inter-American Development Bank
IDCC = International Data Coordinating Centers
IDF = intermediate distribution frame
IDR = intermediate design review
IEA = International Energy Agency
IEC = International Electrotechnical Commission
IEE = Institution of Electrical Engineers
IEEE = Institute of Electrical and Electronic Engineers
IERE = Institution of Electronics and Radio Engineers
IETEL = Instituto Ecuatoriano de Telecomunicaciones
i.f. = intermediate frequency
IFD = International Federation for Documentation
IFIP = International Federation for Information Processing
IFL = interfacility link
IFL = International Frequency List
IFRB = International Frequency Registration Board (ITU)
IFTC = International Film and Television Council
IGU = International Gas Union
IGY = International Geophysical Year
IINA = International Islamic News Agency
IISL = International Institute of Space Law
IIW = International Institute of Welding
ILHUICAHUA = domestic satellite network of Mexico
IM = intermodulation
IMC = international maintenance center
IMO = International Maritime Organization (formerly IMCO)
IMPATT = impact avalanche time transit
INAG = ionospheric network advisory group
INC = inclination
INCATEL = Instituto Centroamericano de Telecomunicaciones

INMARSAT = International Maritime Satellite Organization
INSAT = domestic satellite network of India
INSTEL = Instituto Nacional de Telecomunicaciones (Bolivia)
INTELCAM = Societe des Telecommunications Internationales (Cameroon)
INTELCI = Societe des Telecommunications Internationales de Cote d'Ivoire
INTELCO = Office des Telecommunications Internationales du Congo
INTELPOST = international organization created to conduct experimental electronic mail transmissions
INTELSAT = International Telecommunications Satellite Organization
INTERCOMSA = Intercontinental de Comunicaciones por Satelite, S.A. (Panama)
INTERCOSMOS = international research satellites (USSR)
INTERSPUTNIK = International Space Telecommunication Organization (USSR)
INUKSAT = domestic satellite system (proposed by Denmark)
IOB = Interorganization Board for Information Systems and Related Activities
IOC = INTELSAT Operations Center
IOC = Intergovernmental Oceanographic Commission
IOCTF = IOC TDMA facility
IOR = Indian Ocean region
IOT = In-orbit test antenna
IPA = intermediate power amplifier
IPDC = International Program for Development of Communications
IPTC = International Press Telecommunications Council
IPTG = International Press Telecommunications Council
Ir = power loss from resistive dissipation
i.r. = infrared
IRE = Institute of Radio Engineers
IRIG = Inter-range instrumentation group
ISAS = Institute of Space and Aeronautical Sciences (Japan)
ISCC = International Service Coordination Center
ISCOM = Indian Satellite for Communication Technology
ISD = international subscriber dialing
ISDN = Integrated Services Digital Networks

ISEE = International Sun-Earth Explorers
ISIS = International Satellite for Ionospheric Studies (U.S./Canada)
ISL = intersatellite link
ISMC = International Switching Maintenance Center
ISP = specific impulse
ISPC = International Sound Program Center
ISRO = Indian Space Research Organization
ISS = interrupt safety system
IST = integrated system test
ISTC = integrated system test complex
ITA = Independent Television Authority (U.K.)
ITALSAT = domestic satellite system (Italy)
ITC = International Television Center
ITC = Inter-American Telecommunications Network
ITC = International Teletraffic Congress
ITC = Technological Institute of Electronics and Telecommunications (Colombia)
ITFC = instructional television fixed service (U.S.)
ITMC = international transmission maintenance centers
ITR = INTELSAT test record (form sheets)
ITS = Institute of Telecommunication Sciences (U.S.)
ITT = International Telephone & Telegraph Company
ITU = International Telecommunication Union
IUCAF = Inter-Union Commission of Allocation of Frequencies (for radio)
IUE = International Ultraviolet Explorer
IUGG = International Union of Geodesy and Geophysics
IUPAP = International Union of Pure and Applied Physics
IUS = interim upper stage

JAI = John Andrews International Pty., Ltd.
JAMINTEL = Jamaica International Telecommunications, Ltd.
JASC = Japan sea cable
JFET = junction field effect transistor
JPL = Jet Propulsion Laboratories (U.S.)
JSC = Johnson Space Center

K = Kelvin
K = Boltzmann's constant
kbit/s = kilobits per second
KDD = Kokusai Denshin Denwa Co., Ltd. (Japan)
KENEXTELCOMS = Kenya External Telecommunications Co., Ltd.

kHz = kilohertz
KIKU = engineering test satellite (Japan)
KP = start of pulsing
KSC = Kennedy Space Center
kW = kilowatt

L-SAT = experimental satellite network (France)
LF = line-feed
LAGEOS = laser geodynamic satellite (U.S.)
LANDSAT = Earth resources survey satellite (U.S.)
LAS = Laboratorio de Astronomia Espacial
LCC = launch control center
LCT = Laboratoire Central des Telecommunications (France)
LDM = linear-delta modulation
LDTS = low-density telephony service
LEASAT = U.S. military satellite network
LEC = Lockheed Electronic Company (U.S.)
LED = light emitting diode
LEP = Laboratoire d'Electronique et de Physique (France)
LES = Lincoln Experimental Satellite (U.S.)
LHCP = left-hand circular polarization
LMA = limited motion antenna
LMSC = Lockheed Missiles and Space Corporation (U.S.)
LNA = low-noise amplifier
LNR = low-noise receiver
LO = local oscillator
LOS = line-of-sight
LOUTCH = USSR satellite network
LPF = low-pass filter
LPS = launch process system
LRC = Langley Research Center
LS = link switch
LSE = line signaling equipment
LSI = large scale integration
LSM = line switch marker
LSO = line signaling oscillator
LTOM = less than 1 min
LTOS = less than 1 s
LUXSAT = satellite network (Luxembourg)

M = mega/million
M = modulation index (fm); number of phases (PSK)
MAC = mutual aid committee

MAGSAT = NASA satellite to map magnetic fields of Earth (U.S.)
MAPSAT = mapping satellite (U.S.)
MARECS = maritime experimental comsat system (formerly MAROTS)(ESA)
MARISAT = maritime satellite system (U.S.)
MAROTS = Maritime Orbital Test Satellite (renamed MARECS)(ESA)
MASER = microwave amplification by stimulated emission of radiation
MAWG = mutual aid working group
MBB = Messerschmitt-Bolkow-Blohm (Germany)
Mbit/s = megabits per second
MC = maintenance center
MCC = master control console
MCLF = multichannel loading factor
MCS = Maritime Communications Subsystem
MDA = mechanically despun antenna
MDAC = McDonnell Douglas Astronautics Corporation
MDF = main distribution frame
MDFD = multidestination full duplex
MDHD = multidestination half duplex
MDS = minimum discernible signal
MDTA = modulator-demodulator-translator assembly
MDUS = medium data utilization station (Australia)
Mega = million
MELCO = Mitsubishi Electric Corporation
Mesh = Multi-modal network configuration
MESH = Matra-ERNO-SAAB-Hawker Siddeley Dynamics Aerospace Consortium
Meteor = meteorological satellite (USSR)
MF = multifrequency
MF = medium frequency
MFR = multifrequency register
MFS = multifrequency sender
MG = motor generator
MGC = multigroup controller
MHz = megahertz
MIFR = master international frequency register
MIP = multigroup interference processor
MMIC = microwave-monolithic integrated circuits
Mod = modulator
Modem = modulator-demodulator
MOI = moments of inertia
Molniya = domestic satellite (USSR)
MON = monitoring station
MOS = metal-oxide semiconductor

MS = Meeting of Signatories (INTELSAT)
ms = millisecond
MSC = management services contractor
MSK = minumum shift keying
MSM = microwave switch matrix
MSSF = mate scanner stop failure
MTBF = mean time before/between failures
MTR = meter
MTTF = mean time to failure
MTTR = mean time to repair
MTU = magnetic tape unit
MU = multidestination unidirectional
MUF = maximum usable frequency
Mux = multiplex
Mw = microwave

n = nano (one-billionth)
N = newton
n/a = not applicable
NAAL = North America Aerodynamic Laboratory
NACK = negative acknowledgment
NAM = National Association of Manufacturers (U.S.)
NAS = National Academy of Sciences (U.S.)
NASA = National Aeronautics and Space Administration (U.S.)
NASCOM = NASA Communications Network
NASDA = National Space Development Agency (Japan)
NAT/TFG = North Atlantic Traffic Forecasting Group
NATO = North Atlantic Treaty Organization
NAVSAT = Satellite Navigation Corporation
NB = narrow band
NBC = National Broadcasting Corporation (U.S.)
NBS = National Bureau of Standards (U.S.)
NEC = Nippon Electric Company (Japan)
NESC = National Environmental Satellite Center (U.S.)
NESS = National Environmental Satellite Service (U.S.)
NET = Nigerian External Telecommunications Ltd.
NHK = Japan Broadcasting Corporation
NIC = nearly instantaneous companding
NICATELSAT = Compania Nicaraguense de Telecomunicaciones por Satelite
NIMBUS = meteorological satellite (U.S.)
NMC = National Meteorological Center (U.S.)
NOAA = National Oceanic & Atmospheric Administration (U.S.)

N-on-P = negative on positive
NORAD = North American Air Defense Command (U.S.)
NORDSAT = broadcast television satellite system for Nordic countries
NORSAT = Norwegian domestic system
NPL = National Physical Laboratory (U.K.)
NPR = noise power ratio
NRC = National Research Council (Canada)
NRL = Naval Research Laboratory (U.S.)
NRMS = network reference and monitor station
NRZ = non return-to-zero
ns = nanosecond
NSA = National Standards Association (U.S.)
NSDA = National Space Development Agency (Japan)
NSF = National Science Foundation (U.S.)
NSSDC = National Space Science Data Center (U.S.)
NSTL = National Space Technology Laboratory (U.S.)
NTC = Network Transmission Committee (VITEAC)
NTIA = National Telecommunications Information Administration (U.S.)
NTS = navigation technology satellite
NTSC = National Television System Committee (committee of representatives of the television industry in the U.S. on whose findings and recommendations television standards are established)
NTSC System = system of color television based on the recommendation of the NTSC using 525 lines and 60 fields with a video bandwidth of 4.2 MHz (U.S., Canada, Japan)
NTT = Nippon Telegraph & Telephone Public Corporation (Japan)

O&M = operation and maintenance
O/P = output
OACI = Organizacion de Aviacion Civil Internacional (ICAO)
OAO = Orbiting Astronomical Observatory
OAS = Organization of American States
OAS = orbit adjust subsystem
OAU = Organization of African Unity
OBN = out-of-band noise
OCS = Overseas Communications Service (India)
OERS = Organization of Senegal Riparian States
OFHC = oxygen-free high-conductivity copper
OGO = Orbiting Geophysical Observatory

OIRT = International Radio & Television Organization
OMS = Organizacion Mundial de la Salud (WHO)
OMT = orthomode transducer
ONPTZ = Office National de Postes et Telecommunications du Zaire
ORACLE = Viewdata or teletext system similar to CEEFAX (BBC)
ORBIS = Orbiting Radio Beacon Ionospheric Satellite
ORTF = Office de Radiodiffusion et Television Francaise
OSC = oscillator
OSCAR = amateur radio communications satellite (U.S., Australia)
OSM = miniaturized rigid coaxial cable
OSO = orbiting solar observatory (U.S.)
OSPk = offset keyed quadrature phase shift keying
OSR = optical solar reflector
OTAN = Organizacion del Tratado del Atlantico Norte (NATO)
OTC = Overseas Telecommunications Commission
OTC(A) = Overseas Telecommunications Commission (Australia)
OTE = Hellenic Telecommunications Organization (Greece)
OTEC = ocean thermal energy conversion
OTI = Ibero-American Television Organization
OTP = Office of Telecommunications Policy (U.S.)
OTS = orbital test satellite
OTV = orbital transfer vehicle
OUB = occasional-use bands
OW = orderwire

P = private or special telephone circuit (CCITT)
p = pico (one-trillionth)
P-P = peak-to-peak
PA = power amplifier
PABX = private automatic branch exchange
PAL = phase alternation line
PAL = phase attenuation by line
PAL Color = color television system developed in Germany using 625 lines and 50 fields with a video bandwidth of 5 MHz (Western Europe)

PAL-M = PAL system using NTSC parameters
PALAPA = Indonesian domestic satellite system
PAM = pulse amplitude modulation
PAM = payload assist module
PAMA = pulse-address multiple access
PAML = program authorized materials list
PANA = Pan-African News Agency
PANAFTEL = Pan-African Telecommunication Network
PAPM = pulse amplitude phase modulation
Par = parity
Paramp = parametric low-noise amplifier
PATU = Pan-African Telecommunications Union
PC = printed circuit
PCE = power conditioning electronics
PCM = pulse code modulation
PCU = power control unit
PDM = pulse duration modulation
PDR = preliminary design review
PEM = proto environmental model
PERUMTEL = Perusahaan Umum Telekomunikasi (Indonesia)
pF = picofared
PF = power factor
PFD = power flux density
PFIX = power failure interrupt
PFM = pulse frequency modulation
PGM = program
PH = phase
Phebus = European experimental broadcast television satellite (also, H-SAT)
PHILCOMSAT = Philippine Communications Satellite Corporation
pico = one-trillionth
PIL = pilot
PIP = payload integration plan
PL = parts list
PM = phase modulation
PMA = permanent management arrangements
PMB = proto main body
PMS = picturephone meeting service
PN = pseudonoise
POGO = Polar Orbiting Geophysical Observatory
POL = polarization
Polang = polarization angle
POR = Pacific Ocean region
POTELCO = Philippine Overseas Telecommunications Corporation
POTOK = satellite network of the USSR
PPF = payload processing facility (USAF)

PPM = pulse position modulation
ppm = parts per million
PPS = pulses per second
PQM = proto qualification model
PR = private renter
PRESTEL = trade name for the U.K. Post Office Corporation's viewdata or telex service
PRN = pseudorandom noise
PROC = processor
Progress = modified automatic Soyuz cargo spacecraft (USSR)
PRPM = primary power monitor
PS = power supply
psi = pounds per square inch
PSK = phase shift keying
PTC = Pacific Telecommunications Council
PTLU = pretransmission lineup
PTM = pulse time modulation
PTT = post, telegraph, telephone (general European designation)
PV = present value
pW = picowatt
PWM = pulse width modulation
pWp = picowatt psophometrically weighted
pWpO/pWpP = picowatt psophometrically weighted measured at a point of zero reference level
PWR = power

QA = quality assurance
QAM-PAM = quadrature-amplitude modulation/pulse-amplitude modulation
QM = qualification model
QPSK = quadra-phase shift keyed

R&QA = reliability and quality assurance
RA = right ascension
rad = radian
RAM = random access memory
RARC = Regional Administrative Radio Conference
RAS = remote analysis system
RBV = return beam vidicon
RCA = Radio Corporation of America (now the RCA Corporation)
RCO = restoration control office
Rcvr = receiver
RDF = repeater distribution frame

REF = reference
REL = release
RELAY = early experimental low-medium orbit communications satellite (U.S.)
RET = reliable Earth terminal
rev = revolution
rf = radio frequency
RFI = request for information
RFI = radio frequency interference
RFP = request for proposals
RHCP = right-hand circular polarization
RLO = restoration liaison officer
rms = root mean square
RMTE = remote
RN = reference noise
RO = receive only
rpm = revolutions per minute
rps = revolutions per second
RR = Radio Regulation
RRL = Radio Research Laboratories (Japan)
RSE = register signaling equipment
RSTN = Regional Seismic Test Network
RTE/TSM = TDMA reference and monitor station equipment
RTT = Regie des Telegraphes et des Telephones (Belgium)
Rx = receive
Rz = return to zero

s = seconds
S+DX = speech plus duplex
S/EOS = standard Earth observation satellite
S/N = signal-to-noise ratio
SAC = Strategic Air Command (USAF)
SADA = solar array drive assembly
Sade = solar array drive electronics
SAE = Society of Automotive Engineers (U.S.)
SAGE = stratospheric aerosol & gas experiments satellite (U.S.)
SAGEM = Societe d'Applications Generales d'Electricite et de Mecanique (France)
SAGR = supergroup automatic gain regulator
SAIL = Shuttle Avionics Integration Laboratory (NASA)
SAMOS = USAF reconnaissance satellite
SAMSO = Space and Missile Systems Organization (USAF)
SAPO = South African Post Office

SAR = synthetic aperture radar
SARIT = satellite network of Italy
SATCOL = domestic satellite network of Colombia
SATCOM = U.S. domestic satellite network (owned by RCA American Communications, Inc.)
SBS = Satellite Business Systems (U.S. domestic system owned by IBM, COMSAT General, and Aetna Life)
SBTS = domestic satellite network of Brazil
SCATHA = spacecraft charging at high altitudes scientific satellite (U.S.)
SCC = satellite control center
SCO = subcarrier oscillator
SCORE = signal communications by orbiting relay equipment (U.S. early experimental satellite)
SCPC = single channel per carrier, PCM/PSK telecommunications equipment
SCR = silicon controlled rectifier
SCTO = stalled call timed out
SCUC = Satellite Communications Users Conference
SD = speech detector
SDF = supergroup distribution frame
SDMA = space-division multiple access
SECAM = sequential couleurs a memoire (A color TV system developed in France using 625 lines and 50 fields with a video bandwidth of 6 MHz)
SEL = Space Environment Laboratory (U.S.)
SEN = SENSE command
SEO = satellite for Earth observation (India)
SEOS = synchronous Earth observation satellite
SET = European Telecommunications Working Group of the CEPT
SFER = French Professional Electronic and Engineering Society
SG = Secretary General
SG = supergroup
SGDF = supergroup distribution frame
SHE = spacecraft handling equipment
SHF = super high frequency (3000-30,000 MHz)
Shuttle = U.S. launcher (NASA)
SICC = Spanish International Communications Corporation (U.S.)
SICRAL = satellite network of Italy (planning)
SIRIO = experimental satellite network of Italy
SITA = Sociedad Internacional de Telecomunicaciones Aeronauticas

SKYNET = U.K. defense communications network
SLL = satellite line link
SMA = semimajor access
SMS = Shuttle mission simulator
SMS = synchronous meteorological satellite (U.S.)
SNF = system noise figure
SNIAS = Societe Nationale Aerospatiale (France)
SNR = signal-to-noise ratio
SOLAS = safety of life at sea
SOLWIND = collects data on solar experiments (USAF)
SOM = start of message
SOM = space oblique mercator
SOW = statement of work
Soyuz = two-man spacecraft (USSR)
SPACENET = U.S. domestic satellite network owned by GTE Corp.
SPADE = SCPC/PCM multiple-access demand-assigned equipment
SPCC = Southern Pacific Communications Corporation (U.S.)
SPDT = single-pole double-throw
SPEC = speech predictive encoding system
SPELDA = improved dual-launch system for heavy geostationary payloads
SPM = signal processing modem
Sputnik = experimental satellite for manned space flights (USSR)
SRARQ = selective repeat-automatic repeat request
SRATS = solar radiation and thermospheric satellite (Japan)
SRC = Science Research Council (U.K.)
SRE = speech recognition equipment (or VRE)
SRF = service request flag
SRP = signal reference point
SRTS = supporting RTE/TSM system
SS = satellite switching
SS = spread spectrum
SSMA = spread-spectrum multiple access
SS-TDMA = satellite-switched time-division multiple access
SS/TDMA = space-switched time-division multiple access
SSB-SC-AM = single sideband-suppressed carrier-amplitude
SSB-SC-ASK = single sideband-suppressed carrier-amplitude shift keyed
SSB-SC-PAM = single sideband-suppressed carrier pulse-amplitude modulation

SSLO = solid-state local oscillator
SSM = satellite system monitor
SSMG = satellite system monitoring group
SSOG = satellite system operations guide
SSOP = satellite system operations plan
SSP = signaling and switching processor
SSPA = solid-state power amplifier
SSRA = spread-spectrum random access
SSSF = self-scanner stop failure
SSU = sequential shunt unit
SSU = secondary sampling unit
SSUS = spinning solid upper stage
ST = "end pulsing" command (CCITT format)
STAR = satellite telecommunications with automatic routing (simple form of SPADE system)
STATSIONAR = communications satellite for geostationary orbit operation (USSR)
STBY = standby
STCC = spacecraft technical control center
STDN = space tracking and data network
STE = supergroup translating equipment
STE = spacecraft test equipment
STIMAD = Societe des Telecommunications Internationales de la Republique Malgache
STM = structural thermal model
STOM = SPADE terminal operator's manual
STOX = speech plus duplex (speech and telegraphy in voice channel)
STP = space test program (USAF)
STR = symbol timing recovery
STRATS = solar radiation and thermospheric satellite (Japan)
STS = Space Transportation System (operated by NASA)
STSC = satellite network of Cuba
STSK = Scandanavian Committee for Satellite Telecommunications
STW = satellite network of China (People's Republic of China)
SUDOSAT = Sudan domestic satellite system
SURVSATCOM = survivable communications satellite
SVC = secure voice communications service
SW = switch
SWOF = switchover operation failure
SWR = standing wave radio
SYMCOSAT = consortium for SYMPHONIE satellite

SYMPHONIE = French-German experimental satellite system
SYN = synthesizer
sync = synchronize, synchronizer
SYNCOM = early geostationary satellite (U.S.)
SYNCOM IV = experimental satellite designed by Hughes Aircraft for STS launch (prototype for operational satellites for 1980's)

TACSAT = Tactical Communications Satellite (U.S. Dept. of Defense)
TM = telemetry
T/R = transmit/receive
T/V = thermal vacuum
TA = transistor amplifier
TASI = time-assignment speech interpolation
TASO = television allocations study
TAT = transatlantic telephone (U.S.-Europe)
TBBF = top baseband frequency
TBU = transmit baseband unit
TC&R = telemetry command and ranging
TCE = telemetry and command equipment
TCS = thermal control subsystem
TCTS = Transcanada Telephone System
TDA = terminal diode amplifier
TDB = traffic data base
TDC = telegraphy data channel
TDF = broadcasting satellite network of France (planning)
TDG = transmit data gate
TDM = time-division multiplex
TDMA = time-division multiple access
TDRSS = tracking and data relay satellite system
TE = thermoelectric
TE = threshold extension
TED = threshold extension demodulator
TEL = telephone
TELE-X = experimental satellite network of Sweden
TELECOM = satellite network of France
Teledirektorate = Norwegian telecommunications administration
Teleglobe = Canadian signatory to INTELSAT
Telesat = domestic satellite of Canada
Telespazio = Societa per azioni per le comunicazioni spaziali
Telex = international teleprinter exchange service

Telstar = U.S. domestic experimental satellite network (successor to COMSTAR series)
TEM = transverse electromagnetic
TEMP = temperature
TERA = one trillion
TERLS = Thumba Equatorial Rocket Launching Center (India)
TERM = termination
TES = transportable Earth station
TETR = test and training satellite (U.S.)
TEXTEL = Trinidad and Tobago External Telecommunications Co., Ltd.
TFC = traffic
TFU = timing and frequency unit
Tg = telegraph
TGF = through group filter
TIAS = Treaties and Other International Acts Series (U.S. Dept. of State)
TIG = Societe des Telecommunications Internationales Gabonaises
TIM = Telecommunications Internationales du Mali
TIROS = U.S. meteorological satellite
TIRS = thermal infrared scanner
TIT = Societe des Telecommunications Internationales du Tchad
Titan = U.S. launcher
TIU = terrestrial interface unit
TIUPIL = typical information use per individual
TLP = transmission level point
TNIP = terrestrial network interface processor
TOC = television operating center
TOCC = technical and operational control center
TOD = time of day
TOIRS = transfer orbit infrared Earth sensor
TOPS = telemetry operations
TOS = transfer orbit stage
TOSSA = transfer orbit sun sensor assembly
TOT = time of transmission
TP = test point
TPT = transmission path translator
TRANSAT = U.S. Navy navigational satellite (modified TRANSIT)
TRANSIT = U.S. Navy navigational satellite
Trfc = traffic
TRM = transmit-receive module
TRMA = time random multiple access
TRMS = TDMA reference and monitor station services

TRW = Thompson, Ramo, Wooldridge, Inc.
TS = telegraph service circuit (CCITT)
TSC = technical services contractor
TSESC = telegraph engineering service circuit
TSF = through supergroup filter
TSM = telephony signaling module
TT&C = tracking, telemetry, and command
TTB = trunk test buffer
TTC&M = tracking, telemetry, command and monitoring station
TTP = trunk test panel
TTPB = trunk test panel buffer
TTS = TDMA terminal simulation
TTY = teletypewriter
TV = television (video and assigned audio)
TVRO = television receive-only terminal
TV-SAT = satellite network of Germany (FRG)
TVC = thrust vector control
TWT = traveling wave tube
TWTA = traveling wave tube amplifier
TWX = teletypewriter exchange service
TX = transmit

U = unidirectional
U/C = up-converter
UAMPT = African and Malagasy Postal and Telecommunications Union
UAPT = African Postal and Telecommunications Union
UARTO = United Arab Republic Telecommunication Organization
UBK = unblock
UDEAC = Union Aduanera Economica del Africa Central
UFB = unfit for broadcast
UHF = ultra-high frequency (300-3000 MHz)
UIT = Union Internacional de Telecomunicaciones (ITU)
UITP = International Union of Public Transport
UKPO = United Kingdom Post Office
UNCTAD = United Nations Conference on Trade and Development
UNDP = United Nations Disaster Relief Office
UNESCO = United Nations Educational, Scientific and Cultural Organization
UNIPEDE = International Union of Producers and Distributors of Electrical Energy

UNISAT = domestic satellite network of the U.K.
UNITAR = United Nations Institute for Training and Research
UPS = uninterrupted power supply
URSI = International Union of Radio Science
URTNA = Union of African National Broadcasting and Television
USAF = U.S. Air Force
USB = unified S-band
USGS = U.S. Geological Survey
USISC = U.S. International Service Carriers
USN = United States Navy
USNO = United States Naval Observatory
UST = UNBLOCK SPADE TERMINAL command
UTC = universal coordinated time
UW = unique word

VAN = value added network
VAS = value added service (or VAN)
VCD = valve coil driver
VCO = voltage controlled oscillator
VELCOR = velocity correction
VF = voice frequency
VFO = variable frequency oscillator
VFT = voice frequency telegraph
VHF = very high frequency
VHRR = very high resolution radiometer
VIEWDATA = generic name for a home information system
VITEAC = Video Transmission Engineering Advisory Committee (U.S.)
VLF = very low frequency
VLSL = very large scale integrations
VOLNA = USSR network
VPF = vertical processing facility (KSC)
VRE = voice recognition equipment
VSB = vestigial sideband
VSM = vestigial sideband modulation
VSWR = voltage standing-wave ratio
VTF = via terrestrial facilities
VTR = videotape recording
VU = volume unit

W/G = waveguide
WARC = World Administrative Radio Conference
WARC/MR = World Administrative Radio Conference on Maritime Mobile Telecommunications

WARC/ST	= World Administrative Radio Conference on Space Telecommunication
Wb	= wide-band
WBPA	= wide-band power amplified
WDC	= World Data Center
WESTAR	= U.S. domestic satellite (Western Union)
WF	= weighting function
Wg	= waveguide
WHO	= World Health Organization
WMO	= World Meteorological Organization
WMS	= World Magnetic Survey
WPC	= World Power Conference
WPM	= words per minute
WUI	= Western Union International
WWW	= world weather watch
X	= combined services (CCITT)
XPD	= cross-polarization discrimination
X-POL	= cross-polarized
Xmit	= transmit
XPI	= cross polar interference
XSSF	= scanner sync failure
XTEN	= Xerox Telecommunications Network

Author Index for Volume 93

Alper, J. 363
Astrain, S. 1
Clarke, A. C. 7
Charyk, J. V. 379
Colino, R. R. 55
Durant, F. C. III 21
Edelson, B. I. 39
Goldstein, I. 379
Nosaka, K. 135
Pelton, J. N. 95, 363
Perras, M. 255
Podraczky, E. 95
Quaglione, G. 191
Sachdev, D. K. 331
Schnicke, W. R. 311
Weiss, H. J. 269
Withers, D. J. 269
Wood, H. W. 229

PROGRESS IN ASTRONAUTICS AND AERONAUTICS SERIES VOLUMES

VOLUME TITLE/EDITORS

*1. **Solid Propellant Rocket Research** (1960)
Martin Summerfield
Princeton University

*2. **Liquid Rockets and Propellants** (1960)
Loren E. Bollinger
The Ohio State University
Martin Goldsmith
The Rand Corporation
Alexis W. Lemmon Jr.
Battelle Memorial Institute

*3. **Energy Conversion for Space Power** (1961)
Nathan W. Snyder
Institute for Defense Analyses

*4. **Space Power Systems** (1961)
Nathan W. Snyder
Institute for Defense Analyses

*5. **Electrostatic Propulsion** (1961)
David B. Langmuir
Space Technology Laboratories, Inc.
Ernst Stuhlinger
NASA George C. Marshall Space Flight Center
J.M. Sellen Jr.
Space Technology Laboratories, Inc.

*6. **Detonation and Two-Phase Flow** (1962)
S.S. Penner
California Institute of Technology
F.A. Williams
Harvard University

*7. **Hypersonic Flow Research** (1962)
Frederick R. Riddell
AVCO Corporation

*8. **Guidance and Control** (1962)
Robert E. Roberson
Consultant
James S. Farrior
Lockheed Missiles and Space Company

*9. **Electric Propulsion Development** (1963)
Ernst Stuhlinger
NASA George C. Marshall Space Flight Center

*10. **Technology of Lunar Exploration** (1963)
Clifford I. Cummings and Harold R. Lawrence
Jet Propulsion Laboratory

*11. **Power Systems for Space Flight** (1963)
Morris A. Zipkin and Russell N. Edwards
General Electric Company

*12. **Ionization in High-Temperature Gases** (1963)
Kurt E. Shuler, Editor
National Bureau of Standards
John B. Fenn, Associate Editor
Princeton University

*13. **Guidance and Control—II** (1964)
Robert C. Langford
General Precision Inc.
Charles J. Mundo
Institute of Naval Studies

*14. **Celestial Mechanics and Astrodynamics** (1964)
Victor G. Szebehely
Yale University Observatory

*15. **Heterogeneous Combustion** (1964)
Hans G. Wolfhard
Institute for Defense Analyses
Irvin Glassman
Princeton University
Leon Green Jr.
Air Force Systems Command

*16. **Space Power Systems Engineering** (1966)
George C. Szego
Institute for Defense Analyses
J. Edward Taylor
TRW Inc.

*17. **Methods in Astrodynamics and Celestial Mechanics** (1966)
Raynor L. Duncombe
U.S. Naval Observatory
Victor G. Szebehely
Yale University Observatory

*18. **Thermophysics and Temperature Control of Spacecraft and Entry Vehicles** (1966)
Gerhard B. Heller
NASA George C. Marshall Space Flight Center

*19. **Communication Satellite Systems Technology** (1966)
Richard B. Marsten
Radio Corporation of America

*Out of print.

***20. Thermophysics of Spacecraft and Planetary Bodies: Radiation Properties of Solids and the Electromagnetic Radiation Environment in Space** (1967)
Gerhard B. Heller
NASA George C. Marshall Space Flight Center

***21. Thermal Design Principles of Spacecraft and Entry Bodies** (1969)
Jerry T. Bevans
TRW Systems

***22. Stratospheric Circulation** (1969)
Willis L. Webb
Atmospheric Sciences Laboratory, White Sands, and University of Texas at El Paso

***23. Thermophysics: Applications to Thermal Design of Spacecraft** (1970)
Jerry T. Bevans
TRW Systems

24. Heat Transfer and Spacecraft Thermal Control (1971)
John W. Lucas
Jet Propulsion Laboratory

25. Communication Satellites for the 70's: Technology (1971)
Nathaniel E. Feldman
The Rand Corporation
Charles M. Kelly
The Aerospace Corporation

26. Communication Satellites for the 70's: Systems (1971)
Nathaniel E. Feldman
The Rand Corporation
Charles M. Kelly
The Aerospace Corporation

27. Thermospheric Circulation (1972)
Willis L. Webb
Atmospheric Sciences Laboratory, White Sands, and University of Texas at El Paso

28. Thermal Characteristics of the Moon (1972)
John W. Lucas
Jet Propulsion Laboratory

29. Fundamentals of Spacecraft Thermal Design (1972)
John W. Lucas
Jet Propulsion Laboratory

30. Solar Activity Observations and Predictions (1972)
Patrick S. McIntosh and Murray Dryer
Environmental Research Laboratories, National Oceanic and Atmospheric Administration

31. Thermal Control and Radiation (1973)
Chang-Lin Tien
University of California at Berkeley

32. Communications Satellite Systems (1974)
P.L. Bargellini
COMSAT Laboratories

33. Communications Satellite Technology (1974)
P.L. Bargellini
COMSAT Laboratories

34. Instrumentation for Airbreathing Propulsion (1974)
Allen E. Fuhs
Naval Postgraduate School
Marshall Kingery
Arnold Engineering Development Center

35. Thermophysics and Spacecraft Thermal Control (1974)
Robert G. Hering
University of Iowa

36. Thermal Pollution Analysis (1975)
Joseph A. Schetz
Virginia Polytechnic Institute

37. Aeroacoustics: Jet and Combustion Noise; Duct Acoustics (1975)
Henry T. Nagamatsu, Editor
General Electric Research and Development Center
Jack V. O'Keefe, Associate Editor
The Boeing Company
Ira R. Schwartz, Associate Editor
NASA Ames Research Center

38. Aeroacoustics: Fan, STOL, and Boundary Layer Noise; Sonic Boom; Aeroacoustic Instrumentation (1975)
Henry T. Nagamatsu, Editor
General Electric Research and Development Center
Jack V. O'Keefe, Associate Editor
The Boeing Company
Ira R. Schwartz, Associate Editor
NASA Ames Research Center

39. Heat Transfer with Thermal Control Applications (1975)
M. Michael Yovanovich
University of Waterloo

40. Aerodynamics of Base Combustion (1976)
S.N.B. Murthy, Editor
Purdue University
J.R. Osborn, Associate Editor
Purdue University
A.W. Barrows and J.R. Ward, Associate Editors
Ballistics Research Laboratories

41. Communications Satellite Developments: Systems (1976)
Gilbert E. LaVean
Defense Communications Agency
William G. Schmidt
CML Satellite Corporation

42. Communications Satellite Developments: Technology (1976)
William G. Schmidt
CML Satellite Corporation
Gilbert E. LaVean
Defense Communications Agency

43. Aeroacoustics: Jet Noise, Combustion and Core Engine Noise (1976)
Ira R. Schwartz, Editor
NASA Ames Research Center
Henry T. Nagamatsu, Associate Editor
General Electric Research and Development Center
Warren C. Strahle, Associate Editor
Georgia Institute of Technology

44. Aeroacoustics: Fan Noise and Control; Duct Acoustics; Rotor Noise (1976)
Ira R. Schwartz, Editor
NASA Ames Research Center
Henry T. Nagamatsu, Associate Editor
General Electric Research and Development Center
Warren C. Strahle, Associate Editor
Georgia Institute of Technology

45. Aeroacoustics: STOL Noise; Airframe and Airfoil Noise (1976)
Ira R. Schwartz, Editor
NASA Ames Research Center
Henry T. Nagamatsu, Associate Editor
General Electric Research and Development Center
Warren C. Strahle, Associate Editor
Georgia Institute of Technology

46. Aeroacoustics: Acoustic Wave Propagation; Aircraft Noise Prediction; Aeroacoustic Instrumentation (1976)
Ira R. Schwartz, Editor
NASA Ames Research Center
Henry T. Nagamatsu, Associate Editor
General Electric Research and Development Center
Warren C. Strahle, Associate Editor
Georgia Institute of Technology

47. Spacecraft Charging by Magnetospheric Plasmas (1976)
Alan Rosen
TRW Inc.

48. Scientific Investigations on the Skylab Satellite (1976)
Marion I. Kent and Ernst Stuhlinger
NASA George C. Marshall Space Flight Center
Shi-Tsan Wu
The University of Alabama

49. Radiative Transfer and Thermal Control (1976)
Allie M. Smith
ARO Inc.

50. Exploration of the Outer Solar System (1976)
Eugene W. Greenstadt
TRW Inc.
Murray Dryer
National Oceanic and Atmospheric Administration
Devrie S. Intriligator
University of Southern California

51. Rarefied Gas Dynamics, Parts I and II (two volumes) (1977)
J. Leith Potter
ARO Inc.

52. Materials Sciences in Space with Application to Space Processing (1977)
Leo Steg
General Electric Company

53. Experimental Diagnostics in Gas Phase Combustion Systems (1977)
Ben T. Zinn, Editor
Georgia Institute of Technology
Craig T. Bowman, Associate Editor
Stanford University
Daniel L. Hartley, Associate Editor
Sandia Laboratories
Edward W. Price, Associate Editor
Georgia Institute of Technology
James G. Skifstad, Associate Editor
Purdue University

54. Satellite Communications: Future Systems (1977)
David Jarett
TRW Inc.

55. Satellite Communications: Advanced Technologies (1977)
David Jarett
TRW Inc.

56. Thermophysics of Spacecraft and Outer Planet Entry Probes (1977)
Allie M. Smith
ARO Inc.

57. Space-Based Manufacturing from Nonterrestrial Materials (1977)
Gerard K. O'Neill, Editor
Princeton University
Brian O'Leary, Assistant Editor
Princeton University

58. Turbulent Combustion (1978)
Lawrence A. Kennedy
State University of New York at Buffalo

59. Aerodynamic Heating and Thermal Protection Systems (1978)
Leroy S. Fletcher
University of Virginia

60. Heat Transfer and Thermal Control Systems (1978)
Leroy S. Fletcher
University of Virginia

61. Radiation Energy Conversion in Space (1978)
Kenneth W. Billman
NASA Ames Research Center

62. Alternative Hydrocarbon Fuels: Combustion and Chemical Kinetics (1978)
Craig T. Bowman
Stanford University
Jorgen Birkeland
Department of Energy

63. Experimental Diagnostics in Combustion of Solids (1978)
Thomas L. Boggs
Naval Weapons Center
Ben T. Zinn
Georgia Institute of Technology

64. Outer Planet Entry Heating and Thermal Protection (1979)
Raymond Viskanta
Purdue University

65. Thermophysics and Thermal Control (1979)
Raymond Viskanta
Purdue University

66. Interior Ballistics of Guns (1979)
Herman Krier
University of Illinois at Urbana-Champaign
Martin Summerfield
New York University

67. Remote Sensing of Earth from Space: Role of "Smart Sensors" (1979)
Roger A. Breckenridge
NASA Langley Research Center

68. Injection and Mixing in Turbulent Flow (1980)
Joseph A. Schetz
Virginia Polytechnic Institute and State University

69. Entry Heating and Thermal Protection (1980)
Walter B. Olstad
NASA Headquarters

70. Heat Transfer, Thermal Control, and Heat Pipes (1980)
Walter B. Olstad
NASA Headquarters

71. Space Systems and Their Interactions with Earth's Space Environment (1980)
Henry B. Garrett and Charles P. Pike
Hanscom Air Force Base

72. Viscous Flow Drag Reduction (1980)
Gary R. Hough
Vought Advanced Technology Center

73. Combustion Experiments in a Zero-Gravity Laboratory (1981)
Thomas H. Cochran
NASA Lewis Research Center

74. Rarefied Gas Dynamics, Parts I and II (two volumes) (1981)
Sam S. Fisher
University of Virginia at Charlottesville

75. Gasdynamics of Detonations and Explosions (1981)
J.R. Bowen
University of Wisconsin at Madison
N. Manson
Université de Poitiers
A.K. Oppenheim
University of California at Berkeley
R.I. Soloukhin
Institute of Heat and Mass Transfer, BSSR Academy of Sciences

76. Combustion in Reactive Systems (1981)
J.R. Bowen
University of Wisconsin at Madison
N. Manson
Université de Poitiers
A.K. Oppenheim
University of California at Berkeley
R.I. Soloukhin
Institute of Heat and Mass Transfer, BSSR Academy of Sciences

77. Aerothermodynamics and Planetary Entry (1981)
A.L. Crosbie
University of Missouri-Rolla

78. Heat Transfer and Thermal Control (1981)
A.L. Crosbie
University of Missouri-Rolla

79. Electric Propulsion and Its Applications to Space Missions (1981)
Robert C. Finke
NASA Lewis Research Center

80. Aero-Optical Phenomena (1982)
Keith G. Gilbert and Leonard J. Otten
Air Force Weapons Laboratory

81. Transonic Aerodynamics (1982)
David Nixon
Nielsen Engineering & Research, Inc.

82. Thermophysics of Atmospheric Entry (1982)
T.E. Horton
The University of Mississippi

83. Spacecraft Radiative Transfer and Temperature Control (1982)
T.E. Horton
The University of Mississippi

84. Liquid-Metal Flows and Magnetohydrodynamics (1983)
H. Branover
Ben-Gurion University of the Negev
P.S. Lykoudis
Purdue University
A. Yakhot
Ben-Gurion University of the Negev

85. Entry Vehicle Heating and Thermal Protection Systems: Space Shuttle, Solar Starprobe, Jupiter Galileo Probe (1983)
Paul E. Bauer
McDonnell Douglas Astronautics Company
Howard E. Collicott
The Boeing Company

86. Spacecraft Thermal Control, Design, and Operation (1983)
Howard E. Collicott
The Boeing Company
Paul E. Bauer
McDonnell Douglas Astronautics Company

87. Shock Waves, Explosions, and Detonations (1983)
J.R. Bowen
University of Washington
N. Manson
Université de Poitiers
A.K. Oppenheim
University of California at Berkeley
R.I. Soloukhin
Institute of Heat and Mass Transfer, BSSR Academy of Sciences

88. Flames, Lasers, and Reactive Systems (1983)
J.R. Bowen
University of Washington
N. Manson
Université de Poitiers
A.K. Oppenheim
University of California at Berkeley
R.I. Soloukhin
Institute of Heat and Mass Transfer, BSSR Academy of Sciences

89. Orbit-Raising and Maneuvering Propulsion: Research Status and Needs (1984)
Leonard H. Caveny
Air Force Office of Scientific Research

90. Fundamentals of Solid-Propellant Combustion (1984)
Kenneth K. Kuo
The Pennsylvania State University
Martin Summerfield
Princeton Combustion Research Laboratories, Inc.

91. Spacecraft Contamination: Sources and Prevention (1984)
J.A. Roux
The University of Mississippi
T.D. McCay
NASA Marshall Space Flight Center

92. Combustion Diagnostics by Nonintrusive Methods (1984)
T.D. McCay
NASA Marshall Space Flight Center
J.A. Roux
The University of Mississippi

93. The INTELSAT Global Satellite System (1984)
Joel Alper
COMSAT Corporation
Joseph Pelton
INTELSAT

(Other Volumes are planned.)